T0174666

Handbook of Texture Analysis

The major goals of texture research in computer vision are to understand, model, and process texture and, ultimately, to simulate the human visual learning process using computer technologies. In the last decade, artificial intelligence has been revolutionized by machine learning and big data approaches, outperforming human prediction on a wide range of problems. In particular, deep learning convolutional neural networks (CNNs) are particularly well suited to texture analysis. This volume presents important branches of texture analysis methods which find a proper application in AI-based medical image analysis. This book:

- Discusses first-order and second-order statistical methods, local binary pattern (LBP) methods, and filter bank-based methods.
- Covers spatial frequency-based methods, Fourier analysis, Markov random fields, Gabor filters, and Hough transformation.
- Describes advanced textural methods based on DL as well as BD and advanced applications of texture to medial image segmentation.
- Is aimed at researchers, academics, and advanced students in biomedical engineering, image analysis, cognitive science, and computer science and engineering.

This is an essential reference for those looking to advance their understanding in this applied and emergent field.

Handbook of Texture Analysis

AI-Based Medical Imaging Applications

Volume II

Edited by
Ayman El-Baz,
Mohammed Ghazal and Jasjit S. Suri

CRC Press is an imprint of the
Taylor & Francis Group, an **informa** business

Designed cover image: © Shutterstock

First edition published 2024
by CRC Press
2385 NW Executive Center Drive, Suite 320, Boca Raton FL 33431

and by CRC Press
4 Park Square, Milton Park, Abingdon, Oxon, OX14 4RN

CRC Press is an imprint of Taylor & Francis Group, LLC

ISBN: 978-0-367-48345-6 (hbk)
ISBN: 978-1-032-72743-1 (pbk)
ISBN: 978-0-367-48608-2 (ebk)

DOI: 10.1201/9780367486082

Typeset in Times LT Std
by Apex CoVantage, LLC

With love and affection to my mother and father, whose loving spirit sustains me still.

Ayman El-Baz

To my loving parents.

Mohammed Ghazal

To my late loving parents, immediate family, and children.

Jasjit S. Suri

Contents

Preface

The book covers the state-of-the-art techniques of texture analysis in image processing and their applications. Texture analysis has played a key role in the field of image processing over several decades. It helps in extracting meaningful information from images, and although a lot of research has been done to enhance techniques used to interpret and model texture, it is still very challenging. Being one of the most important topics in image processing and pattern recognition, several state-of-the-art techniques are presented in this book while highlighting points that still need improvement. Among the topics discussed in this book are local binary descriptors for texture classification, precision grading of glioma, liver tumor detection and grading, texture analysis in radiology, texture analysis using a self-organizing feature map, sensor-based human activity recognition analysis using machine learning and topological data analysis, texture analysis in retinal OCT imaging, pneumonia detection, prostatic adenocarcinoma, and texture analysis in cancer prognosis.

In summary, the main aim of this book is to help advance scientific research within the broad field of texture analysis in image processing. The book focuses on major trends and challenges in this area, and it presents work aimed to identify new techniques and their use in artificial intelligence systems.

Contributors

Nahla B. Abdel-Hamid
Computers and Systems Department, Faculty of Engineering
Mansoura University
Mansoura 35511, Egypt

Mohamed Abou El-Ghar
Radiology Department, Urology and Nephrology Center
Mansoura University
Mansoura 35516, Egypt

Hadil Abu Khalifeh
College of Engineering
Abu Dhabi University

Norah Saleh Alghamdi
Department of Computer Sciences, College of Computer and Information Sciences
Princess Nourah Bint Abdulrahman University
PO Box 84428, Riyadh 11671, Saudi Arabia

H. Arafat Ali
Computers and Systems Department, Faculty of Engineering
Mansoura University
Mansoura 35511, Egypt

Ahmed Alksas
BioImaging Laboratory, Bioengineering Department
University of Louisville
Louisville, KY 40292, USA

Emad Alsyed
Department of Nuclear Engineering, Faculty of Engineering
King Abdulaziz University
Jeddah, Saudi Arabia

Arya R.
School of Arts and Sciences, Kochi
Amrita Vishwa Vidyapeetham

Hala Atef
Department of Radiology, Faculty of Medicine
Mansoura University

Sarah M. Ayyad
Computers and Systems Department, Faculty of Engineering
Mansoura University
Mansoura 35511, Egypt

Mohamed A. Badawy
Radiology Department, Urology and Nephrology Center
Mansoura University
Mansoura 35516, Egypt

Hossam Magdy Balaha
Department of Bioengineering
University of Louisville

Rui Bernardes
Coimbra Institute for Biomedical Imaging and Translational Research (CIBIT)/Institute of Nuclear Sciences Applied to Health (ICNAS)
University of Coimbra, Health Sciences Campus, Polo III
3000-548 Coimbra, Portugal

Salvatore Cappabianca
Section of Radiology and Radiotherapy, Department of Precision Medicine
University of Campania "L. Vanvitelli"
80138 Naples, Italy

Miguel Castelo-Branco
Coimbra Institute for Biomedical Imaging and Translational Research (CIBIT)/Institute of Nuclear Sciences Applied to Health (ICNAS)
University of Coimbra, Health Sciences Campus, Polo III
3000-548 Coimbra, Portugal

Ayman El-Baz
BioImaging Laboratory, Bioengineering Department
University of Louisville
Louisville, KY 40292, USA

Moumen El-Melegy
Department of Electrical Engineering
Assiut University
Assiut 71511, Egypt

Asmaa El-Sayed Hassan
Mathematics and Engineering Physics Department, Faculty of Engineering
Mansoura University

Lamiaa Galal El-Serougy
Department of Radiology, Faculty of Medicine
Mansoura University

Mohammed Ghazal
Electrical, Computer, and Biomedical Engineering Department
Abu Dhabi University
UAE

Giuliana Giacobbe
Section of Radiology and Radiotherapy, Department of Precision Medicine
University of Campania "L. Vanvitelli"
80138 Naples, Italy

Daniel Thomas Ginat
University of Chicago

Roberta Grassi
Section of Radiology and Radiotherapy, Department of Precision Medicine
University of Campania "L. Vanvitelli"
80138 Naples, Italy

Noraini Hasan
Universiti Teknologi MARA, Cawangan Melaka, Kampus Jasin
Melaka, Malaysia

Homam Khattab
Department of Bioengineering
University of Louisville

Adel Khelifi
Computer Science and Information Technology
Abu Dhabi University

Mukhit Kulmaganbetov
Centre for Eye and Vision Research (CEVR)
Hong Kong

Labib M. Labib
Computers and Systems Department, Faculty of Engineering
Mansoura University
Mansoura 35511, Egypt

Ali H. Mahmoud
BioImaging Laboratory, Bioengineering Department
University of Louisville
Louisville, KY 40292, USA

Christopher Marshall
Wales Research and Diagnostic PET Imaging Centre
Cardiff, UK

Ahmed Mayel
DuPont Manual High School

Yassin Mohamed-Hassan
DuPont Manual High School

James E. Morgan
Cardiff University
UK

Valerio Nardone
Section of Radiology and Radiotherapy, Department of Precision Medicine
University of Campania "L. Vanvitelli"
80138 Naples, Italy

Ana Nunes
Coimbra Institute for Biomedical Imaging and Translational Research (CIBIT)/Institute of Nuclear Sciences Applied to Health (ICNAS)
University of Coimbra, Health Sciences Campus, Polo III
3000-548 Coimbra, Portugal

Charles Pierce
University of Illinois at Chicago

Alfonso Reginelli
Section of Radiology and Radiotherapy, Department of Precision Medicine
University of Campania "L. Vanvitelli"
80138 Naples, Italy

Nurbaity Sabri
Universiti Teknologi MARA, Cawangan Melaka, Kampus Jasin
Malaysia

Gehad A. Saleh
Department of Radiology, Faculty of Medicine
Mansoura University

Pedro Serranho
Coimbra Institute for Biomedical Imaging and Translational Research (CIBIT)/Institute of Nuclear Sciences Applied to Health (ICNAS)
University of Coimbra, Health Sciences Campus, Polo III
3000-548 Coimbra, Portugal
AND
Mathematics Section, Science and Technology Department
Universidade Aberta
1269-001 Lisboa, Portugal

Ahmed Shaffie
College of Natural Sciences and Mathematics
Louisiana State University at Alexandria

Nur Syafiqah Shaharudin
Universiti Teknologi MARA, Cawangan Melaka, Kampus Jasin
Melaka, Malaysia

Mohamed Shehata
BioImaging Laboratory, Bioengineering Department
University of Louisville
Louisville, KY 40292, USA

Fatma Sherif
Department of Radiology, Faculty of Medicine
Mansoura University

Rhodri Smith
Wales Research and Diagnostic PET Imaging Centre
Cardiff, UK

Ahmed Soliman
Department of Bioengineering
University of Louisville

Asmaa El-Sayed Hassan
Mathematics and Engineering Physics Department, Faculty of Engineering
Mansoura University

Emiliano Spezi
School of Engineering
Cardiff University
Cardiff, UK

Vimina E. R.
School of Arts and Sciences, Kochi
Amrita Vishwa Vidyapeetham

Author Bios

Ayman El-Baz is a distinguished professor at University of Louisville, Kentucky, United States, and University of Louisville at Alamein International University (UofL-AIU), New Alamein City, Egypt. Dr. El-Baz earned his BSc and MSc degrees in electrical engineering in 1997 and 2001, respectively. He earned his PhD in electrical engineering from the University of Louisville in 2006. Dr. El-Baz was named as a fellow for IEEE, Coulter, AIMBE, and NAI for his contributions to the field of biomedical translational research. Dr. El-Baz has almost two decades of hands-on experience in the fields of bio-imaging modeling and non-invasive computer-assisted diagnosis systems. He has authored or co-authored more than 700 technical articles.

Mohammed Ghazal is a professor and chairman of the Department of Electrical, Computer, and Biomedical Engineering at the College of Engineering, Abu Dhabi University, UAE. His research areas are bioengineering, image and video processing, and smart systems. He received his PhD and MASc in electrical and computer engineering (ECE) from Concordia University in Montreal, Canada, in 2010 and 2006, respectively, and his BSc in computer engineering from the American University of Sharjah (AUS) in 2004. He has received multiple awards, including the Distinguished Faculty Award of Abu Dhabi University in 2017 and 2014. Dr. Ghazal has authored or co-authored over 70 publications in recognized international journals and conferences, including *IEEE Transactions in Image Processing*, *IEEE Transactions in Circuits and Systems for Video Technology*, *IEEE Transactions in Consumer Electronics*, Elsevier's *Renewable Energy Reviews*, and Springer's *Multimedia Tools and Applications*.

Jasjit S. Suri is an innovator, scientist, visionary, industrialist, and internationally known world leader in biomedical engineering. Dr. Suri has spent over 25 years in the field of biomedical engineering/devices and its management. He received his PhD from the University of Washington, Seattle, and his business management sciences degree from Weatherhead, Case Western Reserve University, Cleveland, Ohio. Dr. Suri was crowned with the President's Gold medal in 1980 and made a fellow of the American Institute of Medical and Biological Engineering for his outstanding contributions. In 2018, he was awarded the Marquis Lifetime Achievement Award for his outstanding contributions and dedication to medical imaging and its management.

Acknowledgments

The completion of this book could not have been possible without the participation and assistance of so many people, whose names may not all be enumerated. Their contributions are sincerely appreciated and gratefully acknowledged. However, the editors would like to express their deep appreciation and indebtedness, particularly to Dr. Ali H. Mahmoud and Ahmed Alksas, for their endless support.

1 An Exploratory Review on Local Binary Descriptors for Texture Classification

Arya R. and Vimina E. R.

1.1 INTRODUCTION

Computer vision concerns itself with computationally achieving a high-level perception and understanding (approximating the performance of human cognition) of digital images and videos. It attempts the automation of tasks naturally accomplished by the human visual apparatus. Computer vision encompasses scientific methods for the acquisition, processing, analysis, and interpretation of digital images and extraction of multidimensional data from the physical world with a view to produce numerical or symbolic information. For any computer vision applications, the input data is an image or a video, and the objective is to understand the image and its contents. If the goal is to create an application of human vision—like object recognition, defect detection, etc.—then it may be called computer vision.

Texture analysis is a technique for visualizing regions in an image by analyzing the texture information of those regions [1]. *Tactile texture* refers to the physical feel of a surface, and *visual texture* refers to viewing the shape or the contents of an image. In the human vision system, textures diagnosis is relatively feasible, but machine vision and image processing have their own difficulties. Texture feature is a complicated visual design made up of patterns/sub-patterns with extensive characteristics, such as color, intensity, shape, size, slope, etc. It is defined as the changes in pixel intensities and orientation in a spatial domain [2]. Texture information is a useful low-level characteristic that may be utilized to find objects in an image's region of interest. Various textures in an image can be recognized and characterized by considering the distinctive features that define each class of a texture [3]. Since there are several advanced classifiers that exist, the primary difficulty is to generate important features to extract from an image. Transformations ensued due to the rotation, illumination, and other perspective changes in an image can also make texture classification a challenging task. In order to overcome these challenges, we need more powerful feature extraction techniques to get more accurate results [4–6]. Texture analysis is broadly used in the domain of image processing, object recognition, remote sensing applications, surface defect detection, biomedical analysis, document processing, pattern recognition, and so on [7–10].

DOI: 10.1201/9780367486082-1

The process of texture classification consists of two main phases: the phase of learning and recognition. In the first phase, the objective is to construct a texture content model for individual texture classes, which usually includes images with defined class names, in the training data [11]. The texture content of the training images is extracted using the feature extraction method, which provides a group of texture features for each image. The characteristics of these aspects are scalar values or discreet that differentiate textual qualities, such as contrast, spatial structure, roughness, orientation, and so on. In this phase, the test sample image texture is first analyzed using the same technique used in the previous step, and then, using a classification algorithm, the extraction features of the test image are compared with the train imagery, and its class is determined. The general flow chart of texture images classification is shown in Figure 1.1.

An important challenge in texture classification is an efficient texture representation, which can be characterized into statistical, structural, signal processing–based, and model-based methods, and it is shown in Figure 1.2. An image might be captured under different geometric and photometric conditions, and so an excellent method for the categorization of the texture should be sufficiently resilient to prevent change in illuminance, rotation, dimension, and geometry of the surface. In recent years, texture classification approaches, such as texture descriptor–based methods and deep learning–based methods, govern current research. In texture-based approaches, the methods based on locally encoded image features have recently become popular for texture classification tasks, particularly in the existence of large intra-class variation due to changes in illumination, scale, and viewpoint. Deep learning–based approaches are well-known in various applications because of their high performance, however at the increased cost of time complexity, computing power, and data size. These methodologies are also biased by the training data. The traditional learning descriptors are also faced with drawbacks, such as dependency on the training database and its size. But the design aspect of the hand design–based local feature descriptors is very simple and gives very prominent results in various computer vision applications. Some of the key advantages of these descriptors are: (a) they are not based on a database, (b) they have less computational complexity, and (c) lower dimensional descriptors can enhance the time efficiency significantly.

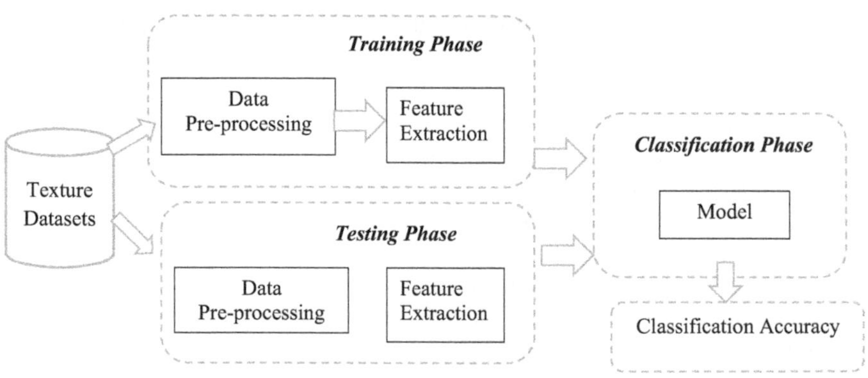

FIGURE 1.1 Block diagram of texture classification.

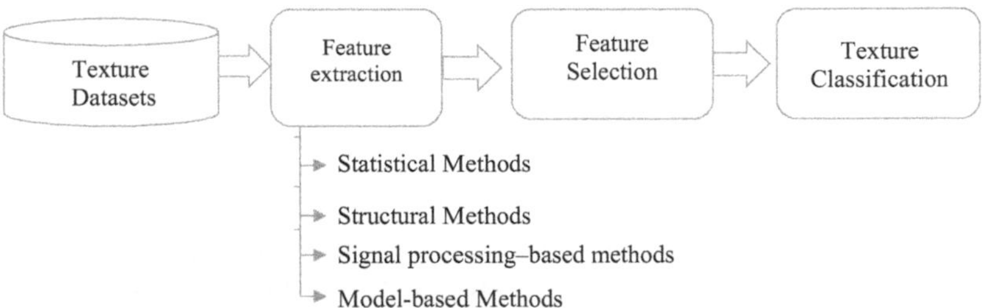

FIGURE 1.2 Different approaches to extract the texture measures for texture representation.

Statistical methodology explores the spatial distribution of gray values, the local features at each pixel position are computed, and statistics are extracted from local feature distributions [12]. In 1994, T. Ojala developed a texture descriptor—local binary pattern (LBP)—for texture classification [13]. Since then, researchers have come up with so many local binary feature extraction approaches for texture classification. In this study, we present a performance analysis of various traditional [14] as well as recent trends, methods, and resources for texture recognition.

1.2 TEXTURE METHODS

Even though deep learning methodologies have gained significant popularity in various image recognition applications in recent years, handcrafted descriptors still have much prominence in this field, generally when there is insufficient data or more computing resources are needed to train a CNN model. In a recent research, it became evident that the use of texture descriptors like LBP might be greater than in the CNN models [15]; in some circumstances, depending on the application, such descriptors can be useful. Hence, we explore many texture descriptors in this section which we consider the most significant for texture classification [16, 17].

1.2.1 Local Binary Pattern (LBP)

LBP (Ojala et al., 1994) is a very simple, robust, and efficient texture descriptor that helps represent images in various real-time applications. LBP is computed by considering the local neighboring intensity values of a pixel. LBP computation is described as follows. Let the intensity value at pixels position (x_c,y_c) be represented by pi_c and the pixel intensities in the 3×3 neighborhood of (x_c,y_c) be represented by pi_n ($n = 0$ to 7) [5]. Then the LBP equivalent of the pixel at (x_c,y_c) is given by:

$$LBP(x_C, y_c) = \sum_{n=0}^{7} T\left(pl_{n-}pl_c\right)2^n \tag{1}$$

$$\text{where, } T\left(v\right) = \begin{cases} 1, if\ v \geq 0 \\ 0, otherwise \end{cases}$$

The computation is pictorially represented in Figure 1.3.

1.2.2 Local Directional Pattern (LDP)

The major issue of LBP is that it suffers from noise, as it depends on the intensity of neighborhood pixels. So a new descriptor, LDP, was introduced by Jabid et al. (2010) [**18**]. Application areas of LDP are facial expression recognition [19], signature verification [20], textural classification [21], etc. LDP is an improved version of LBP descriptor which incorporates a directional component by using Kirsch compass kernels, as shown in Figure 1.4. Here, Kirsch masks are applied on the 3 × 3 neighborhood. The corresponding eight edge responses of Kirsch mask are represented as KM_n in eight different orientations, where *n* varies from 0 to 7.

Applying eight masks, we get eight edge response values, each one representing the edge implication in its respective direction. The response values are not equally important in all directions. The presence of corner or edge shows high response values in particular directions. For each Kirsch mask (KM), corresponding responses are marked

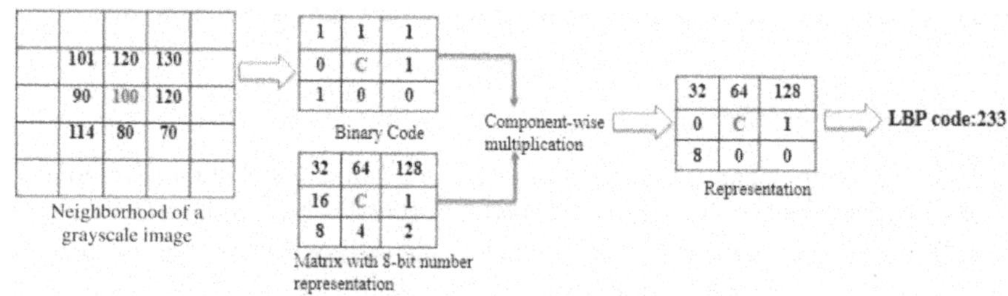

FIGURE 1.3 LBP representation of an image.

East(KM0)

-3	-3	5
-3	0	5
-3	-3	3

North-East(KM1)

-3	5	5
-3	0	5
-3	-3	-3

North(KM3)

5	5	5
-3	0	-3
-3	-3	-3

NorthWest(KM4)

5	5	-3
5	0	-3
-3	-3	-3

West(KM5)

5	-3	-3
5	0	-3
5	-3	-3

SouthWest(KM6)

3	-3	-3
5	0	-3
5	5	-3

South(KM7)

-3	-3	-3
-3	0	-3
5	5	5

SouthEast(KM8)

-3	-3	-3
-3	0	5
-3	5	5

FIGURE 1.4 Kirsch masks.

from M0 to M7, respectively. From the eight edge responses, the most prominent "k" responses are marked as "1"; otherwise, "0." The magnitude of each response is considered. Here, "k" is taken as 3. Then the LDP value for the pixel (x_c, y_c) is given by:

$$LDP_j\left(x_c; y_c\right) = \sum\nolimits_{n=0}^{7} T(KM_n - KM_k).\ 2^n \tag{2}$$

$$\text{where, } T\left(v\right) = \begin{cases} 1, \textit{if } v \geq 0 \\ 0, \textit{otherwise} \end{cases}$$

The LDP computation is pictorially represented in Figure 1.5.

1.2.3 Angular Local Directional Pattern

Though LDP considers the direction element, one of its drawbacks is the selection of "k." Also, while computing LDP, center pixel is not considered. In many of the computer vision applications, center pixel is very important. To address these issues, Abuobayda M. M. et al. have developed angular local directional pattern (ALDP), which is a modified version of LDP. Here also, like LDP, Kirsch masks are used to obtain the edge responses, and only the magnitude of the responses is recorded. The computation of the angular responses in 0°, 45°, 90°, and 135° direction is shown in Table 1.1. Here, *R0*, *R1*, . . . *R7* represent the angular responses of

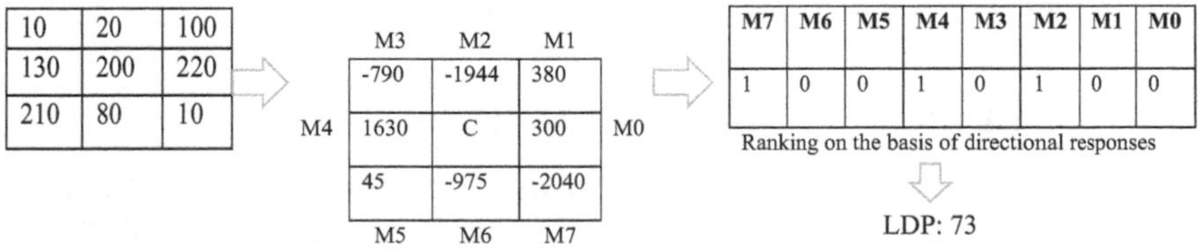

FIGURE 1.5 LDP representation of an image.

TABLE 1.1
Angular Responses of the Eight Neighborhoods of an Image

Angle	Angular Reponses
	R0 = b (M0—C, M4—C)
00	R1 = b (M3—M2,M1—M2)
	R2 = b (M7—M6, M5—M6)
450	R6 = b (M5—C, M1—C)
	R3 = b (M7—M0, M1—M0)
900	R4 = b (M6—C, M2—C)
	R5 = b (M3—M4, M5—M4)
1,350	R7 = b (M3—C, M7—C)

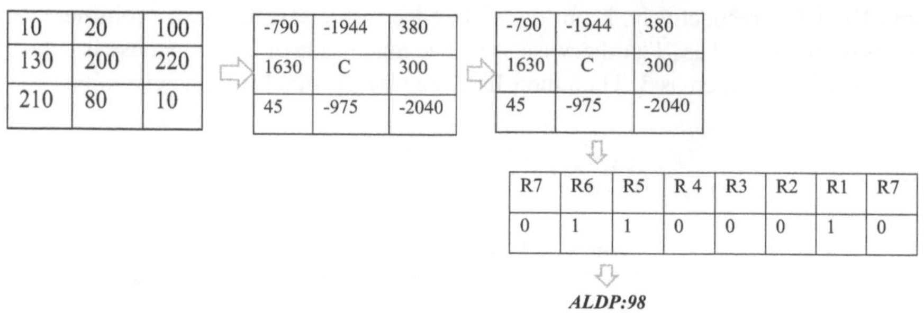

FIGURE 1.6 ALDP representation of an image.

the eight neighborhood positions. The center pixel intensity value "C" (taken from Figure 1.1) is taken as the threshold value for the two neighboring pixels. If the threshold value goes higher than the value of both neighbors, binary code is given the value 1; otherwise, 0 [22].

$$\text{i.e, } R_i = b(i,j)$$

$$\text{where, } \mathrm{b}(i,j) = \begin{cases} 1, \, if \, (i,j) \geq 0 \\ 0, \, otherwise \end{cases}$$

The ALDP computation is pictorially represented in Figure 1.6.

1.2.4 Local Directional Number Pattern (LDNP)

Rivera et al. have introduced a new local binary descriptor known as LDN [23]. It is a 6-bit binary code which is computationally simple. Initially, like LDP, "Kirsch mask" is applied to get the eight edge responses, and only the magnitude of the responses is recorded. The computation of LDN code is obtained by taking the most positive and the least negative three significant bits. The order of the responses is marked on the top right corner of each cell, which is evident in Figure 1.5. The LDN computation is shown in equation 4. The central pixel is denoted as (i, j), while the most positive and the least negative responses are denoted as P_{ij} and N_{ij}, respectively, in equations 4, 5, and 6. The computation is pictorially shown in Figure 1.7.

$$\mathrm{LDN}_{(i,j)} = 8\mathrm{P}_{ij} + \mathrm{N}_{ij} \tag{4}$$

$$\mathrm{P}_{ij} = \arg\max_{\mathrm{P}} \left\{ \mathrm{S}^{\mathrm{P}}(\mathrm{i,j}) \mid 0 \leq \mathrm{P} \leq 7 \right\} \tag{5}$$

$$\mathrm{N}_{ij} = \arg\min_{\mathrm{N}} \left\{ \mathrm{S}^{\mathrm{N}}(\mathrm{i,j}) \mid 0 \leq \mathrm{N} \leq 7 \right\} \tag{6}$$

1.2.5 Local Optimal Oriented Pattern (LOOP)

Recently, Chakraborti et al. (2018) have introduced a novel binary local pattern descriptor known as LOOP, which overcomes some of the drawbacks of LBP and

−790 7	−1944 6	380 5
1630 0	C	300 4
945 1	−975 2	**−2040** 3

⟹ LDN code: **000011**

FIGURE 1.7 LDN representation of an image.

LDP. This descriptor combines rotation invariance into the main formulation of local binary descriptor, thus overcoming drawbacks. LOOP takes the advantages of these descriptors and prevents the non-linear combination of both. As discussed earlier, like in LDP, Kirsch masks are applied in the eight neighboring pixels "pi_n" (where n varies from 0 to 7) in different directions, and further intensity variations are calculated in those directions. The variable "M_C" indicates the eight edge responses of the Kirsch masks, which correspond to intensity of pixels "pi_n," where n varies from 0 to 7. Each of these pixels is assigned an exponential "w_n" (a digit between 0 and 7) according to the rank of the magnitude of M_n among the eight Kirsch mask outputs. Here, the highest Kirsch response is marked as the highest power (2^7), and so on. The LOOP value for the pixel (x_c,y_c) is given in equation 7, and LOOP computation is pictorially represented in Figure 1.8.

$$\text{LOOP}\left(x_c;\, y_c\right) = \sum\nolimits_{n=0}^{7} T\left(pi_n - pi_c\right).\, 2^{w_n} \tag{7}$$

$$\text{where, } T\left(v\right) = \begin{cases} 1, \textit{if } v \geq 0 \\ 0, \textit{otherwise} \end{cases}$$

Based on the principle of LBP, binary patterns are obtained, and using LDP masks, weights are assigned. So the result indicates that for both, LOOP value is $2^7 + 2^6 + 2^5 + 2^4 + 0 = 240$, which demonstrates the rotation invariance, unlike LBP and LDP [24].

1.2.6 Local Line Directional Neighborhood Pattern (LLDNP)

LBP and variants of LBPs have been proved to be an efficient local descriptor for texture analysis. In LLDNP [25], an image is divided into 13 × 13 overlapped regions, and for each region, sign and magnitude patterns are computed for directions, such as 0°, 15°, 30°, 45°, 60°, 75°, 90°, 105°, 120°, 135°, 150°, and 165°. Therefore, the pixel information in 12 directions provides the directional edge information. The computation of sign and magnitude patterns is done using the following equations (7–9).

$$LLDNP(S_{\theta i}) = \sum\nolimits_{n=0}^{N-2} s\left(pi_n - pi_{n+1}\right)2^n \tag{7}$$

Where $\theta = 0°, 15°, 30°, 45°, 60°, 75°, 90°, 105°, 120°, 135°, 150°,$ and $165°$ in 12 directions, and N is the number of neighbors.

$$LLDNP(M_{\theta i}) = \sum\nolimits_{n=0}^{N-2} s\left(\left|pi_n - pi_{n+1}\right| - \mu\right)2^n \tag{8}$$

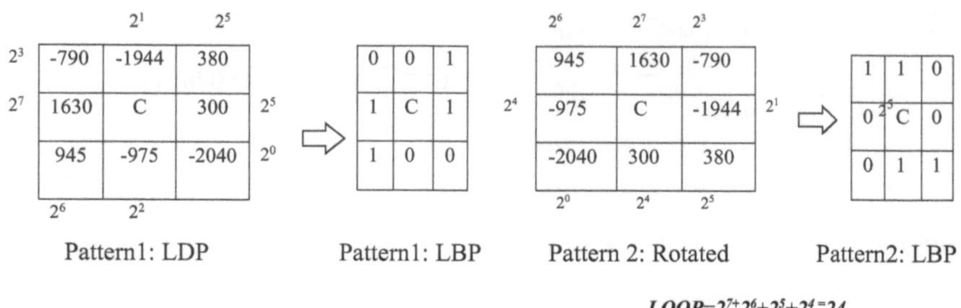

FIGURE 1.8 LOOP representation of an image.

Where $\theta = 0°, 15°, 30°, 45°, 60°, 75°, 90°, 105°, 120°, 135°, 150°,$ and $165°$ in 12 directions, and N indicates the number of neighbors.

$$\mu = \frac{1}{N-1}\sum_{n=0}^{n-2}\left|pi_n - pi_{n+1}\right| \tag{9}$$

Thus, the combination of all the sign and magnitude patterns obtained in 12 directions is represented as $LLDNP(S_{\theta i})$ and $LLDNP(M_{\theta i})$. The equations are shown in the following (10–11).

$$LLDNP(S) = \sum_{i=1}^{i=12} LLDNP(S_{\theta i}) \tag{10}$$

$$LLDNP(M) = \sum_{i=1}^{i=12} LLDNP(M_{\theta i}) \tag{11}$$

Figure 1.9 shows the 13 × 13 pixel region's code generation process for computing $LLDNP(S)$ and $LLDNP(M)$, and Figure 1.10 illustrates the sign and magnitude values obtained from the highlighted pixels considered in those 13 × 13 regions at 13 different directions.

For example, consider 0° direction; the LLDNP sign and magnitude are computed as follows:

LLDNP (S) = {s(138 – 134), s(134 – 130), s(130 – 126), s(126 – 123), s(123 – 123), s(123 – 125), s(125 – 129), s(129 – 134), s(134 – 138), s(138 – 142), s(142 – 145), s(145 – 147)]}
Binary = {1, 1, 1, 1, 1, 0, 0, 0, 0, 0, 0, 0} and decimal = 31

LLDNP (M) = {s|(138 – 134)|–2.83, s|(134 – 130)|–2.83, s|(130 – 126)| –2.83, s|(126 – 123)|–2.83, s|(123 – 123)|–2.83, s|(123 – 125)|–2.83, s|(125 – 129)|–2.83, s|(129 – 134)|–2.83, s|(134 – 138)|–2.83, s|(138 – 142)|–2.83, s|(142 – 145)|–2.83, s|(145 – 147)|–2.83]}
Binary = {1, 1, 1, 1, 0, 0, 1, 1, 1, 1, 1, 0} and decimal = 967

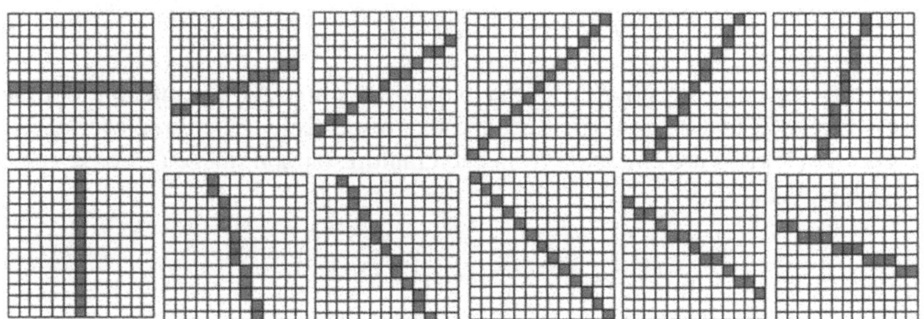

FIGURE 1.9 The code generation process of *LLDNP*(S) and *LLDNP*(M).

Source: [25].

155	152	149	144	138	129	126	125	128	128	124	124	121
152	147	147	143	133	125	122	121	123	123	119	121	122
150	146	143	140	130	125	123	120	121	120	119	123	g_{12} 128
148	143	137	133	128	123	124	127	127	123	g_{10} 124	g_{11} 126	132
144	140	133	127	123	122	128	130	130	g_9 131	129	130	134
143	138	130	126	121	120	126	g_7 128	g_8 133	137	136	137	140
g_0 138	g_1 134	g_2 130	g_3 126	g_4 123	g_5 123	g_6 125	g_7 129	g_8 134	g_9 138	g_{10} 142	g_{11} 145	g_{12} 147
130	125	122	123	g_4 123	g_5 126	129	132	136	141	146	152	152
126	118	118	g_3 121	122	125	133	137	141	146	149	153	155
127	g_1 119	g_2 120	123	124	127	138	145	151	154	154	155	158
g_0 124	120	122	127	130	135	140	149	155	157	158	159	161
122	121	122	126	132	141	144	148	154	158	161	161	161
120	120	118	124	133	143	147	149	154	159	162	162	164

Orientation	Sign	Magnitude
0°	31	967
15°	1561	3523
30°	801	721
45°	2856	848
60°	804	1861
75°	813	2740
90°	1855	2363
105°	191	3098
120°	127	2860
135°	127	2872
150°	127	3885
165°	255	1310

LLDNP_S = 9548

LLDNP_M = 27048

FIGURE 1.10 LLDNP sign and magnitude patterns are calculated for 12 directions.

Source: [25].

1.2.7 Volumetric Local Directional Triplet Patterns (VLDTP)

In most of the traditional 2D approaches, the spatial relationships are encoded by computing the intensity variation between the centers and surrounding pixels. But in VLDTP [26], center pixels are oriented toward their neighbors in three dimensions, in individual directions (0°, 45°, 90°, and 135°). The following circular symmetric Gaussian filter bank is used to generate 3D volume. The equations are shown in the following (12–13).

$$f(x, y, \sigma) = \frac{1}{2\pi\sigma^2} e^{-(x^2 + y^2)/2\sigma^2} \tag{12}$$

$$I_P(\sigma) = g(x, y, \sigma) * I_P(x, y) \tag{13}$$

Where σ = standard deviation, * = convolution operator. As a result, the optimal values of $\sigma 1$, $\sigma 2$, and $\sigma 3$ are 0.5, 0.55, and 0.6, respectively. The three overlapping reference 7 × 7 grids are extracted from each multiresolution image (I_p) of size X × Y as follows (14).

$$I_{Pref(\sigma i)}(a,b) = I_{Pref(\sigma i)}(x+t,\ y+t) \tag{14}$$

Where x = (4, 5, ... X – 3), y = (4, 5, ... Y – 3), i ε (1, 2, 3), t = –3 to 3, a = 1, 2, 3, . . . 7, and b = 1, 2, 3 . . . 7.

In order to obtain the neighbors in direction 0°, 45°, 90°, and 135°, the three multiresolution images must be calculated for the given image. Equations are shown in what follows [15–18].

$$I_P\left(0^0(\mathrm{K})\right) \doteqdot \left[I_{Pref(\sigma 1)}(4,\ \mathrm{x}), I_{Pref(\sigma 2)}(4, 3\mathrm{i}+1), I_{Pref(\sigma 3)}(4,\ \mathrm{x})\right] \tag{15}$$

$$I_P\left(45^0(\mathrm{K})\right) = \left[I_{Pref(\sigma 1)}(8-\mathrm{x},\ \mathrm{x}), I_{Pref(\sigma 2)}\left(3(2-\mathrm{i})+1,\ 3\mathrm{i}+1\right), I_{Pref(\sigma 3)}(8-\mathrm{x},\ \mathrm{x})\right] \tag{16}$$

$$I_P\left(90^0(\mathrm{K})\right) = \left[I_{Pref(\sigma 1)}(\mathrm{x},\ 4), I_{Pref(\sigma 2)}(3\mathrm{i}+1,\ 4), I_{Pref(\sigma 3)}(\mathrm{x},\ 4)\right] \tag{17}$$

$$I_P\left(135^0(\mathrm{K})\right) = \left[I_{Pref(\sigma 1)}(\mathrm{x},\ \mathrm{x}), I_{Pref(\sigma 2)}(3\mathrm{i}+1,\ 3\mathrm{i}+1), I_{Pref(\sigma 3)}(\mathrm{x},\ \mathrm{x})\right] \tag{18}$$

The relationship between a center pixel and its neighbors in all directions is stated as follows in equation 19.

$$I_P\_\left(\theta^0(\mathrm{K})\right) = \left[\left(I_P\_\left(\theta^0(\mathrm{K})\right) - I_P\_\theta^0\left[\frac{K+1}{2}\right]\right]\right. \tag{19}$$

Where **θ** = 0°, 45°, 90°, and 135°.

Thus, three value representation codes for all the direction and the upper LTP and lower LTP are computed using the following equations [20–22].

$$\text{VLDTP}\ (\theta^0) = \left[\left(f\left(I_P\_\left(\theta^0(\mathrm{i})\right) * \mathrm{f}\left(\left(I_P - \theta^0(\mathrm{i})[K-i]\right]\right)\right.\right.\right. \tag{20}$$

$$upper\ LTP = f(x,t) = \begin{cases} 1;\ if\ x \geq t \\ 0;\ else \end{cases} \tag{21}$$

$$lower\ LTP = f(x,t) = \begin{cases} 1;\ if\ x \leq -t \\ 0;\ else \end{cases} \tag{22}$$

$$\text{VLDTP}\ (\theta^0) = \sum\nolimits_{i=1}^{8} (VLDTP_{(upperLTP/lower\ LTP)} * 2^{\mathrm{i}}) \tag{23}$$

The VLDTP code generation process obtained on all the directions is illustrated in Figure 1.11.

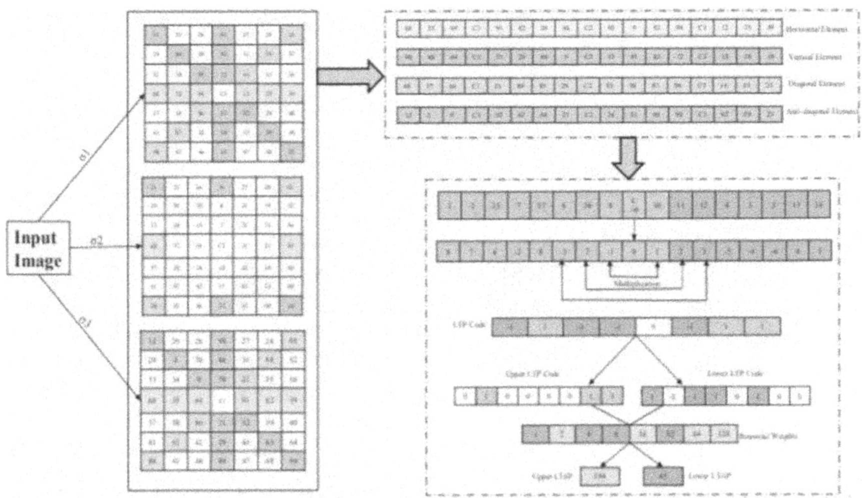

FIGURE 1.11 VLDTP code generation process.

Source: [26].

1.2.8 Local Tri-Directional Patterns (LTriDP)

LTriDP [27] is an improved version of LBP. It explores the relationship of neighboring pixels on the basis of various directions. For a particular radius, each center pixel with their corresponding neighbors is considered. In this method, each pixel's relationship is compared with center pixel and the two most vertical or horizontal adjacent neighbors. The computation of LTriDP is shown in Figure 1.12. Here, the relationship among the neighborhood pixels are compared. Apart from the relationship between center-neighborhood pixels, additional information such as the mutual relationship between adjacent neighbors is also considered, that is, local pattern is obtained based on three direction pixels. Therefore, LTriDP gives more neighborhood relationship information compared to LBP and other traditional methods. Thus, the pattern is obtained from the nearest neighborhood, and also, magnitude patterns are obtained in each pixel. Finally, LTriDP 8-bit pattern and 8-bit magnitude combined to form more efficient feature descriptor. Let's consider pi_c the center pixel and $pi_{1,} pi_{2,} .....pi_{8,}$ the eight-neighbor pixels. Then, compute the intensity variation between the neighbor pixels, that is, each neighbor pixel with their most two adjacent neighbors and the difference between the neighbor pixel and their center pixel. The equation is as follows (24).

$$d1 = p_i - p_{i-1};\ d2 = p_i - p_{i+1};\ d3 = p_i - p_{ic}; \tag{24}$$

The three differences are computes as d1, d2, and d3, and a new pattern is given to three differences, and for each neighborhood pixel "pi = 1, 2, . . . 8," pattern value f (d1, d2, d3) is computed using equation (25), and tri-directional pattern has been obtained using equation (26).

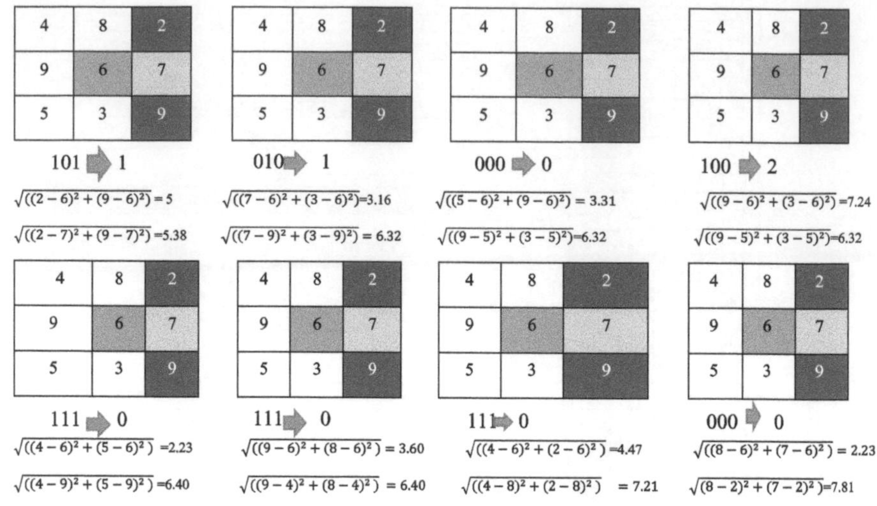

Pattern 1: 11000000; Pattern 2:00010000

FIGURE 1.12 LTriDP code generation process.

$$f\left(d1,\ d2,\ d3\right)=\left\{\#\left(dk<0\right)\right\}\ \text{mod}\ 3\,\forall\,k=1,\ 2,\ 3. \tag{25}$$

$$\text{LTriDP}\ (p_{ic})=\left\{f1,\ f2...\ f8\right\} \tag{26}$$

So a ternary pattern for each center pixel is obtained and converted into two binary patterns.

$$\text{LTriDP}\left(\text{p}_{\text{ic}}\right)\ \text{i}=1,\ 2...7\ \sum\nolimits_{n=0}^{7} 2^{n}\times\text{LTriDPi}\left(p_{ic}\right)\left(\text{n}+1\right) \tag{27}$$

Magnitude pattern is obtained for center pixel, neighborhood pixel, and the two most adjacent pixel using the following equations (28–33).

$$\text{M1}=\sqrt{[(p_{i-1}-p_{ic})^2+\left(p_{i+1}-p_{ic}\right)^2} \tag{28}$$

$$\text{M2}=\sqrt{[(p_{i-1}-p_{ic})^2+\left(p_{i+1}-p_{i}\right)^2},\ \forall\ \text{i}=2,3..,7 \tag{29}$$

$$\text{M1}=\sqrt{[(p_{i8}-p_{ic})^2+\left(p_{i+1}-p_{ic}\right)^2} \tag{30}$$

$$\text{M2}=\sqrt{[(p_{i8-1}-p_{i})^2+\left(p_{i+1}-p_{i}\right)^2},\ \ \forall\ \text{i}=1 \tag{31}$$

$$\text{M1}=\sqrt{[(p_{i-1}-p_{ic})^2+\left(p_{i+1}-p_{ic}\right)^2} \tag{32}$$

$$M2 = \sqrt{[(p_{i-1} - p_i)^2 + (p_{i-1} - p_i)^2}, \ \forall i = 8 \tag{33}$$

Thus, M1 and M2 are computed for each neighborhood pixel, and finally, a magnitude pattern is obtained using the following equations (34–36).

$$Magnitude_i\left(M1, M2\right) = \begin{cases} 1;\ if\ M1 \geq M2 \\ 0; else \end{cases} \tag{34}$$

$$LTriDP_{Magnitude}(p_{ic}) = \left\{Magnitude_1, Magnitude_2 \ldots\ldots Magnitude_8\right\} \tag{35}$$

$$LTriDP_{Magnitude}(p_{ic})_{\text{Magnitude}} = \sum\nolimits_{n=0}^{7} 2^{n.} * LTriDP_{Magnitude}(p_{ic}) \tag{36}$$

LTriDP computation is shown in Figure 1.12. In the figure, center pixel is represented as p_{ic}, and neighborhood pixels $p_{i1,}$ $p_{i2,}$ . . . p_{i7} are shown. The center pixel is marked in blue color, and the first neighbor element is marked in green color, and the corresponding adjacent neighbors are marked in red color. Initially, compare the adjacent pixels and first neighbor pixel with the center pixel and assign either 0 or 1 for all the three values. For example, the first neighborhood pixel is 7, and compare with 2, 9, 6($7 > 2 = 1$; $7 < 9 = 0$; $7 > 6 = 1$) and the pattern 101 is obtained. Likewise, the pattern is computed for all the pixels. Then compute the magnitude of the center pixel with adjacent neighbors and magnitude of 7 with the adjacent neighbors.

Thus, 5 and 5.38 are obtained, and the magnitude of center pixel is less than the magnitude of the neighborhood pixel. The binary pattern 0 is marked; otherwise, 1. Likewise, compute for all the neighbors and LTriDP patterns, and magnitude pattern is obtained.

1.2.9 Local Neighborhood Intensity Pattern (LNIP)

Furthermore, the majority of the local patters, including LBP, focus primarily on the sign information and neglect the magnitude information. Magnitude information provides the additional information of the texture descriptor, which is included in this method by taking into account the mean of absolute deviation from its neighbors. Taking this into consideration, a new texture descriptor, called local neighborhood intensity pattern (LNIP) [28], has developed, which generates a sign and magnitude pattern based on the relative intensity difference between a specific pixel and the center pixel by considering its immediate neighbors. Finally, the sign and magnitude patterns are combined to form a single feature descriptor, yielding a more effective feature descriptor. Each of the eight neighborhood pixels (I_i) in a 3 × 3 region has four adjacent neighbors if i = an odd number, and for two adjacent numbers if i = an even one. It is illustrated in Figure 1.10. Indicate the set of adjacent neighboring pixels w.r.t (I_i) as (S_i). Thus, the number of neighbors in (S_i) is 4 and 2 for i = odd and i = even, respectively. The value corresponding to central pixel (I_c) is computed using equations (37–38).

$$B\left(I_{i\_}I_c\right) = \begin{cases} 1;\ \#D_I = 1 >_2^1 M \\ 0;\ Otherwise \end{cases} \tag{37}$$

$$LNIP_S(\mathrm{I}_C) = \sum\nolimits_{I=1}^{8} B(I_{i-}I_c) * 2^{I-1} \tag{38}$$

Then next is to compute the magnitude pattern using the following equations (39–40).

$$M_i = \frac{1}{M}\sum\nolimits_{k=1}^{M} |S_i(k) - I_i| \tag{39}$$

$$T_c = \frac{1}{8}\sum\nolimits_{k=1}^{8} |I_c - I_i| \tag{40}$$

Where M_i is the mean deviation and T_c is the threshold value, and the magnitude pattern is computed using the following equations (41–42).

$$\mathrm{M}(I_{i-}T_c) = \mathrm{Sign}\ (M_i, T_c) \tag{41}$$

$$LNIP_S(\mathrm{I}_C) = \sum\nolimits_{I=1}^{8} 2^{I-1} * M(I_{i-}T_c) \tag{42}$$

The computation of LNIP is shown in Figure 1.13.

In LNIP, if the concerned pixel is an odd position, then four adjacent neighbors are considered; else, three are considered. The concerned pixel is indicated in green

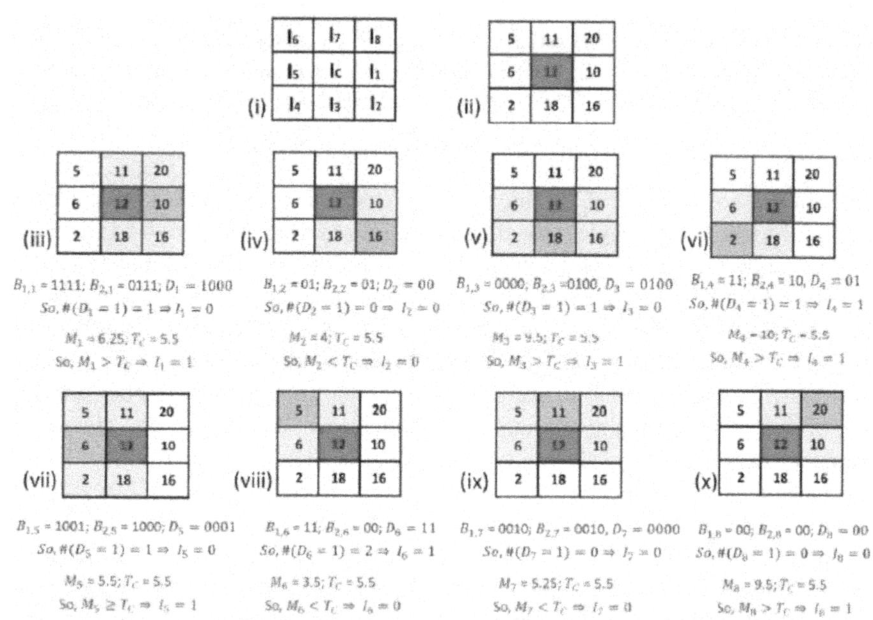

FIGURE 1.13 LNIP representation of an image.

Source: [28].

color, and their surrounding pixels are marked in yellow color; the center pixel is marked green. Initially, compare the center pixel (red) and its corresponding four neighbors (yellow), and if the center value is greater than the surrounding pixels, assign binary as 1, else 0, and compute for the other three ([11 < 12; 20 > 12; 16 > 12; 18 > 12]-binary code B_{11} = 0111). Then, compare the concerned pixel (green) and its corresponding four neighbors (yellow) and assign binary 0 or 1 ([11 > 10; 10 < 20; 10 < 16; 16 < 10)-binary code B_{21} = 1111). An XOR operator is used between these two binary codes, and one 4-bit binary code obtained D1 is 1. The same way, compute for all the neighbors. Then, finally, an 8-bit LNIP sign bit is indicated. Magnitude pattern is computed in all the pixels, and the threshold value T_c is 5.5. The magnitude value is greater than the threshold value. Put binary value as 1; else, 0. Similarly, compute for all the neighbors. Thus obtained another 8-bit LNIP magnitude pattern.

1.2.10 Local Triangular Coded Pattern (LTCP)

Local binary descriptors rely primarily on the intensity difference between neighboring pixels with respect to the center pixel of a region to form the representative value at a particular pixel position. LTCP [3] exploits the relationship between the triangular neighbor pixels in a region to compute the pattern. A set of neighbors in triangular pattern is chosen for comparing with the center pixel instead of choosing one by one with the middle pixel. It doesn't depend on the center pixel (center pixel is not fixed same in all pixel locations). LTCP is made up of 8-bit binary code (256 grayscale values). For each direction, there are four components: three neighbors and one center pixel (threshold). A binary code is generated for that three pixels, instead of replacing one by one. A binary code is generated as 1 if threshold value is greater than any of the two neighboring pixels; otherwise, 0 is generated. The equations are shown in the following (43–46).

$$Cp_0 = \text{binary}\ (\max\ [p_7 - p_c], [p_0 - p_c], [p_1 - p_c]) \tag{43}$$

$$Cp_1 = \text{binary}\ (\max\ [p_1 - p_c], [p_2 - p_c], [p_3 - p_c]) \tag{44}$$

$$Cp_2 = \text{binary}\ (\max\ [p_3 - p_c], [p_4 - p_c], [p_5 - p_c]) \tag{45}$$

$$Cp_3 = \text{binary}\ (\max\ [p_5 - p_c], [p_6 - p_c], [p_7 - p_c]) \tag{46}$$

After computing in four directions, the triangles in each directions are flipped by 180°. The same process is repeated in four directions, and the equations are shown here (47–50).

$$Cp_4 = \text{binary}\ (\max\ [P_6 - p_c], [p_c - p_0], [p_2 - p_0]) \tag{47}$$

$$Cp_5 = \text{binary}\ (\max\ [P_4 - p_c], [p_c\ \ p_2], [p_0 - p_2]) \tag{48}$$

$$Cp_6 = \text{binary}\ (\max\ [P_6 - p_c], [p_c - p_4], [p_2 - p_4]) \tag{49}$$

$$Cp_7 = \text{binary}\ (\max\ [P_4 - p_c], [p_c - p_6], [p_0 - p_6]) \tag{50}$$

Finally, the eight coded pattern values are obtained from the computed using the following equations (51–52):

$$Cp_i = \sum_{i=0}^{n} b(p_i - pc) \geq (n+1)/2 \tag{51}$$

$$LTCP(x_C, y_c) = \sum_{n=0}^{7} T(\text{Cpi} - pc)2^n \tag{52}$$

where, $T(v) = \begin{cases} 1, \text{ if } v \geq 0 \\ 0, \text{ otherwise} \end{cases}$

The code generation of LTCP is shown in Figure 1.14.

Initially, the first triangle in the eastern region is considered, and Cp_0 is the vector which consists of the four components p_7, p_0, p_1, and pc. Here, the center value is fixed as the threshold for the other three values. If any of the two neighborhood pixels are greater than the threshold value, then assign the binary code as 1; otherwise, 0. The same way other three coded patterns are computed. Then each triangular region is rotated by 180^0; p0 is the threshold value, and the other three components are p6, pc, and p2. Likewise, the other three patterns are computed. Thus, LTCP eight coded patterns are obtained.

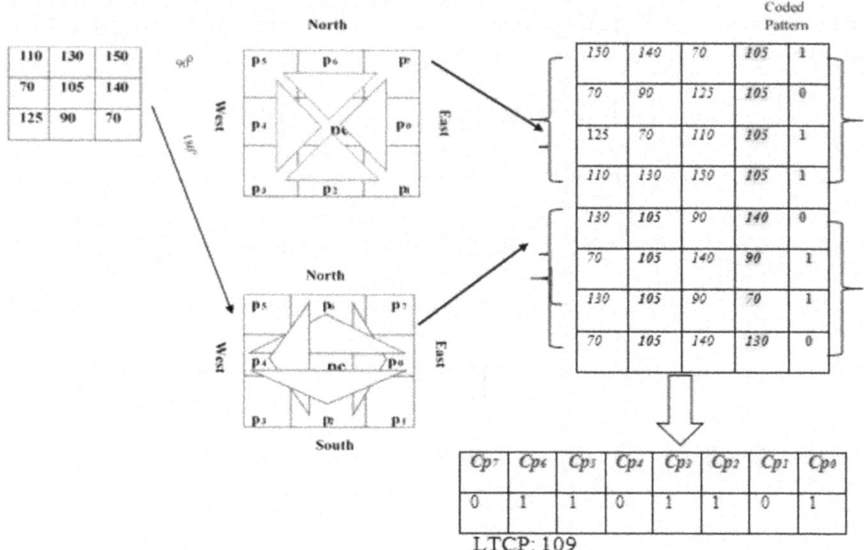

FIGURE 1.14 LTCP code generation process.

1.3 EXPERIMENTAL ANALYSIS AND RESULTS ON LOCAL DESCRIPTORS FOR TEXTURE CLASSIFICATION

1.3.1 Benchmark Texture Datasets

Experimental results of the aforementioned descriptors were evaluated using six different benchmarked texture datasets, namely, Brodatz [29], Outex [30], MIT_Vistex [31], KTH_TIPS 2a [32], KTH-TIPS 2b [32], Kylberg [33], Describable Textures Datasets (DTD) [34], and Salzburg (Stex) [34], which are widely used for texture classification. The database properties are shown in Table 1.2.

- **Brodatz.** Brodatz [29] is a popular multipurpose database that includes natural texture images from Brodatz. It comprises 112 texture images, each of which is 640 × 640 pixels in size. Each image was subdivided into 25 sub-images, each sized 128 × 128 pixels. There are 2,800 images in 112 categories, with 25 images in each category. Sample images from different datasets are shown in Figure 1.12.
- **Outex.** A researcher from Oulue University introduced the Outex [30] dataset in 2002. This is one of the most widely used and the largest datasets. It includes both natural and synthetic textures, which are widely used for image recovery, texture categorization, and other purposes. The selected sets are Tc 000010 and Tc 000012, which are the most distinct intercity and rotation. These two suits share the same 24 classes. OutexTc 000010 has just one light ("Inca"), but Outex Tc 000012 has three illuminates ("Horizon," "Inca," and "TL84") [31]. Each of these two suits has nine different angles

TABLE 1.2
Texture Dataset Properties

	Database	Image Rotation	Illumination	Scale Variation	No. of Classes	Image Size	No. of Samples per Class	Total No. of Images
1	Brodatz	✓	×	✓	112	128 × 128	25	2,800
2	Outex_TC10	✓	×	×	24	128 × 128	180	4,320
3	Outex_TC12	✓	✓	×	24	128 × 128	200	4,800
4	MIT_Vistex	✓	✓	×	40	128 × 128	16	640
5	KTH-TIPS	✓	✓	✓	10	200 × 200	81	810
6	KTH-TIPS2b	✓	×	✓	11	200 × 200	432	4,752
7	Kylberg	✓	✓	×	28	576 × 576	160	4,480
8	DTD	✓	✓	✓	47	600 × 600	120	5,640
9	Stex	×	✓	×	476	128 × 128	16	7616

(0°, 5°, 10°, 15°, 30°, 45°, 60°, 75°, 90°) and 20 textures in each one of the rotation angles. All the images are in RAS format.

- **MIT_Vistex.** The MIT Vision and Modelling Group has developed the MIT-Vistex [31] database. The database contains 40 texture images of size 512 × 512 pixels. Each image is subdivided into 16 sub-images of resolution 128 × 128 pixels. There are 640 images (40 individual texture images × 16 sub-images).
- **KTH-TIPS.** The dataset consists of ten different texture categories. Each category contains 81 images of size 200 × 200 and a total of 810 images. The images are in. png format [32].
- **KTH-TIPS2b.** The database comprises of 11 different categories, and each one consists of four samples in different directions. Around 108 images are there in each samples. All are in. png format of size 200 × 200 [32].
- **Kylberg.** The dataset has 28 categories of 160 images each, with grayscale images of different man-made and natural textured surfaces. All the texture images have a size of 576 × 576 resolution [33].
- **Describable Textures Datasets (DTD).** The dataset contains a total of 5,640 texture images. It has 47 categories of texture images, and each category consists of 120 images of resolution 600 × 600 pixels. All are. jpg format [34].
- **Salzburg Texture Database (STex).** It comprises 7,616 images of 476 categories of size 128 × 128. Each category consists of 16 texture images. The database contains different types of textures, like wood, rubber, etc. [35].

1.3.2 Classifiers

To evaluate the efficacy of these aforementioned descriptors, classifiers like random forest [36], artificial neural network [37], and support vector machine [38] are used.

1.3.2.1 Random Forest

Random forest is one among the classifiers used in this study for analyzing the performance of various descriptors. It is an ensemble learning algorithm and consists of several decision trees. Its highlights include the notion of bagging and the unpredictability of characteristics. We must integrate uncorrelated characteristics of trees, whose prediction results are far better than the individual ones, to create each unique tree [37].

1.3.2.2 Artificial Neural Network

ANNs are well-known classifiers that have been around for a decade and are still frequently employed in recent years. The multilayer perceptron (MLP) is a neural network that is the most basic kind of ANN. During the back-propagation phase, the weights are updated in the training phase. Other neural network variants, such as the probabilistic neural network (PNN) [37, 38], have recently been used in texture classification. The convolutional neural network (CNN) is a self-learning feature extractor that uses convolutional input layers. Consequently, it can perform feature extraction as well as classification in the same architecture [39].

1.3.2.3 Support Vector Machine (SVM)

SVM is the popular machine learning algorithm which has commonly been used in many pattern recognition problems in recent years, including texture classification. SVM is designed to maximize the marginal distance between classes with decision boundaries drawn using different kernels [40]. SVM is designed to work with only two classes by determining the hyperplane to divide two classes. This is done by maximizing the margin from the hyperplane to the two classes. The samples closest to the margin that were selected to determine the hyperplane are known as support vectors [41].

1.3.2.4 Naive Bayes

Another one is Naive Bayes used in this study. One of the statistical approaches for classification is Bayesian theory. Here, a cost function is used, whether there are members of class A incorrectly attributed to class B. Different classes are addressed in this way, each in the form of a probability hypothesis. Each subsequent samples is either increased or decreased from the previous ones, and lastly, the most likely assumptions are grouped together as a class. Thus, error can be minimized [40].

1.3.3 Evaluation Metrics

Evaluation metrics such as classification accuracy and confusion matrix are generated to determine classification effectiveness. Classification accuracy is evaluated, computed using equation (53).

$$\text{Classification Accuracy} = \frac{\text{total no: of correct classification}}{\textit{total}\ \text{no: of samples}} * \mathbf{100} \tag{53}$$

Cross-validation technique is applied on these descriptors with different classifiers to validate the model. The performance is analyzed using five-fold cross-validation. Confusion matrix is used in classification algorithms for summarizing performance and is applied on these descriptors by using different classifiers for summarizing model performance.

1.3.4 Result Analysis

The performance analysis of several descriptors with different classifiers on different texture datasets is shown in the following. Around 70% of the data is used for training, 15% for testing, and 15% for validation purposes. Initially, the features are extracted using different feature extraction methods separately for different datasets. For analyzing the strength of these local descriptors, classifiers such as random forest, ANN (with different activation functions, such sigmoid and tanh), SVM (different polynomial kernels, such as linear and polynomial), and Naive Bayes are used. Ten-fold cross-validation method is applied on all the datasets. Confusion matrix is applied by using different classifiers for summarizing the model performance. The classification accuracy is computed for all the descriptors with the different classifiers on different datasets, as shown in the following table [1–9]. The feature descriptors, such as LBP, LDP, LDNP, ALDP, LOOP, LLDNP, VLDTP, LTriDP, LNIP, and LTCP, use nine texture datasets using classifiers such as random forest, ANN with sigmoid and tanh activation functions, SVM with linear and polynomial kernel functions, and Naive Bayes. The descriptors that give results in all the databases are ALDP, LLDNP, LDRP, and LTCP. The results indicate that all the descriptors, such as ALDP, LLDNP, LDRP, and LTCP, with the combination of random forest, gives the better recognition rate of 98.07%, 99.12%, 98.88%, and 98.17% on the Kylberg dataset. The performance comparison of all the ten descriptors with different datasets and their consolidated results of recognition rate obtained on Brodatz, Outex, MIT_Vistex, KTH_TIPS 2a, KTH-TIPS 2b, Kylberg, DTD, and Stex are shown in Tables 1.3–1.11.

1.4 DISCUSSION

In this chapter, a detailed review and analysis is performed on several local feature extraction methods, namely, LBP, LDP, LDNP, ALDP, LOOP, LLDNP, VLDTP, LTriDP, LNIP, and LTCP, which are commonly used for texture classification. The following are the observations we made while doing the analysis:

- Texture datasets such as on Brodatz, Outex, MIT_Vistex, KTH_TIPS 2a, KTH-TIPS 2b, Kylberg, DTD, and Stex are used for analyzing the classification tasks. Evaluation of these descriptors with different classifiers like random forest, ANN with sigmoid and tanh activation functions, SVM with linear and polynomial kernel functions, and Naive Bayes is done.
- The performance analysis of each descriptor using random forest, ANN with sigmoid and tanh activation functions, SVM with linear and polynomial kernel functions, and Naive Bayes with ten-fold cross-validation is done for better accuracy.

TABLE 1.3
Classification Results of Various Descriptors on Brodatz

Classifiers / Descriptors		ANN		SVM		
	RBF	Sigmoid	Tanh	Linear	Polynomial	Naive Bayes
LBP [5]	87.00	82.88	85.80	82.89	85.66	84.00
LDP [18]	89.66	82.34	85.68	87.66	89.40	84.12
LDNP [23]	87.78	83.89	85.56	80.56	82.11	84.00
ALDP [22]	93.00	84.56	87.45	82.17	88.99	85.12
LOOP [24]	92.99	81.11	88.00	84.00	87.45	86.22
LLDNP [25]	95.12	89.00	94.78	88.90	92.00	90.12
VLDTP [26]	95.00	82.50	89.00	79.80	82.44	88.90
LTriDP [27]	95.22	88.90	86.77	79.88	89.66	84.20
LNIP [28]	93.44	84.44	89.32	81.92	84.48	88.74
LMP [42]	95.98	88.00	91.33	88.16	90.88	90.62
LJP [43]	94.36	89.22	89.00	83.18	87.62	87.42
LDRP [44]	91.84	86.88	88.16	84.48	87.00	85.22
LTCP [3]	92.82	88.44	90.00	85.66	87.22	88.66

TABLE 1.4
Classification Results of Various Descriptors on Outex_TC10

Descriptors \ Classifiers	RBF	ANN		SVM		Naive Bayes
		Sigmoid	Tanh	Linear	Polynomial	
LBP [5]	87.00	79.18	88.12	86.00	88.66	85.52
LDP [18]	93.66	80.32	88.32	85.22	91.12	89.24
LDNP [23]	85.78	80.90	82.12	81.34	84.66	83.00
ALDP [22]	93.10	85.26	88.90	83.88	91.72	89.44
LOOP [24]	91.82	84.00	86.26	82.52	90.61	88.82
LLDNP [25]	96.12	86.00	90.48	87.42	94.33	92.66
VLDTP [26]	94.00	84.66	89.66	88.40	93.56	90.11
LTriDP [27]	96.15	86.30	92.49	94.92	95.81	93.88
LNIP [28]	93.02	86.83	90.34	89.33	93.00	91.96
LMP [42]	94.16	84.12	92.88	90.00	94.00	93.18
LJP [43]	92.88	85.62	89.50	87.26	90.19	89.33
LDRP [44]	94.90	86.00	90.00	88.48	93.00	92.41
LTCP [3]	94.12	85.40	88.66	89.74	92.81	90.44

TABLE 1.5
Classification Results of Various Descriptors on Outex_TC12

Descriptors \ Classifiers	RBF	ANN		SVM		Naive Bayes
		Sigmoid	Tanh	Linear	Polynomial	
LBP [5]	88.30	77.00	85.66	83.10	88.66	85.20
LDP [18]	90.22	78.29	84.50	84.46	89.00	85.56
LDNP [23]	84.12	79.67	83.33	79.38	84.24	80.12
ALDP [22]	92.90	83.78	87.21	85.10	89.33	87.90
LOOP [24]	90.22	82.50	86.00	83.02	89.59	85.16
LLDNP [25]	94.32	85.33	89.56	85.28	90.62	88.52
VLDTP [26]	92.66	84.36	87.42	86.44	90.74	87.64
LTriDP [27]	93.50	84.90	89.66	88.06	92.10	87.08
LNIP [28]	91.12	83.13	88.19	86.13	89.57	87.58
LMP [42]	92.60	82.50	90.55	87.26	90.88	86.00
LJP [43]	90.58	83.33	86.17	87.55	91.21	84.20
LDRP [44]	94.90	85.00	89.02	88.38	91.50	88.72
LTCP [3]	93.66	83.24	87.18	87.26	91.24	87.88

TABLE 1.6
Classification Results of Various Descriptors on MIT_Vistex

Descriptors \ Classifiers	RBF	ANN		SVM		Naive Bayes
		Sigmoid	Tanh	Linear	Polynomial	
LBP [5]	94.00	85.12	88.10	86.00	91.16	90.00
LDP [18]	94.10	87.55	90.28	88.10	93.22	90.12
LDNP [23]	90.22	85.62	87.46	87.82	89.16	86.00
ALDP [22]	96.18	88.70	92.55	90.47	94.59	90.12
LOOP [24]	95.62	87.19	91.70	90.66	93.00	91.22
LLDNP [25]	98.48	90.20	95.82	93.52	97.32	93.12
VLDTP [26]	96.15	89.42	93.50	91.37	94.48	90.90
LTriDP [27]	97.00	89.66	94.51	92.41	95.61	93.20
LNIP [28]	96.28	87.48	93.22	90.17	94.73	93.74
LMP [42]	97.33	87.73	94.64	91.44	96.65	94.62
LJP [43]	97.90	86.99	94.28	92.09	96.22	94.42
LDRP [44]	98.72	88.14	96.72	93.82	97.01	95.22
LTCP [3]	97.66	86.32	95.90	91.22	95.66	94.66

TABLE 1.7
Classification Results of Various Descriptors on KTH_TIPS

Descriptors \ Classifiers	RBF	ANN		SVM		Naive Bayes
		Sigmoid	Tanh	Linear	Polynomial	
LBP [5]	88.12	83.50	85.10	84.40	88.44	84.30
LDP [18]	90.16	84.15	87.28	85.11	89.20	86.52
LDNP [23]	87.72	82.22	85.37	84.24	86.32	84.44
ALDP [22]	96.44	84.00	88.15	87.62	94.88	91.19
LOOP [24]	95.50	83.62	87.00	86.10	94.02	90.76
LLDNP [25]	99.10	87.19	90.82	89.55	98.62	95.55
VLDTP [26]	95.62	85.76	87.55	86.72	93.19	92.67
LTriDP [27]	97.68	89.92	89.09	88.60	96.72	95.90
LNIP [28]	95.44	86.66	88.21	87.77	95.50	94.33
LMP [42]	96.02	85.18	89.52	86.16	95.44	94.82
LJP [43]	96.92	86.40	88.66	88.28	95.63	93.66
LDRP [44]	98.14	87.22	90.52	89.00	96.17	94.50
LTCP [3]	97.00	86.66	90.30	87.99	96.54	93.17

TABLE 1.8
Classification Results of Various Descriptors on KTH_TIPS2b

Classifiers / Descriptors	RBF	ANN		SVM		Naive Bayes
		Sigmoid	Tanh	Linear	Polynomial	
LBP [5]	85.44	80.50	82.00	82.50	86.34	84.22
LDP [18]	88.33	81.33	83.06	84.11	87.65	83.45
LDNP [23]	86.44	83.78	82.57	82.62	84.89.	82.65
ALDP [22]	93.76	85.52	87.15	85.33	90.10	88.58
LOOP [24]	91.18	84.18	86.54	85.75	89.05	86.90
LLDNP [25]	96.62	86.90	90.76	87.88	96.11	93.62
VLDTP [26]	93.33	83.26	88.15	85.40	92.88	90.76
LTriDP [27]	95.24	85.92	90.50	87.12	95.56	93.88
LNIP [28]	92.55	85.44	87.05	84.50	93.50	91.32
LMP [42]	91.70	84.00	88.22	85.90	94.90	90.10
LJP [43]	90.52	84.06	87.11	86.33	93.17	91.04
LDRP [44]	96.90	86.00	90.66	86.56	94.34	92.90
LTCP [3]	95.66	84.62	89.00	85.10	93.66	91.23

TABLE 1.9
Classification Results of Various Descriptors on Kylberg

Classifiers / Descriptors	RBF	ANN		SVM		Naive Bayes
		Sigmoid	Tanh	Linear	Polynomial	
LBP [5]	87.67	80.50	85.18	85.00	87.28	86.17
LDP [18]	91.23	82.99	87.56	89.56	90.36	89.45
LDNP [23]	90.60	83.78	85.67	85.44	89.45	87.65
ALDP [22]	98.17	90.52	92.15	91.32	96.77	94.58
LOOP [24]	96.78	90.18	92.84	92.99	95.10	93.90
LLDNP [25]	99.12	91.90	93.00	92.12	98.50	96.66
VLDTP [26]	96.00	88.50	90.77	88.76	95.45	93.76
LTriDP [27]	98.74	90.32	92.66	91.00	97.62	94.88
LNIP [28]	96.14	87.24	90.17	89.82	95.88	93.32
LMP [42]	97.40	89.99	91.56	88.90	96.92	94.10
LJP [43]	96.72	90.56	92.64	90.33	95.42	93.04
LDRP [44]	98.88	91.00	92.55	92.56	97.01	96.90
LTCP [3]	98.17	90.67	92.10	93.60	97.82	96.00

TABLE 1.10
Classification Results of Various Descriptors on DTD

Classifiers / Descriptors	RBF	ANN		SVM		Naive Bayes
		Sigmoid	Tanh	Linear	Polynomial	
LBP [5]	90.00	85.22	88.60	84.22	89.10	88.66
LDP [18]	93.44	87.80	91.34	83.45	90.28	90.58
LDNP [23]	89.50	85.99	84.72	82.65	87.46	84.22
ALDP [22]	96.66	90.37	91.88	88.58	92.55	91.62
LOOP [24]	95.17	87.66	90.12	86.90	91.70	90.55
LLDNP [25]	96.92	90.12	95.72	93.62	96.12	95.67
VLDTP [26]	93.20	87.22	93.66	90.76	93.50	92.90
LTriDP [27]	96.78	89.12	95.33	93.88	94.51	95.22
LNIP [28]	92.66	88.50	94.50	91.32	93.22	94.32
LMP [42]	94.72	89.57	94.10	90.10	94.64	94.66
LJP [43]	95.56	88.82	93.19	91.04	94.28	93.17
LDRP [44]	96.45	90.66	95.00	92.90	95.72	94.32
LTCP [3]	96.10	90.00	94.72	91.23	95.90	95.00

TABLE 1.11
Classification Results of Various Descriptors on STex

Classifiers / Descriptors	RBF	ANN		SVM		Naive Bayes
		Sigmoid	Tanh	Linear	Polynomial	
LBP [5]	85.18	77.30	82.45	82.11	85.32	Linear
LDP [18]	87.56	78.69	84.21	87.33	89.50	82.89
LDNP [23]	85.67	79.55	82.66	80.72	82.30	87.66
ALDP [22]	92.15	83.29	85.50	82.66	88.55	80.56
LOOP [24]	92.84	82.13	85.12	84.10	87.44	82.17
LLDNP [25]	93.00	85.98	87.90	88.88	88.11	84.00
VLDTP [26]	90.77	84.33	85.33	79.56	82.32	88.90
LTriDP [27]	92.66	84.22	87.66	79.30	89.72	79.80
LNIP [28]	90.17	83.02	84.80	81.00	84.20	79.88
LMP [42]	91.56	82.66	85.62	88.56	90.00	81.92
LJP [43]	92.64	83.72	86.33	83.20	87.10	88.16
LDRP [44]	92.55	85.96	86.92	84.40	87.32	83.18
LTCP [3]	92.10	84.88	85.90	85.33	87.88	84.48

- The results indicate that all the descriptors, namely, ALDP, LLDNP, LDRP, and LTCP, with the combination of random forest, give the better recognition rate of 98.07%, 99.12%, 98.88%, and 98.17% on Kylberg dataset, respectively.

1.5 CONCLUSION

Here, a systematic analysis of several local binary feature descriptors is done. For classification, the performance of ten local binaries, namely, LBP, LDP, LDNP, ALDP, LOOP, LLDNP, VLDTP, LTriDP, LNIP, and LTCP, is evaluated using four classifiers, namely, random forest, ANN with sigmoid and tanh activation functions, SVM with linear and polynomial kernel functions, and Naive Bayes. These descriptors are applied on nine texture datasets—Brodatz, Outex, MIT_Vistex, KTH_TIPS2a, KTH-TIPS 2b, Kylberg, DTD, and Stex—ensuring identical experimental settings. From the results, it is clearly evident that descriptors such as ALDP, LLDNP, LDRP, and LTCP with the combination of random forest give the better recognition rate of 98.07%, 99.12%, 98.88%, and 98.17% on Kylberg dataset, respectively.

REFERENCES

1. M. Ahmed, A. Shaukat, and M. U. Akram. 2016. Comparative analysis of texture descriptors for classification, in *IEEE International Conference on Imaging Systems and Techniques (IST)*, Chania, pp. 24–29, https://doi.org/10.1109/IST.2016.7738192.
2. F. Lehmann. 2011. Turbo segmentation of texture images, *IEEE Transactions on Pattern Analysis and Machine Intelligence*, vol. 33, pp. 16–29.
3. R. Arya and E. R. Vimina. 2021. Local triangular coded pattern: A texture descriptor for image classification, *IETE Journal of Research*, https://doi.org/10.1080/03772063.2021.1919222.
4. L. Armi and S.-F. Ershad. 2019. Texture image analysis and texture classification methods—a review, *International Online Journal of Image Processing and Pattern Recognition*, vol. 2, no. 1, pp. 1–29.
5. M. Tuceryan and A. K. Jain. 1993. Texture analysis, in *Handbook of pattern recognition and computer vision*, World Scientific Publisher, pp. 235–276, https://doi.org/10.1142/9789814343138.
6. E. R. Vimina and M. O. Divya. 2020. Maximal multi-channel local binary pattern with colour information for CBIR, *Multimedia Tools and Applications*, https://doi.org/10.1007/s11042-020-09207-8.
7. M. D. Ansari, S. P. Ghrera, and A. R. Mishra. 2020. Texture feature extraction using intuitionistic fuzzy local binary pattern, *Journal of Intelligent Systems*, vol. 29, no. 1, pp. 19–34, https://doi.org/10.1515/jisys-2016-0155.
8. S. S. Patil and H. Patil. 2013. Study and review of various image texture classification methods, *International Journal of Computer Applications*, vol. 75, pp. 33–38.
9. P. Cavalin and L. Oliveira. 2017. A review of texture classification methods and databases, in *2017 30th SIBGRAPI Conference on Graphics, Patterns and Images Tutorials (SIBGRAPI-T)*, pp. 1–8.
10. C.-C. Hung, E. Song, and Y. Lan. 2019. *Algorithms for image texture classification*, https://doi.org/10.1007/978-3-030–13773–1_3.
11. P. Simon and U. Vijayasundaram. 2018. *Review of texture descriptors for texture classification*, https://doi.org/10.1007/978-981-10-3223-3_15.

12. A. Ramola, A. K. Shakya, and D. Van Pham. 2020. Study of statistical methods for texture analysis and their modern evolutions, *Engineering Reports*, vol. 2, p. e12149, https://doi.org/10.1002/eng2.12149.
13. T. M. Ojala, M. Pietikinen, and D. Harwood. 1994. Performance evaluation of texture measures with classification based on Kullback discrimination of distributions, in *Proceedings of 12th International Conference on Pattern Recognition,* Jerusalem, Israel, vol. 1, pp. 582–585, https://doi.org/10.1109/ICPR.1994.576366.
14. Y. Q. Chen. 1995. *Novel techniques for image texture classification*, PhD Thesis, University of Southampton, United Kingdom.
15. L. Liu, P. Fieguth, Y. Guo, X. Wang, and M. Pietikinen. February 2017. Local binary features for texture classification, *Pattern Recognition*, vol. 62, no. C, pp. 135–160 [Online], https://doi.org/10.1016/j.patcog.2016.08.032.
16. S. Bhosle and P. Khanale. 2019. Texture classification approach and texture datasets: A review, *IJRAR*, vol. 6, pp. 957–968.
17. A. K. Jain and F. Farrokhnia. November 1990. Unsupervised texture segmentation using gabor filters, in *1990 IEEE International Conference on Systems, Man, and Cybernetics Conference Proceedings*, pp. 14–19.
18. T. Jabid, Md. H. Kabir, and O. Chae. 2010. Local directional pattern (LDP) for face recognition, *International Journal of Innovative Computing, Information and Control*, vol. 8, pp. 329–330, https://doi.org/10.1109/ICCE.2010.5418801.
19. F. Zhong and J. Zhang. 2013. Face recognition with enhanced local directional patterns, *Neurocomputing*, vol. 119, pp. 375–384, ISSN 09252312, https://doi.org/10.1016/j.neuron.2013.03.020.
20. T. Jabid, M. H. Kabir, and O. Chae. 2010. Robust facial expression recognition based on local directional pattern, *ETRI Journal*, vol. 32, https://doi.org/10.4218/etrij.10.1510.0132.4.
21. A. M. Shabat and J.-R. Tapamo. 2017. A comparative study of the use of local directional pattern for texture-based informal settlement classification, *Journal of Applied Research and Technology*, vol. 15, no. 3, pp. 250–258, ISSN 1665–6423, https://doi.org/10.1016/j.jart.2016.12.009.7.
22. A. M. M. Shabat and J. Tapamo. 2018. Angled local directional pattern for texture analysis with an application to facial expression recognition, *IET Computer Vision*, vol. 12, no. 5, pp. 603–608, https://doi.org/8.12.10.1049/iet-cvi.2017.0340.
23. A. R. Rivera. J. R. Castillo, and O. O. Chae. 2013. Local directional number pattern for face analysis: Face and expression recognition, *IEEE Transactions on Image Processing: A Publication of the IEEE Signal Processing Society*, vol. 22, https://doi.org/10.1109/TIP.2012.2235848.
24. T. Chakraborti, B. McCane, S. Mills, and U. Pal. May 2018. LOOP descriptor: Local optimal-oriented pattern, *IEEE Signal Processing Letters*, vol. 25, no. 5, pp. 635–639, https://doi.org/10.1109/LSP.2018.2817176.
25. S. Nithya and S. Ramakrishnan. 2018. Local line directional neighborhood pattern for texture classification, *Journal on Image and Video Processing*, vol. 125, https://doi.org/10.1186/s13640-018-0347-x
26. A. B. Gonde, P. W. Patil, G. M. Galshetwar, and L. M. Waghmare. 2017. Volumetric local directional triplet patterns for biomedical image retrieval, in *2017 Fourth International Conference on Image Information Processing (ICIIP)*, pp. 1–6, https://doi.org/10.1109/ICIIP.2017.8313705.
27. M. Verma and B. Raman. 2016. Local tri-directional patterns: A new texture feature descriptor for image retrieval, *Digital Signal Processing*, vol. 51, https://doi.org/10.1016/j.dsp.2016.02.002.
28. P. Banerjee, A. K. Bhunia, A. Bhattacharyya, P. P. Roy, and S. Murala. 2018. Local neighbourhood intensity pattern–a new texture feature descriptor for image retrieval, *Expert Systems with Applications*, vol. 113, pp. 100–115, ISSN 0957–4174, https://doi.org/10.1016/j.eswa.2018.06.044.

29. http://sipi.usc.edu/database/database.php?volume=textures.
30. https://computervisiononline.com/dataset/1105138685.
31. https://vismod.media.mit.edu/vismod/imagery/VisionTexture/vistex.html.
32. P. Mallikarjuna, M. Fritz, A. Tavakoli Targhi, E. Hayman, B. Caputo, and J.-O. Eklundh. *The KTH-TIPS and KTH-TIPS2*, databaseswww.nada.kth.se/cvap/databases/kth-tips.
33. https://kylberg.org/datasets/.
34. M. Cimpoi, S. Maji, I. Kokkinos, S. Mohamed, and A. Vedaldi. 2014. Describing textures in the wild, in *Proceedings of the {IEEE} Conf. on Computer Vision and Pattern Recognition (CVPR).*
35. http://www.wavelab.at/sources/STex/.
36. B. D. Ripley. 1996. *Pattern recognition and neural networks*, United Kingdom: Cambridge University Press.
37. G. Yu and S. V. Kamarthi. 2008. Texture classification using wavelets with a cluster-based feature extraction, in *2008 2nd International Symposium on Systems and Control in Aerospace and Astronautics (ISSCAA).*
38. F. H. C. Tivive and A. Bouzerdoum. 2006. Texture classification using convolutional neural networks, in *IEEE TENCON.*
39. N. Qaiser, M. Hussain, A. Hussain, N. Iqbal, and N. Qaiser. 2006. Dissimilarity analyst of signal processing methods for texture classification, in *IEEE ICEIS.*
40. C. Chen, C. Chen, and C. Chen. 2006. A comparison of texture features based on SVM and SOM, *ICPR*, vol. 2, pp. 630–633.
41. R. Arya and E. R. Vimina. 2020. An evaluation of local binary descriptors for facial emotion classification, in Saini, H., Sayal, R., Buyya, R., and Aliseri, G. (eds.), *Innovations in computer science and engineering*. Lecture notes in networks and systems, vol. 103, Singapore: Springer.
42. S. K. Roy, B. Chanda, B. B. Chaudhuri, D. K. Ghosh, and S. R. Dubey. 2018. Local morphological pattern: A scale space shape descriptor for texture classification, *Digital Signal Processing*, vol. 82, pp. 152–165, ISSN 1051–2004, https://doi.org/10.1016/j.dsp.2018.06.016.
43. S. K. Roy, B. Chanda, B. B. Chaudhuri, D. K. Ghosh, & S. R. Dubey. 2018. Local jet pattern: A robust descriptor for texture classification, *Multimedia Tools and Applications*, https://doi.org/10.1007/s11042-018-6559-3.
44. S. R. Dubey. 2019. Local directional relation pattern for unconstrained and robust face retrieval, *Multimedia Tools and Applications*, vol. 78, pp. 28063–28088, https://doi.org/10.1007/s11042-019-07908-3.
45. L. Armi and S. Fekri Ershad. 2019. Texture image analysis and texture classification methods—a review, *International Online Journal of Image Processing and Pattern Recognition*, vol. 2, no. 1, pp. 1–29.

AUTHORS' BIOGRAPHIES

Arya R. received her BCA from Mahatma Gandhi University, Kerala, India, and her MCA and MPhil in computer science from Amrita Vishwa Vidyapeetham University, India. She is currently working toward a PhD degree in the area of image processing at Amrita Vishwa Vidyapeetham University, Kochi. Her areas of research interests are image classification, content-based image retrieval, data mining, and video analytics. Corresponding author's email: arya.arya.88@gmail.com.

Vimina E. R. received her BTech degree in electrical and electronics engineering from Mahatma Gandhi University, Kerala, India; ME degree in computer science and engineering from Bharathiyar University, Tamil Nadu, India; and PhD degree from Cochin University of Science and Technology, Kerala, India. She is currently working as an assistant professor in the Department of Computer Science and IT of Amrita Vishwa Vidyapeetham, Kochi Campus, India. Her major fields of interests are content-based image retrieval, biomedical imaging, and video analysis. *Email*: vimina.er@gmail.com.

2 Precision Grading of Glioma

A System for Accurate Diagnosis and Treatment Planning

Asmaa El-Sayed Hassan, Mohamed Shehata, Hossam Magdy Balaha, Hala Atef, Ahmed Alksas, Ali H. Mahmoud, Fatma Sherif, Norah Saleh Alghamdi, Mohammed Ghazal, Ahmed Mayel, Lamiaa Galal El-Serougy, and Ayman El-Baz

2.1 INTRODUCTION

Primary cancerous tumors of the brain and spinal cord are a type of cancer that originates in these regions. In young adults along with children, they are the second most frequently occurring cancer after leukemia [1]. According to the American Brain Tumor Association, new cases, over 87,000, of primary brain and central nervous system (CNS) tumors are found each year in the United States alone [2]. Based on the tumor's position, histological characteristics, and genetic features, primary brain and spinal cord tumors can be categorized. The tumor's type and grade determine the survival rate for brain and spinal cord tumors. The five-year survival rate for all primary brain tumors is roughly 34% [2]. Adult mortality from cancer of the CNS is significant in the USA, where 18,990 deaths are predicted to occur in 2023, making it the tenth highest cause of death [3]. Gliomas are a specific kind of brain tumor that arises in the central nervous system's (CNS) glial cells, which are in charge of supporting and nourishing the brain's neurons. The three types of glial cells that can give rise to gliomas are oligodendrocytes, ependymal cells, and astrocytes, with astrocytes being the most prevalent [4, 5]. Gliomas are divided into four classes by the World Health Organization (WHO), with grade IV being the most aggressive and malignant type. Both grades I and II gliomas are regarded as low-grade; however, grade I have benign tendencies, while grade II gliomas have a high recurrence rate. Both glioblastoma (GBM) and anaplastic glioma (grade III and grade IV, respectively) are categorized as high-grade malignant gliomas with a poor prognosis [6, 7]. The most malignant form of GBM has a 0.068 survival rate [8]. In adults, the

DOI: 10.1201/9780367486082-2

majority of primary malignant tumors of central nervous are gliomas, which make up around 0.8 of all malignant brain tumors and about 0.45 of all primary brain tumors [2]. Gliomas also have a high rate of morbidity and mortality, with an incidence of five to ten cases per 100,000 people annually [6]. The prognosis is still poor despite improvements in medical therapies that particularly target glioma-related biological pathways. After the first diagnosis, the median survival rate is less than 15 months, and the five-year survival rate is less than 0.1 [9, 10].

The symptoms of gliomas can differ depending on the location and size of the tumor. The most popular symptoms include seizures, uncontrolled headaches, cognitive and behavioral changes, motor dysfunction, and problems with vision or speech. The symptoms may develop slowly or rapidly, depending on the tumor's aggressiveness and location [11]. To diagnose gliomas, a combination of imaging tests, including computed tomography (CT), positron emission tomography (PET), and magnetic resonance imaging (MRI) scans, is usually used, along with a biopsy to determine the tumor's histological type and grade [12]. Additionally, molecular testing can help identify any genetic mutations in the tumor that could guide the choice of treatment. The available treatment options for gliomas are determined by various factors, such as the tumor's location, size, grade, and genetic characteristics. Surgery is the primary approach for treating gliomas, and the goal is to eliminate as much of the tumor as possible. However, complete resection may not be feasible in some cases, especially if the tumor is located in critical regions of the brain. After surgery, radiation therapy is often employed to destroy any remaining malignant cells and minimize the risk of recurrence. Chemotherapy may also be used either in combination with radiation therapy or as a standalone treatment. Targeted therapies that aim to inhibit specific molecular pathways involved in tumor growth may be utilized in certain instances. However, the tumors' resistance to traditional surgical surgery and other therapeutic techniques is a result of the extensive infiltration of malignant gliomas into the brain parenchyma and the distinctive tumor microenvironment that usually encourages the development of the glioma [13].

The malignancy of the tumor, which is evaluated by histology of tissue acquired from surgical biopsy or resection, determines the grade of gliomas. The WHO published updated grading standards for gliomas in 2016 that take into account both histological characteristics and molecular and genetic data. The improved prognosis, treatment response, and tumor behavior predictions made possible by this updated technique [14, 15] are essential for assisting neuro-oncologists in creating efficient treatment regimens and enhancing patient outcomes [16, 17]. Although biopsy is still the gold standard for determining the level of the glioma, it is intrusive, is expensive, and can have unfavorable effects. For the early and accurate grading of gliomas, numerous researchers have investigated non-invasive imaging approaches [18, 19]. The most widely used imaging technique for gliomas and other brain tumors is the MRI. The location, multiplicity, morphology, and mass-related consequences of the glioma can all be evaluated using conventional MRI. Advanced MRI methods, such as diffusion-weighted imaging, perfusion-weighted imaging, and magnetic resonance spectroscopy (MRS), can reveal vital physiological details concerning brain malignancies. They provide a more accurate evaluation of tumor behavior both before and after therapy because they provide precise assessments of tumor cellularity, vascularization, and metabolism [20]. Diffusion-weighted MRI (DW-MRI) is especially valuable for distinguishing gliomas by their grade and predicting prognosis. Generally, higher grades are linked to lower apparent diffusion coefficients (ADC), since increased tumor cellularity restricts water diffusion [17]. Perfusion MRI techniques, such as arterial spin labeling (ASL), can determine tumor vascularity, which is linked to neoangiogenesis [21]. By determining the quantity of numerous tumoral cell metabolites, MRS assesses the metabolic environment of the tumor. Creatinine (Cr), choline (Cho), and N-acetyl aspartate (NAA) are markers of cell membrane breakdown and turnover, neuronal integrity, and metabolism, respectively, in glioma grading. Higher Cho, lower NAA and Cr, and higher Cho/NAA and Cho/Cr ratios are associated with higher glioma grades [22].

The early prediction of several medical diseases has benefited greatly from recent developments in artificial intelligence [23–27], notably in the machine and deep learning algorithms [28, 29]. This has led to the use of these algorithms in several studies to aid in the clinical diagnosis of gliomas [30]. To correctly classify a glioma as grade I, grade II, grade III, or grade IV, this chapter introduces a comprehensive glioma grading computer-aided diagnostic (GG-CAD) system that combines 3D appearance features, volumetric features, functional features, and 3D first- and second-order textural features. This information is essential in developing an appropriate medical management plan.

The following is a summary of the book chapter's contributions:

1. Presenting a thorough glioma grading computer-aided design (GG-CAD) system.
2. Accurately classifying the glioma grade as grade, I, grade II, grade III, or grade IV by fusing functional features retrieved from multimodal MR images with 3D first- and second-order textural features, volumetric features, and 3D appearance features.
3. Presenting the latest performance metrics and benchmarking them against various relevant studies.

The remainder of this chapter is structured as follows: The associated literature is reviewed and summarized in Section 2.2. The used materials are covered in Section 2.3. It describes the input data and discusses the imaging protocols, reference standard diagnosis, and glioma tumor preprocessing. The approach, engineering features, classification and hyperparameter tuning, engineering features selection, and experiments are all covered in Section 2.4. The details and discussions of the experiments' results are presented in Section 2.5. The book chapter is concluded, and future work is discussed, in Section 2.6.

2.2 RELATED STUDIES

Recent advancements in artificial intelligence, specifically machine learning algorithms, have had a profound impact on healthcare, revolutionizing the early prediction and diagnosis of medical conditions [31, 32]. In the field of gliomas, a type of brain tumor, machine learning algorithms have shown immense potential in facilitating clinical diagnosis. Integrating these algorithms into glioma diagnosis studies has proven invaluable, enhancing the accuracy and efficiency of glioma detection [33]. Researchers have dedicated their efforts to developing innovative and non-invasive approaches that assist healthcare professionals in diagnosing and managing gliomas effectively. This integration represents a significant milestone in medical imaging and personalized medicine, as machine learning (ML) algorithms can analyze extensive medical data (e.g., radiological images, patient histories, and biomarkers) to generate accurate predictions. This transformative potential in glioma diagnosis stems from the algorithms' ability to process complex data and provide insights for personalized treatment plans. Ongoing research and refinement of machine learning approaches are expected to lead to further breakthroughs in understanding brain tumors and improving glioma diagnosis. The research was done by Zhang et al. [34] to determine how well multimodal MR images for glioma grading were performed when texture analysis and machine learning methods were combined. The study comprised 120 individuals with glioma, including 28 cases of low-grade gliomas (LGGs) and 92 cases of high-grade gliomas (HGGs), distributed as follows: grades I, II, III, and IV instances fall into the following categories: 3, 25, 29, and 63, respectively. Data was balanced using an oversampling approach after texture characteristics were taken from the data. Twenty-five ML classifiers were used for classification. While obtaining an accuracy of 0.786 on the original data, the best model graded gliomas (grades II, III, and IV) with an accuracy of 0.961 using oversampling. Functional or aesthetic characteristics, which may enhance the model's performance, especially in light of the unbalanced data, were not included in the research. A strategy for predicting the grades of gliomas using radiomics imaging characteristics was put out by Cho and Park [35]. Using the data (i.e., multimodal MRI) retrieved from the MICCAI BRATs 2015 challenge [36], they assessed the precision of glioma categorization. To measure the characteristics of gliomas, they collected 45 radiomics features from each fluid-attenuated inversion recovery (FLAIR), T1, T1-contrast, and T2 picture using the shape, histogram, and gray level co-occurrence matrix (GLCM). Significant characteristics were chosen using L1-norm regularization (LASSO) from among the 180 features. Gliomas were categorized into LGG or HGG using logistic regression using the LASSO coefficient and a few chosen feature values. The output of the classification was validated using ten-fold cross-validation. The accuracy, sensitivity, specificity, and area under the curve (AUC) of the suggested approach were 0.8981, 0.8889, 0.9074, and 0.8870, respectively.

Using texture analysis applied to standard brain MRI, Suarez-Garcia et al. [37] sought to develop a straightforward and reasonably priced classification algorithm that can discriminate between LGG and HGG. They looked at several MRI contrast combinations (T1Gd and T2) as well as a segmented glioma area (NCR/NET, or necrotic and non-enhancing tumor core). They acquired a total of 285 individuals (LGG = 75 and HGG = 210) from the BRATs 2018 challenge, and they used the gray level size zone matrix (GLSZM) to compute texture characteristics. They divided the data into several training subsets using an undersampling technique, extracting complementary information, and then creating unique classification models. To discover the optimal model for categorizing gliomas, they then submitted these variables to several linear regression models. The sensitivity, specificity, and accuracy of this model, which only included three texture characteristics, were 0.9412, 0.8824, and 0.9118, respectively. According to the selected features, when the NCR/NET region was examined, LGGs had a more heterogeneous texture than HGGs in the T2 images and vice versa in the T1Gd images. The study presented by Banerjee et al. [38] investigated the effectiveness of convolutional neural networks (CNNs) and multi-sequence MR images in distinguishing between HGG and LGG. With the use of MRI slices, patches, and multi-planar volumetric slices, they created new ConvNet models. A total of 746 patients (472 HGG and 274 LGG) from BRATs and the Cancer Imaging Archive (TCIA) was included in the research. Through the optimization of two pre-existing ConvNet models (i.e., VGGs and ResNet), the researchers also investigated the application of transfer learning. Utilizing holdout dataset testing as well as leave-one-patient-out testing, ConvNets performance was assessed. The outcomes showed that the suggested ConvNets had superior accuracy in all cases. The model outperformed traditional models by achieving an accuracy of 0.95 for classifying LGG and HGG.

In a recent study, Alis et al. [39] employed ANNs to distinguish between LGG and HGG. They assessed the diagnostic efficacy of analyzing the texture of conventional MR images. Additionally, regions of interest (ROIs) for 181 participants (HGG = 97 and LGG = 84) were manually placed, and first-order and second-order texture characteristics were retrieved from those ROIs. These data comprised high-order texture features and histogram parameters from contrast-enhanced T1-weighted images and T2-weighted FLAIR images that encompassed the tumor's volume. The evaluation of interobserver reliability was done to assess the repeatability of the features. Training and test divisions of the cohort (121 and 60, respectively) were created. While the training set was used to select features and construct the model, the test set was used to perform the evaluation of the diagnostic performance of the trained ANNs in distinguishing between the two categories of glioma (i.e., HGG and LGG). For the ANN models employing texture data from T2-weighted FLAIR and contrast-enhanced T1-weighted images, AUCs of 87% and 86% were achieved, respectively. The ANN model with the selected texture features has an AUC of 92% and a maximum diagnosis accuracy of 88.3%. Hsieh et al. [40] presented a CAD system that utilizes intensity-invariant MR imaging features to classify glioma grades. The imaging of 107 individuals with glioma (34 HGG and 73 LGG) was examined. The MR images' local texture was transformed into an intensity-invariant local binary pattern (LBP), and thereafter, quantitative image features (e.g., textures and histogram moments) were retrieved. These data were integrated to create a malignancy prediction model using a logistic regression classifier, which was then evaluated against traditional features to show the conducted improvement. The LBP-based CAD system produced the greatest results, with accuracy, sensitivity, negative predictive value, and AUC of 93%, 97%, 99%, and 94%, respectively. Kalaiselvi et al. [41] developed six different CNN models to classify HGG and LGG lesions from volumetric MR scans. These models were developed by modifying various hyperparameters using the standard CNN architecture. The used CNN models were built as follows: (1) five layers with dropout, (2) two layers with five epochs, (3) five layers with dropout and batch normalization (FLSCBN), (4) five layers with stopping criteria (FLSC), and (5) FLSCBN with dropout. Two datasets (i.e., BRATs 2013 [30 volumes] and WBA [i.e.,

Whole Brain Atlas]) were used to randomly choose 4,500 images to test the models. With a score of 88.91%, the FLSCBN model had the best accuracy for detecting brain tumors.

Zhuge et al. [23] put forward two new methods to differentiate LGG and HGG non-invasively on MRI images conducted from 315 patients (LGG = 105 and HGG = 210) from BRATs 2018 and TCIA. They employed deep CNNs and a modified U-Net model to segment brain tumors in 3D. Following that, researchers divided tumors into two segments for use in tumor classification. In the first phase, they identified the slice with the biggest tumor area and graded the tumor using the state-of-the-art mask region-based CNN (R-CNN) model. Additionally, they used a 2D data augmentation strategy to improve the model's efficiency by diversifying the training photos. In the subsequent stage, they classified the segmented tumors by immediately applying a 3D volumetric CNN, known as 3DConvNet, to the bounding image areas. This method took advantage of the 3D spatial contextual information of volumetric image data. The suggested model had the best diagnostic performance, with a 97.1% accuracy rate. Cho et al. [24] evaluated the ability of radiomic features to discriminate between HGG and LGG using 285 multimodal MR images retrieved from the BRATs 2017 dataset (LGG = 75 and HGG = 210). To distinguish between the training and test sets, they used a five-fold cross-validation procedure. The radiomics features (468 features) were computed using three different types of ROIs. In order to choose meaningful features for categorizing glioma grades in the training cohort, the minimum redundancy maximum relevance method was used. Several classifiers (i.e., support vector machines [SVM], logistics, and random forest [RF]) were constructed using the set features. Accuracy, sensitivity, specificity, and AUC were used to assess the performance. The ML classifiers identified five significant features, and the used classifiers achieved an average AUC of 94% and 90.30% for training and testing cohorts, respectively. The logistic regression, SVM, and RF classifiers scored accuracies of 90.10%, 88.66%, and 92.13%, respectively, for the test cohorts.

2.2.1 Research Gap

Addressing these limitations and exploring additional features can significantly enhance the accuracy and clinical utility of glioma grading using MRIs and CAD systems. Many previous studies [23, 24, 35, 37–41] focused solely on differentiating between LGGs and HGGs. However, it is crucial to go beyond this binary classification and adopt a comprehensive grading approach, encompassing specific grades I, II, III, or IV. This refined grading strategy enables a more precise assessment of tumor aggressiveness and facilitates personalized treatment planning tailored to the individual needs of patients.

Furthermore, the incorporation of functional features into the grading process can provide valuable insights into the biological behavior of gliomas. Functional features, such as perfusion parameters or diffusion characteristics, capture the dynamic nature of tumor physiology and offer a more comprehensive understanding of tumor grade and aggressiveness. Integrating these functional features into CAD systems has the potential to improve diagnostic performance and support clinicians in making well-informed decisions regarding treatment strategies.

Even while some researchers [34, 37, 39, 40] have included texture elements in their work, they haven't explored appearance or shape features, which may have improved the system's overall performance. Shape features, including tumor volume, irregularity, and surface characteristics, can provide additional quantitative information about tumor morphology [42], enabling differentiation between different grades. Appearance features, such as contrast-enhancement patterns or signal intensity variations, reveal distinct imaging characteristics associated with specific tumor grades. By integrating these shape and appearance features into CAD systems, a more comprehensive set of imaging biomarkers can be harnessed to improve the reliability of the grading process.

2.3 MATERIALS

Description of the Entry Data. The study comprised 99 glioma tumors in total (49 male and 50 female), all of which had biopsies confirming their presence. The participants were aged from 1 to 79 years, with a 40.15 ± 19.94 years average. The sample had the following distribution of tumor grades: 13, 22, 22, and 42 for grade I, grade II, grade III, and grade IV, respectively. All participants, including children under the age of 18, gave their informed consent to take part in the study, either directly or through their legal guardians. The study used multimodal MR images from the Mansoura University Hospital in Egypt, including T2-MR (FLAIR), DW-MR, and T1-MR with pre-contrast and post-contrast phases.

Protocols for Imaging. A Philips Ingenia 1.5 Tesla magnetic resonance imaging (MRI) scanner was used for the MRI scans. Participants were positioned in the supine position, and a standard eight-channel head coil was used. The imaging settings were as follows: a matrix size and a slice thickness of 256 × 256 and 3 mm, respectively. Axial FLAIR (TR/TE/TI = 8,000/140/2000 ms), axial T2 (TR/TE = 1,250/100 ms), and axial T1 (TR/TE = 475/15 ms) sequences were all carried out. Axial images for the contrast-enhanced T1-MR were obtained after 0.1 mmol/kg of gadolinium-based contrast agents were administered. The axial DW-MR was performed utilizing a multi-section single-shot spin-EPI (echo-planar imaging) procedure (TE\TR\NEX = 88\3,000\1) with two b-values (0 and 1,000 s/mm2). It is important to note that for each modality, multiple axial cross sections were obtained to include the entire volume; further, the saving format was DICOM.

Reference Standard Diagnosis. The biopsy, which entails pathologically examining aberrant brain tissues, is the gold standard for conclusively detecting gliomas and establishing a prognosis. Numerous cores were taken from the solid-enhancing region of the tumor, which varies in size and position according to the kind of tumor and its prognosis, during an excisional biopsy, which was the procedure used in the majority of patients. A stereotactic needle biopsy was carried out on patients with gliomas in difficult-to-reach or delicate locations that may be harmed by a more invasive treatment. In order to remove tiny fragments of aberrant tissue, a neurosurgeon had to make a tiny hole in the skull and insert a thin needle under the direction of radiological scans. The collected samples from both types of biopsies were sent for pathological analysis, and the highest pathological glioma grade was given to the entire tumor. Additionally, it is significant to highlight that the current study adhered to the principles of the Declaration of Helsinki and was given approval by Mansoura University's Institutional Review Board (MD.20.01.278).

Preprocessing of Glioma Tumors. The precision of feature extraction relies on the accuracy of segmentation [**43**]. Multiple gray level images recorded in DICOM format comprised the initial data for each participant, which were subsequently transferred to a workstation (extended MR Workspace release 2.6, Philips Medical Systems, BV, Eindhoven, the Netherlands). For each patient, binary masks were produced by manually segmenting the ROIs. For T1-MR and DW-MR, the ROIs were used to particularly recognize the tumor, while for T2-MR (FLAIR), the ROIs included the tumor and any edema presented around it. Two experienced radiologists with more than ten years of practical expertise in the analysis of medical pictures carried out the segmentation procedure while remaining blind to one another's segmentations. The overlapping regions between the two observers were removed in order to construct the ground truth segmentation; however, when it comes to medical image analysis, a radiologist with more than 25 years of experience examined whether the changes should be viewed as being a part of the tumor or normal tissues. The agreement between the two segmentations was also evaluated and quantified using Bland–Altman analysis.

2.4 METHODOLOGY

The proposed system, called GG-CAD, offers the capability to distinguish the various grades of gliomas through a series of steps. First, it creates a histogram of oriented gradients (HOG) to extract higher-order 3D appearance information from the segmented tumor and then calculates the volume of the tumor utilizing both T2-MR (FLAIR) and contrast-enhanced T1-MR imaging. Second, it creates histograms, GLCM, and GLRLM based on T2-MR (FLAIR) and contrast-enhanced T1-MR images to derive first-order and second-order textural properties from the segments. Thirdly, it analyzes the enhancement degree among the pre-contrast and post-contrast phases of T1-MR imaging and generates 3D ADC maps for the segments using DW-MR images which had been obtained at a b-value = 1,000 s/mm^2. Fourthly, it chooses the most important set of features out of all the collected features utilizing feature selection using the Gini impurity technique. To determine the final categorization of the tumor (i.e., grade I, II, III, or IV), the proposed MLP-ANN classification model is fed with the optimal set of features.

2.4.1 Features Engineering

After preprocessing the glioma tumors, it becomes crucial to transform the structured objects representing the different subjects into a set of features that are not only distinguishable but also standardized and easily comprehensible by ML algorithms.

These features serve as the key to discriminating between various subjects by providing the learning algorithm with valuable insights into the unique characteristics of each object. The quality and relevance of these features play a pivotal role in defining and enhancing the predictive capabilities of the presented ML model [44]. To ensure a comprehensive and well-rounded approach, we actively engaged input obtained from the medical team. Through collaborative discussions, a consensus on multiple categories of distinctive features that are well-aligned with the specific nature of our problem is reached. These categories encompass a wide range of attributes that capture the essence of the glioma tumors, allowing for a more comprehensive analysis. In the subsequent sections, we will delve into an in-depth exploration of the extracted imaging features. These features encompass various aspects, including but not limited to structural, functional, and textural characteristics. By examining these features closely, we aim to uncover hidden patterns, correlations, and dependencies within the glioma tumors' imaging data. This deeper understanding will not only contribute to the overall knowledge of glioma classification but also pave the way for the development of an effective and accurate ML model.

2.4.1.1 Higher-Order 3D-Appearance Features

Identifying and incorporating specific parametric higher-order appearance features is essential to develop the GG-CAD system with high sensitivity and specificity, capable of accurately differentiating between different glioma grades [45]. It is believed that higher-grade gliomas have more aggressive development rates and complicated, irregular, and rough forms than lower-grade gliomas, which is the basis for using these 3D-appearance qualities. Therefore, obtaining a correct diagnosis depends on precisely detecting, modeling, and extracting such characteristics. According to the suggested framework, the HOG analysis is used to identify gliomas along with accounting for the overall tumor volume. By employing HOG, the system can capture intricate details and patterns related to the texture and shape of the glioma tumor [46]. This enables the system to discern specific characteristics that are indicative of different grades of glioma tumors. Furthermore, the inclusion of total tumor volume as a feature adds dimension to the analysis. By considering the overall size of the tumor, the system gains a deeper understanding of the tumor's extent and spread, which can be valuable in distinguishing between grades of glioma tumors, as higher-grade tumors typically exhibit larger sizes compared to lower-grade tumors [47].

2.4.1.2 3D-HOG

The HOG descriptor is a method used to capture the morphological structure and appearance of an object by creating a simplified representation of the image that focuses on the most significant details [48, 49]. It quantifies the frequency of gradient orientations within specific areas of an object. In this study, the volumes obtained from the pre- and post-contrast T1- and T2-MR (FLAIR) images are treated using a 3D-HOG technique. The recovered ROIs are scaled to a consistent shape of (32 × 32 × slices per volume) to ensure uniformity and accelerate the analysis. The length (X) and width (Y) of each volume were uniform thanks to the resizing procedure; however, the depth (Z) varied according to the number of 2D slices contained in each volume. Then, algorithm 1 is fed with the enlarged ROIs, with the number of cells and bins set to 4 and 9, respectively. Through this process, the HOG descriptor was generated. With four cells, each containing eight histograms, and each histogram comprising nine bins, the overall number of features obtained was calculated as $4 \times 8 \times 9 = 288$ features for each volume. This means that each volume was represented by 288 distinctive features derived from the 3D-HOG analysis. By employing the 3D-HOG approach, the relevant information related to the morphological characteristics of glioma tumors can be effectively captured and represented. These features serve as valuable inputs for our classification and prediction algorithms, enabling accurate and comprehensive analysis of the tumor volumes.

Algorithm 1 Histogram of Oriented Gradients (HOG)

Input: Region of interest (ROI), denoted as V, with dimensions $m \times n \times p$ and voxel size $\Delta x \times \Delta y \times \Delta z$, as well as a specified cell size (C) and the number of bins (B).

Output: The HOG H_{ijk}

For each voxel located at (i, j, k), perform the following:

$$M_{ijk} = \sqrt{\left(V_{i+1,jk} - V_{i-1,jk}\right)^2 + \left(V_{ij+1,k} - V_{ij-1,k}\right)^2 + \left(V_{ijk+1} - V_{ijk-1}\right)^2}$$

$$\Theta_{ijk} := \cos^{-1}\left(\frac{V_{i+1,jk}V_{i-1,jk} + V_{ij+1,k}V_{ij-1,k} + V_{ijk+1}V_{ijk-1}}{\sqrt{V_{i+1,jk}^2 + V_{ij+1,k}^2 + V_{ijk+1}^2}\sqrt{V_{i-1,jk}^2 + V_{ij-1,k}^2 + V_{ijk-1}^2}}\right)$$

For each non-overlapping cell i of size $C \times C \times p$, perform the following:

For each bin j perform:

$$D_j := \left[(j-1)\pi/B, j\pi/B\right]$$

$$H_{ij} := \Sigma\left\{M_{rst} \mid (r,s,t) \in i \wedge \Theta_{rst} \in D_j\right\}$$

For each k in the 8-neighborhood of i perform:

$$H_{ijk} := H_{ij} / \Sigma\left\{M_{rst} \mid (r,s,t) \in k\right\}$$

2.4.1.3 Textural Features

In our quest to improve the early distinction between different glioma grades, an approach that involved extracting a comprehensive set of first-order and second-order texture features was employed [50]. These features provided detailed descriptions that captured the heterogeneity or uniformity (i.e., homogeneity) of the detected gliomas with precision. The motivation behind this stemmed from the theory that HGG tumors exhibit greater heterogeneity in their textural appearance compared to low-grade gliomas, as supported by previous studies [51, 52]. By analyzing the textural characteristics of the tumors, we aimed to uncover subtle variations in their internal structure, such as patterns, irregularities, and spatial distribution of intensities. These features offer valuable insights into the underlying tumor biology, highlighting the complexity and diverse composition of high-grade gliomas. By contrast, low-grade gliomas tend to exhibit more uniform and homogeneous textural patterns.

2.4.1.4 GLCM

The estimation of first-order textural features involved the use of a normalized empirical histogram, which provided valuable information about the distribution of intensity levels within the ROI. However, it is important to note that first-order textural features can be sensitive to noise and may not fully capture the heterogeneity between different grades of tumors. To address this limitation and achieve a more robust quantification of heterogeneity, we incorporated second-order textural features, specifically the GLCM [53, 54]. We used second-order textural characteristics, notably the gray level co-occurrence matrix (GLCM), to overcome this drawback and produce a more reliable assessment of heterogeneity [53, 54]. In order to assess the spatial relationships between voxels inside a certain neighborhood block, GLCM is often used. It represents a bivariate histogram that contains information about the frequency of particular intensity pairings between adjacent voxels. We initially established the amount of quantization and the intensity range inside the ROI before building the GLCM. The neighborhood system was then created by defining the spatial connection between neighboring voxel pairs. The likelihood that a voxel with intensity "i" has a nearby voxel with intensity "j" is proportional to each element (i, j) of the GLCM. The GLCM matrix is then normalized to ensure that all elements sum up to unity, facilitating meaningful comparisons [55, 56]. In our study, voxel intensities were quantized to 8-bit precision, resulting in a 256 × 256 GLCM. We defined *voxel neighbors* as those within a distance of $\leq \sqrt{2}$ mm, which yielded an approximately symmetric GLCM, excluding the impact of boundary voxels. By incorporating second-order textural features like GLCM, we aimed to capture and quantify the spatial relationships and co-occurrence patterns of intensity values within the tumor region. This information is crucial for differentiating between various grades of gliomas and providing valuable insights for clinical decision-making.

2.4.1.5 GLRLM

Along with analyzing the frequency of voxel pairings represented by GLCM, gray level run-length matrix (GLRLM) was applied to analyze voxel runs and assess voxel connectivity within the region of interest. GLRLM counts the occurrences of a run of successive voxels with the same intensity. Similar to GLCM, establishing the range of the gray level and its quantization, generally spanning from 0 to 255 in 8-bit precision, is required for GLRLM construction. The fixed range sets the rows number in the GLRLM, and the greatest extent of the ROI in the particular dimension where voxel runs are being recorded determines the column dimension. Each of the GLRLM's elements (i, j) reflects the relative occurrence of a run made up of *j* successive voxels with an intensity value of *i*. By analyzing voxel runs, we gain insights into the spatial characteristics and patterns of voxel connectivity within the region of interest [54, 56]. Due to the discrepancy between the spacing of MRI slices and the pixel spacing, we constructed two GLRLMs for our analysis. The first GLRLM captures voxel runs in both directions (x and y directions) within the same slice, while the second GLRLM considers runs in the third direction (z-direction), specifically across slices at the same (x, y) location in each slice. This approach allows us to account for the varying spatial orientations and connectivity patterns present in glioma tumors across different dimensions of the MRI data. By incorporating GLRLM alongside GLCM, we aim to capture a comprehensive range of textural features that effectively represent the heterogeneity and connectivity of glioma tumors, enabling accurate differentiation between different tumor grades.

2.4.1.6 Functional Features

The glioma functionality has an immediate impact on their imaging characteristics, which can significantly enhance the model's capabilities in accurately specifying the glioma grade. With this understanding, two distinct features derived from both T1- and DW-MR imaging modalities are explored to observe the functional aspects of various gliomas. By incorporating functional features into our analysis, we aim to capture the dynamic nature of gliomas and gain insights into their underlying functional properties. These features provide valuable information about the tumor's behavior and can serve as important indicators for determining the grade of glioma tumors.

2.4.1.7 Contrast-Enhancement Slopes (T1-MR)

The glioma functionality can be assessed through its *hyperenhancement*, which refers to the degree of contrast enhancement observed in the tumor. By analyzing the temporal changes in gray values between the pre-contrast and post-contrast phases of T1-MR, the contrast-enhancement slope can be estimated. The contrast-enhancement slope captures the rate of change in gray level intensities between the two phases [57, 58]. This slope serves as a quantitative measure of the speed and magnitude of contrast enhancement within the tumor. By calculating the slope, we can identify the differences in enhancement patterns between HGG and LGG. HGGs tend to exhibit higher and faster slopes, indicating more intense and rapid contrast enhancement compared to LGGs. These differences in contrast-enhancement slopes reflect the functional characteristics of the tumors and provide valuable insights into their grade and aggressiveness.

2.4.1.8 ADCs (DW-MR)

Gliomas, being pathological tissues, exhibit unique diffusivity properties compared to healthy tissue. Diffusion-weighted MR imaging provides valuable information about the movement of water molecules within the tissue, reflecting the tissue's microstructural characteristics. In addition to water diffusion, capillary perfusion can also affect the diffusion-weighted MR signal. To quantify the diffusion properties of gliomas, we utilize the ADC functional parameter. The values of this parameter can be estimated by acquiring diffusion-weighted images with different gradient field strengths and durations, commonly referred to as b-values. Higher b-values cause the DW-MR signal intensity to weaken more, allowing for a more accurate evaluation of the tissue's diffusivity [59]. In this study, a 1,000 s/mm^2 b-value is selected to calculate the ADC values for each voxel within the segmented tumor. However, using the raw ADC measurements as descriptive features can be problematic due to variations in the number of voxels within different segmented regions. To standardize the length of the descriptor and ensure comparability between different regions, a technique called binning is employed. The ADC values are binned, and the cumulative distribution function (CDF) of those measurements is calculated; after that, a standardized representation of the diffusion characteristics within the tumor is obtained. This standardized ADC descriptor provides a compact and informative summary of the diffusion properties of gliomas [60], facilitating comparisons and analysis across different regions and cases.

2.4.2 Classification and Hyperparameters Tuning

After glioma tumors from three different MR modalities were chosen for the discriminative criteria, such as appearance, textural, and functional characteristics, the classification procedure was carried out. Using MLP-ANN, our goal was to successfully distinguish between various grades of gliomas. The selected characteristics revealed important information on the visual appearance, functional aspects, and textural qualities of glioma tumors.

MLP-ANNs. MLP-ANNs are highly regarded and widely used in machine learning as robust classifiers. These networks comprise multiple layers, including an input layer, one or more hidden layers, and an output layer. Within each layer, numerous nodes, also known as activation units, perform intricate computations. The architecture of an MLP-ANN is carefully designed to ensure that there are connections between every pair of adjacent layers. This interconnectedness enables the network to efficiently process and analyze the input data. By incorporating non-linear activation functions within the neurons, the MLP-ANN can effectively divide the input space into complex and flexible regions, enabling the representation of intricate patterns and relationships within the data.

During the training process, the primary objective of the MLP-ANN is to minimize a loss function, which quantifies the discrepancy between the predicted outputs and the actual targets. To achieve this, the network employs a supervised learning technique known as backpropagation. Through backpropagation, the connection weights and biases in the network are iteratively adjusted using gradient descent methods [61]. This iterative optimization procedure fine-tunes the parameters of the network to enhance its predictive accuracy and reduce errors.

During the training process of MLP-ANNs, labeled examples are provided to the network, where the expected outputs are already known. By learning from these examples, the network gradually improves its capacity to generalize and make accurate predictions on new, unseen data. This learning process involves iteratively adjusting the connection weights and biases of the network based on the gradients calculated during backpropagation. Through this iterative adjustment, the MLP-ANN learns how to effectively map the input features to the desired outputs [62]. The strength of MLP-ANNs lies in their ability to learn and capture intricate relationships within the data, allowing them to tackle a wide range of classification problems. They excel in modeling non-linear relationships and are capable of handling high-dimensional data, making them a suitable choice for tasks such as differentiating glioma grades [63].

Hyperparameters Tuning. Optimizing hyperparameters is crucial as it governs the overall performance and behavior of a learning model [64]. In the suggested GG-CAD system, an intensive grid search technique was used to find the best set of hyperparameters. The grid search involved setting up a grid with various combinations of hyperparameters [65, 66] and then training and testing the MLP-ANN on each combination to search for the best performance. The accuracy served as the optimization metric for the whole search procedure. Utilizing LOSO cross-validation, the hyperparameter optimization was carried out to make sure that the chosen hyperparameters would generalize well to unobserved data. This method entailed progressively removing one subject from the dataset before using the other subjects to train and test the model and assess how well it performed with data that had not yet been seen. After careful optimization, the following set of parameters was determined as the optimal configuration for the MLP-ANN: training function, trainlm; maximum epochs, 500; hidden layers, the first, second, and third hidden layers are set to 200, 100, and 50; goal, 0; maximum validation failure, 6; minimum gradient, 10^{-7}; and training gain (μ), initial $\mu = 0.001$; μ decrease factor = 0.1; μ increase factor = 10; and $\mu_{max} = 10^{10}$. By fine-tuning these hyperparameters through the grid search and LOSO cross-validation, the MLP-ANN in the GG-CAD system is expected to achieve optimal performance in differentiating glioma grades.

2.4.3 Engineering Features Selection

In the typical data science workflow, feature selection plays a crucial role in identifying the most relevant and significant features from a large pool of prospective options. From a collection of n alternatives, this procedure seeks to extract a subset of m features, where m is typically smaller than n and represents the most informative and meaningful features for the task at hand [67]. To achieve this, a Gini impurity–based selection technique is employed [68, 69].

Selection Based on Gini Impurity. Along with being one of the most reliable ML classifiers, RFs have been used in the data science pipeline to choose features and markers. The inherent tree-based nature of random forests enables them to naturally assess the enhancement in node purity, referred to as Gini impurity. Nodes with the greatest reduction in impurity are emphasized throughout the tree development, while those nodes with the least impurity reduction are positioned at the end. By pruning the trees below a specified node, we can create a subset of the most important markers.

The feature selection algorithm is implemented using a combined marker selection approach. In this approach, the Gini impurity–based technique was applied to the complete set of markers to identify and thus utilize the optimal subset of markers. To determine the optimal impurity threshold value, three alternative scenarios are conducted by using three different threshold values. Each scenario involved performing the selection process and reducing the feature set based on the specified impurity threshold. To assess the performance of the reduced feature sets, we employed the LOSO cross-validation method. Each reduced feature set was used to train and test the presented MLP classifier. Then, utilizing the three condensed sets of features, we evaluated the diagnostic effectiveness of the presented system. Our investigation showed that the best performance was obtained with an impurity threshold of 0.001. This threshold value resulted in a set of 332 selected features, which we identified as the optimal setting for our proposed model. By employing this comprehensive approach, the most relevant markers from various categories are captured successfully, thus enhancing the overall performance of the model.

2.4.4 Experiments

A range of experiments was carried out to assess the performance of the proposed system (i.e., GG-CAD), with the objective of comprehensively evaluating its effectiveness and dependability. These experiments were designed to evaluate the system's performance across different dimensions and provide comprehensive insights into its capabilities. The subsequent experiments took place:

Experiment I. The tuned MLP-ANN model was used, with an optimal set of features chosen using the Gini impurity–based approach, to evaluate the performance of the presented system. The effectiveness of this system was evaluated using the LOSO and k-fold stratified cross-validation techniques, which are both considered two commonly used cross-validation approaches. In the LOSO approach, the classification model was trained using all available data except for one subject, which was reserved for testing. For a total of 99 iterations, this technique was done repeatedly for each participant in the dataset. Ninety-eight participants were used as the training sample size for each iteration, while a single individual was used for testing. With regard to the k-fold stratified cross-validation, a subset of the data $\left(i.e., \frac{1}{k}\times 100\%\right)$ was randomly selected and set aside for testing, while the remaining data was used for training. The process was repeated *k* times, with *k* being either 5 or 10, depending on the specific experiment. The classification model was reinitialized in each iteration of the k-fold cross-validation, the training set consisting of the subjects from the previous iteration was added, while a new group $\frac{1}{k}\times 100\%$ of subjects was selected for testing.

Stratification was employed to ensure that the distribution of subjects across different glioma grades was preserved in both the training and testing sets. This approach mitigated potential biases and variances that could arise during the cross-validation process. With 0.13, 0.22, 0.22, and 0.43 of the training/testing sets coming from grade I, grade II, grade III, and grade IV, respectively, the stratification strategy preserved the initial distribution of individuals in the entire dataset. By utilizing the LOSO and k-fold stratified cross-validation approaches, the proposed model's robustness was thoroughly evaluated, ensuring that it is not prone to overfitting and performs consistently across different subsets of the data.

Experiment II. To demonstrate the contribution and significance of each group of features, as well as emphasize the importance of integrating them, the monothetic classifier performance was evaluated using each group of features individually. The goal behind this was to evaluate the discriminative power and information content of each feature group in isolation. To achieve this, the monothetic classifier was trained and tested using only one group of features at a time. The performance metrics were computed for each feature group separately. By examining the performance of the classifier when using only morphological features, textural features, and functional features independently, we could assess their individual impact on the classification outcome.

Experiment III. Two established methods from the literature [**39, 41**] were applied to our dataset of 99 samples to assess and validate the diagnostic performance attained by the presented system. These methods were chosen specifically to handle the same glioma grading classification issue (grades I, II, III, and IV), allowing a fair and relevant comparison. By applying these established approaches to our dataset, we aimed to estimate the significance of our GG-CAD system in comparison to the existing methods. This comparison would provide valuable insights into the strengths and improvements offered by our system. After executing the two comparison approaches on the dataset, we obtained diagnostic results for glioma grading. Subsequently, these results were carefully analyzed and compared with the final diagnostic outcomes achieved by our GG-CAD system.

Evaluation Metrics. The evaluation of the classification performance was done using two of the performance metrics (i.e., accuracy [**70, 71**] and quadratic-weighted Cohen's kappa [**72**]). Quadratic-weighted Cohen's kappa is a statistical measure utilized to evaluate the level of agreement among two rates. It extends the concept of Cohen's kappa coefficient by considering the agreement beyond what could occur by chance alone. In classification evaluation, quadratic-weighted Cohen's kappa considers not only the accuracy of the classification but also the proximity between the predicted and actual labels. It incorporates a weighting scheme that assigns different weights to agreement levels based on their severity or importance. This weighting system enables a more nuanced assessment of the agreement, emphasizing cases where the disagreement has a greater impact. On the other hand, accuracy is considered the most commonly applied metric in classification tasks that quantify the overall correctness of a model's predictions. It provides a straightforward measure of the model's performance by determining the ratio of correctly classified instances to the total number of instances in the dataset [**73**]. Accuracy is computed by comparing the model's predicted labels with the true labels of the instances [**74, 75**]. Each prediction that matches the true label is considered correct, while any prediction that differs from the true label is deemed incorrect. To ensure robustness and reliability, the experimental results, which will be presented in the following section, were recorded based on the mean value along with the standard deviation. This process was repeated 15 times to account for variations and provide a comprehensive understanding of performance consistency and variability.

2.5 RESULTS AND DISCUSSION

2.5.1 Results

2.5.1.1 Experiment I

By combining the selected features set with MLP-ANN, some diagnostic findings were achieved in order to evaluate the classification abilities of the tuned MLP-ANN classifier. Further, a positive comparison using the three validation methods previously described was performed in order to demonstrate the model's capacity to generalize and reproduce outcomes. The findings, which are shown in Table 2.1,

TABLE 2.1
Cross-Validation, LOSO and k-fold Stratified Cross-Validation (with k = 5 and 10), Results for Evaluating the GG-CAD System in Glioma Diagnosis Was Provided. Glioma Grades I, II, III, and IV Were Designated as GG-I, GG-II, GG-III, and GG-IV, Respectively

	Accuracy					
Approach	**GG-I**	**GG-II**	**GG-III**	**GG-IV**	**Overall**	**Kappa**
10-Fold	0.95 ± 0.06	0.93 ± 0.03	0.87 ± 0.06	0.95 ± 0.02	0.93± 0.02	0.91 ± 0.02
5-Fold	0.90 ± 0.06	0.94 ± 0.06	0.81 ± 0.05	0.94 ± 0.02	0.90 ± 0.02	0.90 ± 0.03
LOSO	0.96 ± 0.04	0.97 ± 0.04	0.91 ± 0.04	0.98 ± 0.03	0.96 ± 0.02	0.96 ± 0.02

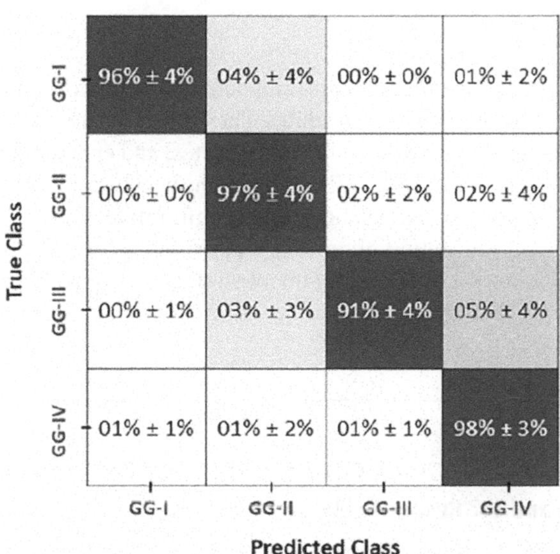

FIGURE 2.1 The confusion matrix of the ultimate MLP-ANN model illustrates the accuracy of classification for the various glioma grades. In this matrix, the values depict the degree of confusion between different grades. The glioma grades are denoted as GG-I, GG-II, GG-III, and GG-IV.

show that the suggested GG-CAD system consistently demonstrates a strong diagnostic performance across all validation techniques. Figure 2.1 depicts the confusion matrix, with an emphasis on accuracy, for the proposed model, which employs MLP-ANN with LOSO cross-validation.

2.5.1.2 Experiment II

As mentioned before, the effectiveness of the presented system was evaluated by combining specific features with the tuned MLP-ANN classification models. We aimed to emphasize the benefits of merging these distinct characteristics by comparing the diagnostic performance of the proposed model, which incorporates the merged extracted features with the performance of the individual models. The integrated model demonstrated superior performance, as indicated by an overall accuracy of 0.958 ± 0.019 and a kappa value of 0.96 ± 0.02. Table 2.2 clearly shows that the performance of the presented system outperformed that of all other individual models. This improvement is due to the incorporation of several quantitative features that represent many different aspects of the tumor.

2.5.1.3 Experiment III

In this experiment, comprehensive comparisons with various methodologies were made in order to highlight the benefits of the presented system. As shown in Table 2.3, the CAD system's diagnostic performance for glioma grading outperformed that of the methods previously stated.

2.5.2 Discussion

Mortality and morbidity are significantly impacted by gliomas, which are the most prevalent and common brain tumors in adults. They typically arise in the glial tissue of the CNS, which includes astrocytes, oligodendrocytes, and ependymal cells. Gliomas are divided into four classes by the WHO based on their histological characteristics. Pilocytic astrocytomas, which are solid and non-infiltrative tumors, are classified as grade I, while diffuse infiltrating gliomas are classified as grades II–IV. Since each glioma grade necessitates a different treatment strategy, accurate grading is essential. Despite being the gold standard for grading, surgical biopsy has certain drawbacks.

TABLE 2.2
The Abbreviations GG-I, GG-II, GG-III, and GG-IV Referred to Glioma Grades I, II, III, and IV, Respectively

	Accuracy						
MR Modality	GG-I	GG-II	GG-III	GG-IV	Overall	Kappa	MLP-ANN
Higher order 3D Appearance Features							
T2-MR (FLAIR)	0.93 ± 0.07	0.85 ± 0.08	0.73 ± 0.07	0.94 ± 0.02	0.87 ± 0.02	0.85 ± 0.03	(50, 25)
T1-MR (pre)	0.84 ± 0.07	0.84 ± 0.06	0.68 ± 0.04	0.93 ± 0.02	0.83 ± 0.02	0.71 ± 0.06	(50, 25)
T1-MR (post)	0.88 ± 0.06	0.87 ± 0.06	0.66 ± 0.05	0.93 ± 0.01	0.83 ± 0.02	0.78 ± 0.05	(50, 25)
Textural Features							
T2-MR (FLAIR)	0.88 ± 0.07	0.72 ± 0.06	0.61 ± 0.05	0.94 ± 0.02	0.81 ± 0.01	0.82 ± 0.02	(25, 10)
T1-MR (pre)	0.80 ± 0.09	0.77 ± 0.06	0.64 ± 0.06	0.92 ± 0.02	0.81 ± 0.02	0.79 ± 0.04	(25, 10)
T1-MR (post)	0.86 ± 0.04	0.81 ± 0.02	0.70 ± 0.06	0.92 ± 0.01	0.84 ± 0.03	0.78 ± 0.05	(25, 10)
Functional Features							
DW- & T1-MR	0.83 ± 0.08	0.79 ± 0.05	0.65 ± 0.07	0.92 ± 0.02	0.82 ± 0.03	0.79 ± 0.05	(50, 25)
Integrated Features							
GG-CAD (All)	0.96 ± 0.04	0.97 ± 0.04	0.91 ± 0.04	0.098 ± 0.03	0.958 ± 0.19	0.96 ± 0.02	(200, 100, 50)

TABLE 2.3
The Diagnostic Performance of the Proposed CAD System Was Compared With the Approaches Provided by Alis et al. [39] and Kalaiselvi et al. [41]

	Accuracy					
Model	**GG-I**	**GG-II**	**GG-III**	**GG-IV**	**Overall**	**Kappa**
GG-CAD	0.96 ± 0.04	0.97 ± 0.04	0.91 ± 0.04	0.98 ± 0.03	0.96 ± 0.02	0.96 ± 0.02
Alis [39]	0.62 ± 0.17	0.77 ± 0.06	0.67 ± 0.02	0.91 ± 0.02	0.79 ± 0.05	0.79 ± 0.05
Kalaiselvi [41]	0.77 ± 0.15	0.79 ± 0.05	0.67 ± 0.08	0.91 ± 0.02	0.81 ± 0.04	0.82 ± 0.05

Consequently, there has been considerable interest in exploring the potential of MRI-based grading using computer-aided diagnosis (CAD) systems as a non-invasive alternative to biopsy, as evidenced by previous studies [23, 24, 34, 35, 37–41].

This study demonstrated that the suggested GG-CAD system has a high level of diagnostic precision in identifying various glioma tumor grades. This precise tumor grading facilitated by the system empowers neuro-oncologists to make well-informed decisions regarding treatment planning and prognosis. While more sophisticated MRI modalities allow for quantitative evaluations of metabolic characteristics, tumor cellularity, and vascularity, conventional MRI offers insights into tumor architecture. Three different MRI modalities (i.e., DW, contrast-enhanced T1, and T2 [FLAIR]) were used to record different elements of tumor characteristics. By leveraging the most significant sets of extracted discriminatory features in combination with a powerful ML classifier such as MLP-ANN, the accurate determination of glioma tumor malignancy grade has been proven highly effective.

In this research, the study focused on combining and visualizing regions of interest (ROIs) extracted from images of different modalities as 3D objects representing the subjects. In contrast-enhanced T1 and DW, these 3D objects include several voxels that represent the tumor by itself, while in T2 (FLAIR), the tumor plus the surrounding edema are included. Different histopathological factors affect each voxel's signal intensity, which results in particular grayscale values. Despite being visually undetectable, the 3D arrays of grayscale values in the ROIs can show detailed geometric patterns that are specific to individual tumor grades. In this work, texture analysis was used to find these patterns.

Texture analysis offers a useful explanation of how the distribution of voxel values is influenced by the grayscale of each voxel inside a given region. Previous research [24, 34, 35, 37, 39, 40] has demonstrated the important influence of these texture data on the effectiveness of categorization algorithms. The generated model in this work used both first- and second-order texture features, utilizing a variety of texture analysis techniques and algorithms. First-order features, sometimes referred to as histogram-based features, use a histogram to convey the image intensity distribution (i.e., the voxel intensity signals are dispersed across the tumor). The spatial orientation and connections between voxels are not taken into consideration by these features. On the other hand, second-order features record statistical associations based on intensity levels between neighboring voxels or groups of voxels. By using matrices that precisely record the spatial interactions between the voxels inside certain regions of the tumor, these attributes quantitatively reflect the intratumoral heterogeneity. The GLCM and GLRLM were both used in the study. While GLRLM provides details about the size of homogeneous runs within the same object, GLCM offers the likelihood that two intensity levels will occur in adjacent voxels inside a single object.

Texture analysis plays a vital role in determining the glioma grade; however, the glioma severity also impacts the shape of the tumor along with the surrounding edema. In comparison to lower-grade gliomas, higher-grade gliomas have a more complex architecture as well as sharper borders. As a result, we included 3D appearance attributes to include the differences in shape across the various classes of gliomas. In addition, the tumor and the tumor combined with the edema volumes were computed.

Furthermore, it was crucial in our study to include the functionality of gliomas of varying grades, as it could significantly contribute to our objective. The functionality of gliomas influences their imaging characteristics, thereby enhancing the model's ability to identify the glioma grade. To assess functionality, we acquired 3D diffusion-weighted magnetic resonance images of each glioma using $b = 0$ and $b = 1,000$ s/mm^2. Subsequently, we calculated the apparent diffusion coefficients (ADCs) at the non-zero b-value to discern functional differences between subjects with different grades. In contrast-enhanced T1, we observed changes in contrast enhancement that varied according to the tumor's severity. To quantify these distinctions, we estimated contrast-enhancement slopes, which capture variations in the distribution of gray levels between the pre- and post-contrast phases. This measurement provides insight into the variations in enhancement among the different grades.

There are significant differences in the characteristics of glioma grades, but there is still some overlap between them. This overlap makes it challenging to accurately identify the grade of glioma using a single feature class, even with the most appropriate magnetic resonance (MR) sequence. However, we may get a better representation of features for identifying various glioma grades by selecting an appropriate feature set from the combination of all features. The GG-CAD system we created has great classification performance, correctly identifying the different glioma grades using a suitable feature set. Our results show that the suggested approaches, when combined with MR imaging, have the ability to diagnose and are clinically helpful. Tables 2.1–3 outline the findings, whereas Figure 2.1 shows the confusion matrix of the final suggested model.

2.6 CONCLUSIONS AND FUTURE WORK

In summary, the GG-CAD system has demonstrated outstanding diagnostic performance by effectively integrating and selecting optimal appearance, textural, and functional features. Utilizing an MLP-ANN classification model, the system achieved a high level of accuracy, with a kappa coefficient of 0.96 ± 0.02 and an overall accuracy of $95.8\% \pm 1.9\%$. These remarkable results underscore the potential and effectiveness of combining diverse and significant features that capture various aspects of glioma tumor characteristics, including their appearance, texture, and functionality.

Nevertheless, it is crucial to acknowledge the limitations associated with the relatively small size of the dataset used in this study. To overcome this limitation and further advance our research, we are actively working on collecting a larger and more comprehensive dataset. This expanded dataset will enable us to explore the capabilities of different deep learning techniques, such as convolutional neural networks (CNN) and stacked autoencoders, for more precise segmentation of glioma tumors, extraction of discriminating features, and automated identification of glioma grades. By leveraging these advanced techniques on a larger dataset, we can enhance the robustness and reliability of our grading system, ensuring its effectiveness across diverse patient populations.

Furthermore, our future investigations will focus on establishing correlations between pathological diagnoses and treatment responses, encompassing various outcomes such as no response, partial response, complete response, or progressive response. This critical aspect of our research aims to develop a computer-aided prediction (CAP) system capable of accurately forecasting treatment responses reliably and objectively. Such a personalized medicine approach will empower clinicians to devise the most suitable treatment plans for individual glioma patients, considering their unique characteristics and predicted responses to therapy. Ultimately, this integration of advanced technologies and personalized treatment strategies holds tremendous potential to improve patient outcomes and revolutionize the field of glioma diagnosis and treatment. By pursuing these avenues of research, we are dedicated to advancing our understanding of glioma pathophysiology, refining diagnostic accuracy, and tailoring treatment decisions to each patient's specific needs. Through these efforts, we aspire to make significant contributions to the field of glioma research, ultimately leading to improved patient care, enhanced treatment outcomes, and a brighter future for individuals affected by gliomas. In addition to the brain [76–98], this work could also be applied to various other applications in medical imaging, such as the prostate [99–103], the kidney [104–122], the heart [123–140], the lung [141–190], the vascular system [191–201], the retina [202–211], the bladder [212–216], the liver [217, 218], the head and neck [219–221], and injury prediction [222], as well as several non-medical applications [223–229].

REFERENCES

1. National Cancer Institute. *Brain and spinal cord tumors in adults*, 2021. Available online: https://www.cancer.gov/types/brain/hp/adult-brain-treatment-pdq
2. American Brain Tumor Association. *Key statistics*, 2021. Available online: https://www.abta.org/brain-tumor-facts-statistics/
3. Cancer.Net Editorial Board. *Brain tumor: Statistics*, 2022. Available online: https://www.cancer.net/cancer-types/brain-tumor/statistics
4. S. Ahmed, A. Gull, T. Khuroo, M. Aqil, and Y. Sultana, "Glial cell: a potential target for cellular and drug based therapy in various CNS diseases," *Current Pharmaceutical Design*, vol. 23, no. 16, 2017, pp. 2389–2399.
5. Y. Jiang, and L. Uhrbom, "On the origin of glioma," *Upsala Journal of Medical Sciences*, vol. 117, no. 2, 2012, pp. 113–121.
6. S. Gutta, J. Acharya, M. S. Shiroishi, D. Hwang, and K. S. Nayak, "Improved glioma grading using deep convolutional neural networks," *American Journal of Neuroradiology*, vol. 42, no. 2, 2021, pp. 233–239.

7. Q. Wang, D. Lei, Y. Yuan, and H. Zhao, "Accuracy of magnetic resonance imaging texture analysis in differentiating low-grade from high-grade gliomas: systematic review and meta-analysis," *BMJ Open*, vol. 9, no. 9, 2019, pp. e027144.
8. National Brain Tumor Society. *About brain Tumor: Types and statistics*, 2022. Available online: https://braintumor.org/braintumors/about-brain-tumors/brain-tumor-types/astrocytoma/
9. M. Qian, Z. Chen, X. Guo, S. Wang, Z. Zhang, W. Qiu,. . .and G. Li, "Exosomes derived from hypoxic glioma deliver miR-1246 and miR-10b-5p to normoxic glioma cells to promote migration and invasion," *Laboratory Investigation*, vol. 101, no. 5, 2021, pp. 612–624.
10. J. Ferlay, I. Soerjomataram, R. Dikshit, S. Eser, C. Mathers, M. Rebelo,. . .and F. Bray, "Cancer incidence and mortality worldwide: sources, methods and major patterns in GLOBOCAN 2012," *International Journal of Cancer*, vol. 136, no. 5, 2015, pp. E359–E386.
11. A. Sharma, and J. J. Graber, "Overview of prognostic factors in adult gliomas," *Annals of Palliative Medicine*, vol. 10, no. 1, 2021, pp. 863–874.
12. Y. Yang, M. Z. He, T. Li, and X. Yang, "MRI combined with PET-CT of different tracers to improve the accuracy of glioma diagnosis: a systematic review and meta-analysis," *Neurosurgical Review*, vol. 42, 2019, pp. 185–195.
13. R. Goldbrunner, M. Ruge, M. Kocher, C. W. Lucas, N. Galldiks, and S. Grau, "The treatment of gliomas in adulthood," *Deutsches Ärzteblatt International*, vol. 115, no. 20–21, 2018, p. 356.
14. P. Wesseling, and D. W. H. O. Capper, "WHO 2016 classification of gliomas," *Neuropathology and Applied Neurobiology*, vol. 44, no. 2, 2018, pp. 139–150.
15. J. J. Miller, and W. Wick, "What's new in grade II and grade III gliomas?," in *Seminars in neurology*. Thieme Medical Publishers, vol. 38, no. 1, 2018, February, pp. 41–49.
16. F. Citak-Er, Z. Firat, I. Kovanlikaya, U. Ture, and E. Ozturk-Isik, "Machine-learning in grading of gliomas based on multi-parametric magnetic resonance imaging at 3T," *Computers in Biology and Medicine*, vol. 99, 2018, pp. 154–160.
17. A. K. A. Razek, L. El-Serougy, M. Abdelsalam, G. Gaballa, and M. Talaat, "Multi-parametric arterial spin labelling and diffusion-weighted magnetic resonance imaging in differentiation of grade II and grade III gliomas," *Polish Journal of Radiology*, vol. 85, no. 1, 2020, pp. 110–117.
18. H. H. Cho, S. H. Lee, J. Kim, and H. Park, "Classification of the glioma grading using radiomics analysis," *PeerJ*, vol. 6, 2018, p. e5982.
19. A. Alksas, M. Shehata, H. Atef, F. Sherif, M. Yaghi, M. Alhalabi,. . .and A. El-Baz, "A comprehensive non-invasive system for early grading of gliomas," in *2022 26th International Conference on Pattern Recognition (ICPR)*. IEEE, 2022, August, pp. 4371–4377.
20. M. Iv, and S. Bisdas, "Neuroimaging in the era of the evolving WHO classification of brain tumors, from the AJR special series on cancer staging," *American Journal of Roentgenology*, no. 1, 2021, pp. 3–15.
21. A. A. K. A. Razek, M. Talaat, L. El-Serougy, G. Gaballa, and M. Abdelsalam, "Clinical applications of arterial spin labeling in brain tumors," *Journal of Computer Assisted Tomography*, vol. 43, no. 4, 2019, pp. 525–532.
22. A. Aggarwal, P. K. Das, A. Shukla, S. Parashar, M. Choudhary, A. Kumar,. . .and S. Dutta, "Role of multivoxel intermediate TE 2D CSI MR spectroscopy and 2D echoplanar diffusion imaging in grading of primary glial brain tumours," *Journal of Clinical and Diagnostic Research: JCDR*, vol. 11, no. 6, 2017, p. TC05.
23. Y. Zhuge, H. Ning, P. Mathen, J. Y. Cheng, A. V. Krauze, K. Camphausen, and R. W. Miller, "Automated glioma grading on conventional MRI images using deep convolutional neural networks," *Medical Physics*, vol. 47, no. 7, 2020, pp. 3044–3053.

24. H. H. Cho, S. H. Lee, J. Kim, and H. Park, "Classification of the glioma grading using radiomics analysis," *PeerJ*, vol. 6, 2018, p. e5982.
25. H. M. Balaha, E. M. El-Gendy, and M. M. Saafan, "A complete framework for accurate recognition and prognosis of COVID-19 patients based on deep transfer learning and feature classification approach," *Artificial Intelligence Review*, vol. 55, no. 6, 2022, pp. 5063–5108.
26. N. R. Yousif, H. M. Balaha, A. Y. Haikal, and E. M. El-Gendy, "A generic optimization and learning framework for Parkinson disease via speech and handwritten records," *Journal of Ambient Intelligence and Humanized Computing*, 2022, pp. 1–21.
27. A. T. Shalata, M. Shehata, E. Van Bogaert, K. M. Ali, A. Alksas, A. Mahmoud,. . .and A. El-Baz, "Predicting recurrence of non-muscle-invasive bladder cancer: current techniques and future trends," *Cancers*, vol. 14, no. 20, 2022, p. 5019.
28. H. M. Balaha, A. O. Shaban, E. M. El-Gendy, and M. M. Saafan, "A multi-variate heart disease optimization and recognition framework," *Neural Computing and Applications*, vol. 34, no. 18, 2022, pp. 15907–15944.
29. A. A. K. Abdel Razek, A. Alksas, M. Shehata, A. AbdelKhalek, K. Abdel Baky, A. El-Baz, and E. Helmy, "Clinical applications of artificial intelligence and radiomics in neuro-oncology imaging," *Insights into Imaging*, vol. 12, 2021, pp. 1–17.
30. K. Clark, B. Vendt, K. Smith, J. Freymann, J. Kirby, P. Koppel,. . .and F. Prior, "The cancer imaging archive (TCIA): maintaining and operating a public information repository," *Journal of Digital Imaging*, vol. 26, 2013, pp. 1045–1057.
31. K. V. Chang, and L. Özçakar, "What can artificial intelligence do for pain medicine?," *Asia Pacific Journal of Pain*, vol. 1, 2022, p. 1.
32. W. Gao, X. Zhang, J. Yuan, D. Li, Y. Sun, Z. Chen, and Z. Gu, "Application of medical imaging methods and artificial intelligence in tissue engineering and organ-on-a-chip," *Frontiers in Bioengineering and Biotechnology*, 2022, p. 1557.
33. M. Tabatabaei, A. Razaei, A. H. Sarrami, Z. Saadatpour, A. Singhal, and H. Sotoudeh, "Current status and quality of machine learning-based radiomics studies for glioma grading: a systematic review," *Oncology*, vol. 99, no. 7, 2021, pp. 433–443.
34. X. Zhang, L. F. Yan, Y. C. Hu, G. Li, Y. Yang, Y. Han,. . .and G. B. Cui, "Optimizing a machine learning based glioma grading system using multi-parametric MRI histogram and texture features," *Oncotarget*, vol. 8, no. 29, 2017, pp. 47816.
35. H. H. Cho, and H. Park, "Classification of low-grade and high-grade glioma using multi-modal image radiomics features," in *2017 39th Annual International Conference of the IEEE Engineering in Medicine and Biology Society (EMBC)*. IEEE, 2017, July, pp. 3081–3084.
36. D. G. Altman, and J. M. Bland, "Measurement in medicine: the analysis of method comparison studies," *Journal of the Royal Statistical Society: Series D (The Statistician)*, vol. 32, no. 3, 1983, pp. 307–317.
37. J. G. Suárez-García, J. M. Hernández-López, E. Moreno-Barbosa, and B. de Celis-Alonso, "A simple model for glioma grading based on texture analysis applied to conventional brain MRI," *PLoS One*, vol. 15, no. 5, 2020, p. e0228972.
38. S. Banerjee, S. Mitra, F. Masulli, and S. Rovetta, "Glioma classification using deep radiomics," *SN Computer Science*, vol. 1, no. 4, 2020, p. 209.
39. D. E. N. İ. Z. Alis, O. Bagcilar, Y. D. Senli, C. Isler, M. Yergin, N. Kocer,. . .and O. Kizilkilic, "The diagnostic value of quantitative texture analysis of conventional MRI sequences using artificial neural networks in grading gliomas," *Clinical Radiology*, vol. 75, no. 5, 2020, pp. 351–357.
40. K. L. C. Hsieh, C. Y. Chen, and C. M. Lo, "Quantitative glioma grading using transformed gray-scale invariant textures of MRI," *Computers in Biology and Medicine*, vol. 83, 2017, pp. 102–108.

41. T. Kalaiselvi, T. Padmapriya, P. Sriramakrishnan, and V. Priyadharshini, "Development of automatic glioma brain tumor detection system using deep convolutional neural networks," *International Journal of Imaging Systems and Technology*, vol. 30, no. 4, 2020, pp. 926–938.
42. I. Sharaby, A. Alksas, A. Nashat, H. M. Balaha, M. Shehata, M. Gayhart,. . .and A. El-Baz, "Prediction of wilms' tumor susceptibility to preoperative chemotherapy using a novel computer-aided prediction system," *Diagnostics*, vol. 13, no. 3, 2023, p. 486.
43. H. M. Balaha, M. H. Balaha, and H. A. Ali, "Hybrid COVID-19 segmentation and recognition framework (HMB-HCF) using deep learning and genetic algorithms," *Artificial Intelligence in Medicine*, vol. 119, 2021, p. 102156.
44. H. Patel, "What is feature engineering—importance, tools and techniques for machine learning," *Towards Data Science*, 2021. Available online: https://towardsdatascience.com/what-is-feature-engineering-importance-tools-and-techniques-formachine-learning-2080b0269f10 (visited on 08/31/2022).
45. A. Alksas, M. Shehata, H. Atef, F. Sherif, N. S. Alghamdi, M. Ghazal,. . .and A. El-Baz, "A novel system for precise grading of glioma," *Bioengineering*, vol. 9, no. 10, 2022, p. 532.
46. H. Khan, P. M. Shah, M. A. Shah, S. ul Islam, and J. J. Rodrigues, "Cascading handcrafted features and convolutional neural network for IoT-enabled brain tumor segmentation," *Computer Communications*, vol. 153, 2020, pp. 196–207.
47. W. Chen, B. Liu, S. Peng, J. Sun, and X. Qiao, "Computer-aided grading of gliomas combining automatic segmentation and radiomics," *International Journal of Biomedical Imaging*, vol. 2018, 2018.
48. R. K. McConnell, *Method of and apparatus for pattern recognition* (No. US 4567610), 1986.
49. N. Dalal, and B. Triggs, "Histograms of oriented gradients for human detection," in *2005 IEEE Computer Society Conference on Computer Vision and Pattern Recognition (CVPR'05)*, vol. 1. IEEE, 2005, June, pp. 886–893.
50. H. M. Balaha, and A. E. S. Hassan, "A variate brain tumor segmentation, optimization, and recognition framework," *Artificial Intelligence Review*, 2022, pp. 1–54.
51. N. Soni, S. Priya, and G. Bathla, "Texture analysis in cerebral gliomas: a review of the literature," *American Journal of Neuroradiology*, vol. 40, no. 6, 2019, pp. 928–934.
52. K. Skogen, A. Schulz, J. B. Dormagen, B. Ganeshan, E. Helseth, and A. Server, "Diagnostic performance of texture analysis on MRI in grading cerebral gliomas," *European Journal of Radiology*, vol. 85, no. 4, 2016, pp. 824–829.
53. R. M. Haralick, "Statistical and structural approaches to texture," *Proceedings of the IEEE*, vol. 67, no. 5, 1979, pp. 786–804.
54. M. M. Galloway, "Texture analysis using gray level run lengths," *Computer Graphics and Image Processing*, vol. 4, no. 2, 1975, pp. 172–179.
55. A. Alksas, M. Shehata, G. A. Saleh, A. Shaffie, A. Soliman, M. Ghazal,. . .and A. El-Baz, "A novel computer-aided diagnostic system for accurate detection and grading of liver tumors," *Scientific Reports*, vol. 11, no. 1, 2021, pp. 1–18.
56. A. Alksas, M. Shehata, G. A. Saleh, A. Shaffie, A. Soliman, M. Ghazal,. . .and A. El-Baz, "A novel computer-aided diagnostic system for early assessment of hepatocellular carcinoma," in *2020 25th International Conference on Pattern Recognition (ICPR)*. IEEE, 2021, January, pp. 10375–10382.
57. M. Shehata, A. Alksas, R. T. Abouelkheir, A. Elmahdy, A. Shaffie, A. Soliman,. . .and A. El-Baz, "A comprehensive computer-assisted diagnosis system for early assessment of renal cancer tumors," *Sensors*, vol. 21, no. 14, 2021, p. 4928.
58. M. Shehata, A. Alksas, R. T. Abouelkheir, A. Elmahdy, A. Shaffie, A. Soliman,. . .and A. El-Baz, "A new computer-aided diagnostic (cad) system for precise identification of renal tumors," in *2021 IEEE 18th International Symposium on Biomedical Imaging (ISBI)*. IEEE, 2021, April, pp. 1378–1381.

59. S. M. Ayyad, M. A. Badawy, M. Shehata, A. Alksas, A. Mahmoud, M. Abou El-Ghar,. . .and A. El-Baz, "A new framework for precise identification of prostatic adenocarcinoma," *Sensors*, vol. 22, no. 5, 2022, p. 1848.
60. M. A. Vermoolen, T. C. Kwee, and R. A. J. Nievelstein, "Apparent diffusion coefficient measurements in the differentiation between benign and malignant lesions: a systematic review," *Insights into Imaging*, vol. 3, 2012, pp. 395–409.
61. I. A. Basheer, and M. Hajmeer, "Artificial neural networks: fundamentals, computing, design, and application," *Journal of Microbiological Methods*, vol. 43, no. 1, 2000, pp. 3–31.
62. T. Hastie, R. Tibshirani, J. H. Friedman, and J. H. Friedman, *The elements of statistical learning: data mining, inference, and prediction.* Springer, vol. 2, 2009, pp. 1–758.
63. S. Lee, and J. Y. Choeh, "Predicting the helpfulness of online reviews using multilayer perceptron neural networks," *Expert Systems with Applications*, vol. 41, no. 6, 2014, pp. 3041–3046.
64. W. M. Bahgat, H. M. Balaha, Y. AbdulAzeem, and M. M. Badawy, "An optimized transfer learning-based approach for automatic diagnosis of COVID-19 from chest x-ray images," *PeerJ Computer Science*, vol. 7, 2021, p. e555.
65. N. A. Baghdadi, A. Malki, S. F. Abdelaliem, H. M. Balaha, M. Badawy, and M. Elhosseini, "An automated diagnosis and classification of COVID-19 from chest CT images using a transfer learning-based convolutional neural network," *Computers in Biology and Medicine*, vol. 144, 2022, p. 105383.
66. H. M. Balaha, H. A. Ali, E. K. Youssef, A. E. Elsayed, R. A. Samak, M. S. Abdelhaleem,. . .and M. M. Mohammed, "Recognizing arabic handwritten characters using deep learning and genetic algorithms," *Multimedia Tools and Applications*, vol. 80, 2021, pp. 32473–32509.
67. D. Fahmy, A. Alksas, A. Elnakib, A. Mahmoud, H. Kandil, A. Khalil,. . .and A. El-Baz, "The role of radiomics and AI technologies in the segmentation, detection, and management of hepatocellular carcinoma," *Cancers*, vol. 14, no. 24, 2022, p. 6123.
68. C. Albon, *Machine learning with python cookbook: practical solutions from preprocessing to deep learning*. O'Reilly Media, Inc., 2018.
69. H. M. Balaha, and A. E. S. Hassan, "Comprehensive machine and deep learning analysis of sensor-based human activity recognition," *Neural Computing and Applications*, 2023, pp. 1–39.
70. H. M. Balaha, H. A. Ali, M. Saraya, and M. Badawy, "A new Arabic handwritten character recognition deep learning system (AHCR-DLS)," *Neural Computing and Applications*, vol. 33, 2021, pp. 6325–6367.
71. Y. Abdulazeem, H. M. Balaha, W. M. Bahgat, and M. Badawy, "Human action recognition based on transfer learning approach," *IEEE Access*, vol. 9, 2021, pp. 82058–82069.
72. J. Cohen, "Weighted kappa: nominal scale agreement provision for scaled disagreement or partial credit," *Psychological Bulletin*, vol. 70, no. 4, 1968, p. 213.
73. H. M. Balaha, E. M. El-Gendy, and M. M. Saafan, "CovH2SD: a COVID-19 detection approach based on Harris Hawks optimization and stacked deep learning," *Expert Systems with Applications*, vol. 186, 2021, p. 115805.
74. H. M. Balaha, M. Saif, A. Tamer, and E. H. Abdelhay, "Hybrid deep learning and genetic algorithms approach (HMB-DLGAHA) for the early ultrasound diagnoses of breast cancer," *Neural Computing and Applications*, vol. 34, no. 11, 2022, pp. 8671–8695.
75. H. M. Balaha, and A. E. S. Hassan, "Skin cancer diagnosis based on deep transfer learning and sparrow search algorithm," *Neural Computing and Applications*, vol. 35, no. 1, 2023, pp. 815–853.
76. A. A. K. Abdel Razek, A. Alksas, M. Shehata, A. AbdelKhalek, K. Abdel Baky, A. El-Baz, and E. Helmy, "Clinical applications of artificial intelligence and radiomics in neuro-oncology imaging," *Insights into Imaging*, vol. 12, no. 1, 2021, pp. 1–17.

77. Y. ElNakieb, M. T. Ali, O. Dekhil, M. E. Khalefa, A. Soliman, A. Shalaby, A. Mahmoud, M. Ghazal, H. Hajjdiab, A. Elmaghraby et al., "Towards accurate personalized autism diagnosis using different imaging modalities: sMRI, fMRI, and DTI," in *2018 IEEE International Symposium on Signal Processing and Information Technology (ISSPIT)*. IEEE, 2018, pp. 447–452.
78. Y. ElNakieb, A. Soliman, A. Mahmoud, O. Dekhil, A. Shalaby, M. Ghazal, A. Khalil, A. Switala, R. S. Keynton, G. N. Barnes et al., "Autism spectrum disorder diagnosis framework using diffusion tensor imaging," in *2019 IEEE International Conference on Imaging Systems and Techniques (IST)*. IEEE, 2019, pp. 1–5.
79. R. Haweel, O. Dekhil, A. Shalaby, A. Mahmoud, M. Ghazal, R. Keynton, G. Barnes, and A. El-Baz, "A machine learning approach for grading autism severity levels using task-based functional MRI," in *2019 IEEE International Conference on Imaging Systems and Techniques (IST)*. IEEE, 2019, pp. 1–5.
80. O. Dekhil, M. Ali, R. Haweel, Y. Elnakib, M. Ghazal, H. Hajjdiab, L. Fraiwan, A. Shalaby, A. Soliman, A. Mahmoud et al., "A comprehensive framework for differentiating autism spectrum disorder from neurotypicals by fusing structural MRI and resting state functional mri," in *Seminars in Pediatric Neurology*. Elsevier, 2020, p. 100805.
81. R. Haweel, O. Dekhil, A. Shalaby, A. Mahmoud, M. Ghazal, A. Khalil, R. Keynton, G. Barnes, and A. El-Baz, "A novel framework for grading autism severity using task-based fMRI," in *2020 IEEE 17th International Symposium on Biomedical Imaging (ISBI)*. IEEE, 2020, pp. 1404–1407.
82. A. El-Baz, A. Elnakib, F. Khalifa, M. A. El-Ghar, P. McClure, A. Soliman, and G. Gimel'farb, "Precise segmentation of 3-D magnetic resonance angiography," *IEEE Transactions on Biomedical Engineering*, vol. 59, no. 7, 2012, pp. 2019–2029.
83. A. El-Baz, A. Farag, A. Elnakib, M. F. Casanova, G. Gimel'farb, A. E. Switala, D. Jordan, and S. Rainey, "Accurate automated detection of autism related corpus callosum abnormalities," *Journal of Medical Systems*, vol. 35, no. 5, 2011, pp. 929–939.
84. A. El-Baz, G. Gimel'farb, R. Falk, M. A. El-Ghar, V. Kumar, and D. Heredia, "A novel 3D joint Markov-Gibbs model for extracting blood vessels from PC–MRA images," in *Medical Image Computing and Computer-Assisted Intervention–MICCAI 2009*, vol. 5762. Springer, 2009, pp. 943–950.
85. A. Elnakib, A. El-Baz, M. F. Casanova, G. Gimel'farb, and A. E. Switala, "Image-based detection of corpus callosum variability for more accurate discrimination between dyslexic and normal brains," in *Proceedings of IEEE International Symposium on Biomedical Imaging: From Nano to Macro (ISBI'2010)*. IEEE, 2010, pp. 109–112.
86. A. Elnakib, M. F. Casanova, G. Gimel'farb, A. E. Switala, and A. El-Baz, "Autism diagnostics by centerline-based shape analysis of the corpus callosum," in *Proceedings of IEEE International Symposium on Biomedical Imaging: From Nano to Macro (ISBI'2011)*. IEEE, 2011, pp. 1843–1846.
87. A. Elnakib, M. Nitzken, M. Casanova, H. Park, G. Gimel'farb, and A. El-Baz, "Quantification of age-related brain cortex change using 3D shape analysis," in *Pattern Recognition (ICPR), 2012 21st International Conference on*. IEEE, 2012, pp. 41–44.
88. M. Nitzken, M. Casanova, G. Gimel'farb, A. Elnakib, F. Khalifa, A. Switala, and A. El-Baz, "3D shape analysis of the brain cortex with application to dyslexia," in *Image Processing (ICIP), 2011 18th IEEE International Conference on*, Brussels, Belgium. IEEE, 2011, September, pp. 2657–2660. (Selected for oral presentation. Oral acceptance rate is 10 percent and the overall acceptance rate is 35 percent).
89. F. E.-Z. A. El-Gamal, M. M. Elmogy, M. Ghazal, A. Atwan, G. N. Barnes, M. F. Casanova, R. Keynton, and A. S. El-Baz, "A novel cad system for local and global early diagnosis of alzheimer's disease based on pib-pet scans," in *2017 IEEE International Conference on Image Processing (ICIP)*. IEEE, 2017, pp. 3270–3274.

90. M. M. Ismail, R. S. Keynton, M. M. Mostapha, A. H. ElTanboly, M. F. Casanova, G. L. Gimel'farb, and A. El-Baz, "Studying autism spectrum disorder with structural and diffusion magnetic resonance imaging: a survey," *Frontiers in Human Neuroscience*, vol. 10, 2016, p. 211.
91. A. Alansary, M. Ismail, A. Soliman, F. Khalifa, M. Nitzken, A. Elnakib, M. Mostapha, A. Black, K. Stinebruner, M. F. Casanova et al., "Infant brain extraction in t1-weighted MR images using bet and refinement using LCDG and MGRF models," *IEEE Journal of Biomedical and Health Informatics*, vol. 20, no. 3, 2016, pp. 925–935.
92. E. H. Asl, M. Ghazal, A. Mahmoud, A. Aslantas, A. Shalaby, M. Casanova, G. Barnes, G. Gimel'farb, R. Keynton, and A. El-Baz, "Alzheimer's disease diagnostics by a 3D deeply supervised adaptable convolutional network," *Frontiers in Bioscience* (Landmark Edition), vol. 23, 2018, pp. 584–596.
93. O. Dekhil, M. Ali, Y. El-Nakieb, A. Shalaby, A. Soliman, A. Switala, A. Mahmoud, M. Ghazal, H. Hajjdiab, M. F. Casanova, A. Elmaghraby, R. Keynton, A. El-Baz, and G. Barnes, "A personalized autism diagnosis cad system using a fusion of structural MRI and resting-state functional MRI data," *Frontiers in Psychiatry*, vol. 10, 2019, p. 392. [Online]. Available online: https://www.frontiersin.org/article/10.3389/fpsyt.2019.00392
94. O. Dekhil, A. Shalaby, A. Soliman, A. Mahmoud, M. Kong, G. Barnes, A. Elmaghraby, and A. El-Baz, "Identifying brain areas correlated with ADOS raw scores by studying altered dynamic functional connectivity patterns," *Medical Image Analysis*, vol. 68, 2021, p. 101899.
95. Y. A. Elnakieb, M. T. Ali, A. Soliman, A. H. Mahmoud, A. M. Shalaby, N. S. Alghamdi, M. Ghazal, A. Khalil, A. Switala, R. S. Keynton et al., "Computer aided autism diagnosis using diffusion tensor imaging," *IEEE Access*, vol. 8, 2020, pp. 191298–191308.
96. M. T. Ali, Y. A. Elnakieb, A. Shalaby, A. Mahmoud, A. Switala, M. Ghazal, A. Khelifi, L. Fraiwan, G. Barnes, and A. El-Baz, "Autism classification using sMRI: a recursive features selection based on sampling from multi-level high dimensional spaces," in *2021 IEEE 18th International Symposium on Biomedical Imaging (ISBI)*. IEEE, 2021, pp. 267–270.
97. M. T. Ali, Y. ElNakieb, A. Elnakib, A. Shalaby, A. Mahmoud, M. Ghazal, J. Yousaf, H. Abu Khalifeh, M. Casanova, G. Barnes et al., "The role of structure MRI in diagnosing autism," *Diagnostics*, vol. 12, no. 1, 2022, p. 165.
98. Y. ElNakieb, M. T. Ali, A. Elnakib, A. Shalaby, A. Soliman, A. Mahmoud, M. Ghazal, G. N. Barnes, and A. El-Baz, "The role of diffusion tensor MR imaging (DTI) of the brain in diagnosing autism spectrum disorder: promising results," *Sensors*, vol. 21, no. 24, 2021, p. 8171.
99. I. Reda, M. Ghazal, A. Shalaby, M. Elmogy, A. AbouEl-Fetouh, B. O. Ayinde, M. AbouEl-Ghar, A. Elmaghraby, R. Keynton, and A. El-Baz, "A novel ADCs-based CNN classification system for precise diagnosis of prostate cancer," in *2018 24th International Conference on Pattern Recognition (ICPR)*. IEEE, 2018, pp. 3923–3928.
100. I. Reda, A. Khalil, M. Elmogy, A. Abou El-Fetouh, A. Shalaby, M. Abou El-Ghar, A. Elmaghraby, M. Ghazal, and A. El-Baz, "Deep learning role in early diagnosis of prostate cancer," *Technology in Cancer Research & Treatment*, vol. 17, 2018, p. 1533034618775530.
101. I. Reda, B. O. Ayinde, M. Elmogy, A. Shalaby, M. El-Melegy, M. A. El-Ghar, A. A. El-fetouh, M. Ghazal, and A. El-Baz, "A new cnn-based system for early diagnosis of prostate cancer," in *2018 IEEE 15th International Symposium on Biomedical Imaging (ISBI 2018)*. IEEE, 2018, pp. 207–210.
102. S. M. Ayyad, M. A. Badawy, M. Shehata, A. Alksas, A. Mahmoud, M. Abou El-Ghar, M. Ghazal, M. El-Melegy, N. B. Abdel-Hamid, L. M. Labib, H. A. Ali, and A. El-Baz, "A new framework for precise identification of prostatic adenocarcinoma," *Sensors*, vol. 22, no. 5, 2022. [Online]. Available online: https://www.mdpi.com/1424-8220/22/5/1848

103. K. Hammouda, F. Khalifa, M. El-Melegy, M. Ghazal, H. E. Darwish, M. A. El-Ghar, and A. El-Baz, "A deep learning pipeline for grade groups classification using digitized prostate biopsy specimens," *Sensors*, vol. 21, no. 20, 2021, p. 6708.
104. M. Shehata, A. Shalaby, A. E. Switala, M. El-Baz, M. Ghazal, L. Fraiwan, A. Khalil, M. A. El-Ghar, M. Badawy, A. M. Bakr et al., "A multimodal computer-aided diagnostic system for precise identification of renal allograft rejection: preliminary results," *Medical Physics*, vol. 47, no. 6, 2020, pp. 2427–2440.
105. M. Shehata, F. Khalifa, A. Soliman, M. Ghazal, F. Taher, M. Abou El-Ghar, A. C. Dwyer, G. Gimel'farb, R. S. Keynton, and A. El- Baz, "Computer-aided diagnostic system for early detection of acute renal transplant rejection using diffusion-weighted MRI," *IEEE Transactions on Biomedical Engineering*, vol. 66, no. 2, 2018, pp. 539–552.
106. E. Hollis, M. Shehata, M. Abou El-Ghar, M. Ghazal, T. El-Diasty, M. Merchant, A. E. Switala, and A. El-Baz, "Statistical analysis of ADCs and clinical biomarkers in detecting acute renal transplant rejection," *The British Journal of Radiology*, vol. 90, no. 1080, 2017, p. 20170125.
107. M. Shehata, A. Alksas, R. T. Abouelkheir, A. Elmahdy, A. Shaffie, A. Soliman, M. Ghazal, H. Abu Khalifeh, R. Salim, A. A. K. Abdel Razek et al., "A comprehensive computer-assisted diagnosis system for early assessment of renal cancer tumors," Sensors, vol. 21, no. 14, 2021, p. 4928.
108. F. Khalifa, G. M. Beache, M. A. El-Ghar, T. El-Diasty, G. Gimel'farb, M. Kong, and A. El-Baz, "Dynamic contrast-enhanced MRI- based early detection of acute renal transplant rejection," IEEE Transactions on Medical Imaging, vol. 32, no. 10, 2013, pp. 1910–1927.
109. F. Khalifa, M. A. El-Ghar, B. Abdollahi, H. Frieboes, T. El-Diasty, and A. El-Baz, "A comprehensive non-invasive framework for automated evaluation of acute renal transplant rejection using DCE-MRI," *NMR in Biomedicine*, vol. 26, no. 11, 2013, pp. 1460–1470.
110. F. Khalifa, A. Elnakib, G. M. Beache, G. Gimel'farb, M. A. El-Ghar, G. Sokhadze, S. Manning, P. McClure, and A. El-Baz, "3D kidney segmentation from CT images using a level set approach guided by a novel stochastic speed function," in *Proceedings of International Conference Medical Image Computing and Computer-Assisted Intervention, (MICCAI'11)*, Toronto, Canada, 2011, September 18–22, pp. 587–594.
111. M. Shehata, F. Khalifa, E. Hollis, A. Soliman, E. Hosseini-Asl, M. A. El-Ghar, M. El-Baz, A. C. Dwyer, A. El-Baz, and R. Keynton, "A new non-invasive approach for early classification of renal rejection types using diffusion-weighted MRI," in *IEEE International Conference on Image Processing (ICIP)*. IEEE, 2016, pp. 136–140.
112. F. Khalifa, A. Soliman, A. Takieldeen, M. Shehata, M. Mostapha, A. Shaffie, R. Ouseph, A. Elmaghraby, and A. El-Baz, "Kidney segmentation from CT images using a 3D NMF-guided active contour model," in *IEEE 13th International Symposium on Biomedical Imaging (ISBI)*. IEEE, 2016, pp. 432–435.
113. M. Shehata, F. Khalifa, A. Soliman, A. Takieldeen, M. A. El-Ghar, A. Shaffie, A. C. Dwyer, R. Ouseph, A. El-Baz, and R. Keynton, "3D diffusion MRI-based CAD system for early diagnosis of acute renal rejection," in *Biomedical Imaging (ISBI), 2016 IEEE 13th International Symposium on*. IEEE, 2016, pp. 1177–1180.
114. M. Shehata, F. Khalifa, A. Soliman, R. Alrefai, M. A. El-Ghar, A. C. Dwyer, R. Ouseph, and A. El-Baz, "A level set-based framework for 3D kidney segmentation from diffusion MR images," in *IEEE International Conference on Image Processing (ICIP)*. IEEE, 2015, pp. 4441–4445.
115. M. Shehata, F. Khalifa, A. Soliman, M. A. El-Ghar, A. C. Dwyer, G. Gimel'farb, R. Keynton, and A. El-Baz, "A promising non-invasive cad system for kidney function assessment," in *International Conference on Medical Image Computing and Computer-Assisted Intervention*. Springer, 2016, pp. 613–621.

116. F. Khalifa, A. Soliman, A. Elmaghraby, G. Gimel'farb, and A. El-Baz, "3D kidney segmentation from abdominal images using spatial-appearance models," *Computational and Mathematical Methods in Medicine*, vol. 2017, 2017, pp. 1–10.
117. E. Hollis, M. Shehata, F. Khalifa, M. A. El-Ghar, T. El-Diasty, and A. El-Baz, "Towards non-invasive diagnostic techniques for early detection of acute renal transplant rejection: a review," *The Egyptian Journal of Radiology and Nuclear Medicine*, vol. 48, no. 1, 2016, pp. 257–269.
118. M. Shehata, F. Khalifa, A. Soliman, M. A. El-Ghar, A. C. Dwyer, and A. El-Baz, "Assessment of renal transplant using image and clinical-based biomarkers," in *Proceedings of 13th Annual Scientific Meeting of American Society for Diagnostics and Interventional Nephrology (ASDIN'17)*, New Orleans, LA, USA, 2017, February 10–12.
119. M. Shehata, F. Khalifa, A. Soliman, M. A. El-Ghar, A. C. Dwyer, and A. El-Baz, "Early assessment of acute renal rejection," in *Proceedings of 12th Annual Scientific Meeting of American Society for Diagnostics and Interventional Nephrology (ASDIN'16)*, Pheonix, AZ, USA, 2016, February 19–21.
120. A. Eltanboly, M. Ghazal, H. Hajjdiab, A. Shalaby, A. Switala, A. Mahmoud, P. Sahoo, M. El-Azab, and A. El-Baz, "Level sets-based image segmentation approach using statistical shape priors," *Applied Mathematics and Computation*, vol. 340, 2019, pp. 164–179.
121. M. Shehata, A. Mahmoud, A. Soliman, F. Khalifa, M. Ghazal, M. A. El-Ghar, M. El-Melegy, and A. El-Baz, "3D kidney segmentation from abdominal diffusion MRI using an appearance-guided deformable boundary," *PLoS One*, vol. 13, no. 7, 2018, p. e0200082.
122. H. Abdeltawab, M. Shehata, A. Shalaby, F. Khalifa, A. Mahmoud, M. A. El-Ghar, A. C. Dwyer, M. Ghazal, H. Hajjdiab, R. Keynton et al., "A novel cnn-based cad system for early assessment of transplanted kidney dysfunction," *Scientific Reports*, vol. 9, no. 1, 2019, p. 5948.
123. K. Hammouda, F. Khalifa, H. Abdeltawab, A. Elnakib, G. Giridharan, M. Zhu, C. Ng, S. Dassanayaka, M. Kong, H. Darwish et al., "A new framework for performing cardiac strain analysis from cine MRI imaging in mice," *Scientific Reports*, vol. 10, no. 1, 2020, pp. 1–15.
124. H. Abdeltawab, F. Khalifa, K. Hammouda, J. M. Miller, M. M. Meki, Q. Ou, A. El-Baz, and T. Mohamed, "Artificial intelligence based framework to quantify the cardiomyocyte structural integrity in heart slices," *Cardiovascular Engineering and Technology*, 2021, pp. 1–11.
125. F. Khalifa, G. M. Beache, A. Elnakib, H. Sliman, G. Gimel'farb, K. C. Welch, and A. El-Baz, "A new shape-based framework for the left ventricle wall segmentation from cardiac first-pass perfusion MRI," in *Proceedings of IEEE International Symposium on Biomedical Imaging: From Nano to Macro, (ISBI'13)*, San Francisco, CA, 2013, April 7–11, pp. 41–44.
126. F. Khalifa, G. M. Beache, A. Elnakib, H. Sliman, G. Gimel'farb, K. C. Welch, and A. El-Baz, "A new nonrigid registration framework for improved visualization of transmural perfusion gradients on cardiac first–pass perfusion MRI," in *Proceedings of IEEE International Symposium on Biomedical Imaging: From Nano to Macro, (ISBI'12)*, Barcelona, Spain, 2012, May 2–5, pp. 828–831.
127. F. Khalifa, G. M. Beache, A. Firjani, K. C. Welch, G. Gimel'farb, and A. El-Baz, "A new nonrigid registration approach for motion correction of cardiac first-pass perfusion MRI," in *Proceedings of IEEE International Conference on Image Processing, (ICIP'12)*, Lake Buena Vista, Florida, 2012, September 30–October 3, pp. 1665–1668.
128. F. Khalifa, G. M. Beache, G. Gimel'farb, and A. El-Baz, "A novel CAD system for analyzing cardiac first-pass MR images," in *Proceedings of IAPR International Conference on Pattern Recognition (ICPR'12)*, Tsukuba Science City, Japan, 2012, November 11–15, pp. 77–80.

129. F. Khalifa, G. M. Beache, G. Gimel'farb, and A. El-Baz, "A novel approach for accurate estimation of left ventricle global indexes from short-axis cine MRI," in *Proceedings of IEEE International Conference on Image Processing, (ICIP'11)*, Brussels, Belgium, 2011, September 11–14, pp. 2645–2649.
130. F. Khalifa, G. M. Beache, G. Gimel'farb, G. A. Giridharan, and A. El-Baz, "A new image-based framework for analyzing cine images," in *Handbook of multi modality state-of-the-art medical image segmentation and registration methodologies*, A. El-Baz, U. R. Acharya, M. Mirmedhdi, and J. S. Suri, Eds. Springer, vol. 2, ch. 3, 2011, pp. 69–98.
131. F. Khalifa, G. M. Beache, G. Gimel'farb, G. A. Giridharan, and A. El-Baz, "Accurate automatic analysis of cardiac cine images," *IEEE Transactions on Biomedical Engineering*, vol. 59, no. 2, 2012, pp. 445–455.
132. F. Khalifa, G. M. Beache, M. Nitzken, G. Gimel'farb, G. A. Giridharan, and A. El-Baz, "Automatic analysis of left ventricle wall thickness using short-axis cine CMR images," in *Proceedings of IEEE International Symposium on Biomedical Imaging: From Nano to Macro, (ISBI'11)*, Chicago, Illinois, 2011, March 30–April 2, pp. 1306–1309.
133. M. Nitzken, G. Beache, A. Elnakib, F. Khalifa, G. Gimel'farb, and A. El-Baz, "Accurate modeling of tagged CMR 3D image appearance characteristics to improve cardiac cycle strain estimation," in *Image Processing (ICIP), 2012 19th IEEE International Conference on*, Orlando, Florida, USA. IEEE, 2012, September, pp. 521–524.
134. M. Nitzken, G. Beache, A. Elnakib, F. Khalifa, G. Gimel'farb, and A. El-Baz, "Improving full-cardiac cycle strain estimation from tagged CMR by accurate modeling of 3D image appearance characteristics," in *Biomedical Imaging (ISBI), 2012 9th IEEE International Symposium on*, Barcelona, Spain. IEEE, 2012, May, pp. 462–465. (Selected for oral presentation).
135. M. J. Nitzken, A. S. El-Baz, and G. M. Beache, "Markov-Gibbs random field model for improved full-cardiac cycle strain estimation from tagged CMR," *Journal of Cardiovascular Magnetic Resonance*, vol. 14, no. 1, 2012, pp. 1–2.
136. H. Sliman, A. Elnakib, G. Beache, A. Elmaghraby, and A. El-Baz, "Assessment of myocardial function from cine cardiac MRI using a novel 4D tracking approach," *Journal of Computer Science and Systems Biology*, vol. 7, 2014, pp. 169–173.
137. H. Sliman, A. Elnakib, G. M. Beache, A. Soliman, F. Khalifa, G. Gimel'farb, A. Elmaghraby, and A. El-Baz, "A novel 4D PDE-based approach for accurate assessment of myocardium function using cine cardiac magnetic resonance images," in *Proceedings of IEEE International Conference on Image Processing (ICIP'14)*, Paris, France, 2014, October 27–30, pp. 3537–3541.
138. H. Sliman, F. Khalifa, A. Elnakib, G. M. Beache, A. Elmaghraby, and A. El-Baz, "A new segmentation-based tracking framework for extracting the left ventricle cavity from cine cardiac MRI," in *Proceedings of IEEE International Conference on Image Processing, (ICIP'13)*, Melbourne, Australia, 2013, September 15–18, pp. 685–689.
139. H. Sliman, F. Khalifa, A. Elnakib, A. Soliman, G. M. Beache, A. Elmaghraby, G. Gimel'farb, and A. El-Baz, "Myocardial borders segmentation from cine MR images using bi-directional coupled parametric deformable models," *Medical Physics*, vol. 40, no. 9, 2013, pp. 1–13.
140. H. Sliman, F. Khalifa, A. Elnakib, A. Soliman, G. M. Beache, G. Gimel'farb, A. Emam, A. Elmaghraby, and A. El-Baz, "Accurate segmentation framework for the left ventricle wall from cardiac cine MRI," in *Proceedings of International Symposium on Computational Models for Life Science, (CMLS'13)*, vol. 1559, Sydney, Australia, 2013, November 27–29, pp. 287–296.
141. A. Sharafeldeen, M. Elsharkawy, N. S. Alghamdi, A. Soliman, and A. El-Baz, "Precise segmentation of covid-19 infected lung from CT images based on adaptive first-order appearance model with morphological/anatomical constraints," *Sensors*, vol. 21, no. 16, 2021, p. 5482.

142. M. Elsharkawy, A. Sharafeldeen, F. Taher, A. Shalaby, A. Soliman, A. Mahmoud, M. Ghazal, A. Khalil, N. S. Alghamdi, A. A. K. A. Razek et al., "Early assessment of lung function in coronavirus patients using invariant markers from chest x-rays images," *Scientific Reports*, vol. 11, no. 1, 2021, pp. 1–11.
143. B. Abdollahi, A. C. Civelek, X.-F. Li, J. Suri, and A. El-Baz, "PET/CT nodule segmentation and diagnosis: a survey," in *Multi detector CT imaging*, L. Saba, and J. S. Suri, Eds. Taylor, Francis, ch. 30, 2014, pp. 639–651.
144. B. Abdollahi, A. El-Baz, and A. A. Amini, "A multi-scale non-linear vessel enhancement technique," in *Engineering in Medicine and Biology Society, EMBC, 2011 Annual International Conference of the IEEE*. IEEE, 2011, pp. 3925–3929.
145. B. Abdollahi, A. Soliman, A. Civelek, X.-F. Li, G. Gimel'farb, and A. El-Baz, "A novel gaussian scale space-based joint MGRF framework for precise lung segmentation," in *Proceedings of IEEE International Conference on Image Processing, (ICIP'12)*. IEEE, 2012, pp. 2029–2032.
146. B. Abdollahi, A. Soliman, A. Civelek, X.-F. Li, G. Gimel'farb, and A. El-Baz, "A novel 3D joint MGRF framework for precise lung segmentation," in *Machine learning in medical imaging*. Springer, 2012, pp. 86–93.
147. A. M. Ali, A. S. El-Baz, and A. A. Farag, "A novel framework for accurate lung segmentation using graph cuts," in *Proceedings of IEEE International Symposium on Biomedical Imaging: From Nano to Macro, (ISBI'07)*. IEEE, 2007, pp. 908–911.
148. A. El-Baz, G. M. Beache, G. Gimel'farb, K. Suzuki, and K. Okada, "Lung imaging data analysis," *International Journal of Biomedical Imaging*, vol. 2013, 2013, pp. 1–2.
149. A. El-Baz, G. M. Beache, G. Gimel'farb, K. Suzuki, K. Okada, A. Elnakib, A. Soliman, and B. Abdollahi, "Computer-aided diagnosis systems for lung cancer: Challenges and methodologies," *International Journal of Biomedical Imaging*, vol. 2013, 2013, pp. 1–46.
150. A. El-Baz, A. Elnakib, M. Abou El-Ghar, G. Gimel'farb, R. Falk, and A. Farag, "Automatic detection of 2D and 3D lung nodules in chest spiral CT scans," *International Journal of Biomedical Imaging*, vol. 2013, 2013, pp. 1–11.
151. A. El-Baz, A. A. Farag, R. Falk, and R. La Rocca, "A unified approach for detection, visualization, and identification of lung abnormalities in chest spiral CT scans," in *International Congress Series*, vol. 1256. Elsevier, 2003, pp. 998–1004.
152. A. El-Baz, A. A. Farag, R. Falk, and R. La Rocca, "Detection, visualization and identification of lung abnormalities in chest spiral CT scan: Phase-I," in *Proceedings of International Conference on Biomedical Engineering*, vol. 12, no. 1, Cairo, Egypt, 2002.
153. A. El-Baz, A. Farag, G. Gimel'farb, R. Falk, M. A. El-Ghar, and T. Eldiasty, "A framework for automatic segmentation of lung nodules from low dose chest CT scans," in *Proceedings of International Conference on Pattern Recognition, (ICPR'06)*, vol. 3. IEEE, 2006, pp. 611–614.
154. A. El-Baz, A. Farag, G. Gimel'farb, R. Falk, and M. A. El-Ghar, "A novel level set-based computer-aided detection system for automatic detection of lung nodules in low dose chest computed tomography scans," *Lung Imaging and Computer Aided Diagnosis*, vol. 10, 2011, pp. 221–238.
155. A. El-Baz, G. Gimel'farb, M. Abou El-Ghar, and R. Falk, "Appearance-based diagnostic system for early assessment of malignant lung nodules," in *Proceedings of IEEE International Conference on Image Processing, (ICIP'12)*. IEEE, 2012, pp. 533–536.
156. A. El-Baz, G. Gimel'farb, and R. Falk, "A novel 3D framework for automatic lung segmentation from low dose CT images," in *Lung imaging and computer aided diagnosis*, A. El-Baz, and J. S. Suri, Eds. Taylor, Francis, ch. 1, 2011, pp. 1–16.
157. A. El-Baz, G. Gimel'farb, R. Falk, and M. El-Ghar, "Appearance analysis for diagnosing malignant lung nodules," in *Proceedings of IEEE International Symposium on Biomedical Imaging: From Nano to Macro (ISBI'10)*. IEEE, 2010, pp. 193–196.

158. A. El-Baz, G. Gimel'farb, R. Falk, and M. A. El-Ghar, "A novel level set-based CAD system for automatic detection of lung nodules in low dose chest CT scans," in *Lung imaging and computer aided diagnosis*, A. El-Baz, and J. S. Suri, Eds. Taylor, Francis, vol. 1, ch. 10, 2011, pp. 221–238.
159 A. El-Baz, G. Gimel'farb, R. Falk, and M. A. El-Ghar, "A new approach for automatic analysis of 3D low dose CT images for accurate monitoring the detected lung nodules," in *Proceedings of International Conference on Pattern Recognition, (ICPR'08)*. IEEE, 2008, pp. 1–4.
160. A. El-Baz, G. Gimel'farb, R. Falk, and M. A. El-Ghar, "A novel approach for automatic follow-up of detected lung nodules," in *Proceedings of IEEE International Conference on Image Processing, (ICIP'07)*, vol. 5. IEEE, 2007, pp. V–501.
161. A. El-Baz, G. Gimel'farb, R. Falk, and M. A. El-Ghar, "A new CAD system for early diagnosis of detected lung nodules," in *Image Processing, 2007, ICIP 2007, IEEE International Conference on*, vol. 2. IEEE, 2007, pp. II–461.
162. A. El-Baz, G. Gimel'farb, R. Falk, M. A. El-Ghar, and H. Refaie, "Promising results for early diagnosis of lung cancer," in *Proceedings of IEEE International Symposium on Biomedical Imaging: From Nano to Macro, (ISBI'08)*. IEEE, 2008, pp. 1151–1154.
163. A. El-Baz, G. L. Gimel'farb, R. Falk, M. Abou El-Ghar, T. Holland, and T. Shaffer, "A new stochastic framework for accurate lung segmentation," in *Proceedings of Medical Image Computing and Computer-Assisted Intervention, (MICCAI'08)*, 2008, pp. 322–330.
164. A. El-Baz, G. L. Gimel'farb, R. Falk, D. Heredis, and M. Abou El-Ghar, "A novel approach for accurate estimation of the growth rate of the detected lung nodules," in *Proceedings of International Workshop on Pulmonary Image Analysis*, 2008, pp. 33–42.
165. A. El-Baz, G. L. Gimel'farb, R. Falk, T. Holland, and T. Shaffer, "A framework for unsupervised segmentation of lung tissues from low dose computed tomography images," in *Proceedings of British Machine Vision, (BMVC'08)*, 2008, pp. 1–10.
166. A. El-Baz, G. Gimel'farb, R. Falk, and M. A. El-Ghar, "3D MGRF-based appearance modeling for robust segmentation of pulmonary nodules in 3D LDCT chest images," in *Lung imaging and computer aided diagnosis*. CRC Press, ch. 3, 2011, pp. 51–63.
167. A. El-Baz, G. Gimel'farb, R. Falk, and M. A. El-Ghar, "Automatic analysis of 3D low dose CT images for early diagnosis of lung cancer," *Pattern Recognition*, vol. 42, no. 6, 2009, pp. 1041–1051.
168. A. El-Baz, G. Gimel'farb, R. Falk, M. A. El-Ghar, S. Rainey, D. Heredia, and T. Shaffer, "Toward early diagnosis of lung cancer," in *Proceedings of Medical Image Computing and Computer-Assisted Intervention, (MICCAI'09)*. Springer, 2009, pp. 682–689.
169. A. El-Baz, G. Gimel'farb, R. Falk, M. A. El-Ghar, and J. Suri, "Appearance analysis for the early assessment of detected lung nodules," in *Lung imaging and computer aided diagnosis*. CRC Press, ch. 17, 2011, pp. 395–404.
170. A. El-Baz, F. Khalifa, A. Elnakib, M. Nitkzen, A. Soliman, P. McClure, G. Gimel'farb, and M. A. El-Ghar, "A novel approach for global lung registration using 3D Markov Gibbs appearance model," in *Proceedings of International Conference Medical Image Computing and Computer-Assisted Intervention, (MICCAI'12)*, Nice, France, 2012, October 1–5, pp. 114–121.
171. A. El-Baz, M. Nitzken, A. Elnakib, F. Khalifa, G. Gimel'farb, R. Falk, and M. A. El-Ghar, "3D shape analysis for early diagnosis of malignant lung nodules," in *Proceedings of International Conference Medical Image Computing and Computer-Assisted Intervention, (MICCAI'11)*, Toronto, Canada, 2011, September 18–22, pp. 175–182.
172. A. El-Baz, M. Nitzken, G. Gimel'farb, E. Van Bogaert, R. Falk, M. A. El-Ghar, and J. Suri, "Three-dimensional shape analysis using spherical harmonics for early assessment of detected lung nodules," in *Lung imaging and computer aided diagnosis*. CRC Press, ch. 19, 2011, pp. 421–438.

173. A. El-Baz, M. Nitzken, F. Khalifa, A. Elnakib, G. Gimel'farb, R. Falk, and M. A. El-Ghar, "3D shape analysis for early diagnosis of malignant lung nodules," in *Proceedings of International Conference on Information Processing in Medical Imaging, (IPMI'11)*, Monastery Irsee, Germany (Bavaria), 2011, July 3–8, pp. 772–783.
174. A. El-Baz, M. Nitzken, E. Vanbogaert, G. Gimel'Farb, R. Falk, and M. Abo El-Ghar, "A novel shape-based diagnostic approach for early diagnosis of lung nodules," in *Biomedical Imaging: From Nano to Macro, 2011 IEEE International Symposium on*. IEEE, 2011, pp. 137–140.
175. A. El-Baz, P. Sethu, G. Gimel'farb, F. Khalifa, A. Elnakib, R. Falk, and M. A. El-Ghar, "Elastic phantoms generated by microfluidics technology: validation of an imaged-based approach for accurate measurement of the growth rate of lung nodules," *Biotechnology Journal*, vol. 6, no. 2, 2011, pp. 195–203.
176. A. El-Baz, P. Sethu, G. Gimel'farb, F. Khalifa, A. Elnakib, R. Falk, and M. A. El-Ghar, "A new validation approach for the growth rate measurement using elastic phantoms generated by state-of-the-art microfluidics technology," in *Proceedings of IEEE International Conference on Image Processing, (ICIP'10)*, Hong Kong, 2010, September 26–29, pp. 4381–4383.
177. A. El-Baz, P. Sethu, G. Gimel'farb, F. Khalifa, A. Elnakib, R. Falk, and M. A. E.-G. J. Suri, "Validation of a new imaged-based approach for the accurate estimating of the growth rate of detected lung nodules using real CT images and elastic phantoms generated by state-of-the-art microfluidics technology," in *Handbook of lung imaging and computer aided diagnosis*, A. El-Baz, and J. S. Suri, Eds. Taylor & Francis, vol. 1, ch. 18, 2011, pp. 405–420.
178. A. El-Baz, A. Soliman, P. McClure, G. Gimel'farb, M. A. El-Ghar, and R. Falk, "Early assessment of malignant lung nodules based on the spatial analysis of detected lung nodules," in *Proceedings of IEEE International Symposium on Biomedical Imaging: From Nano to Macro, (ISBI'12)*. IEEE, 2012, pp. 1463–1466.
179. A. El-Baz, S. E. Yuksel, S. Elshazly, and A. A. Farag, "Non-rigid registration techniques for automatic follow-up of lung nodules," in *Proceedings of Computer Assisted Radiology and Surgery, (CARS'05)*, vol. 1281. Elsevier, 2005, pp. 1115–1120.
180. A. S. El-Baz, and J. S. Suri, *Lung imaging and computer aided diagnosis*. CRC Press, 2011.
181. A. Soliman, F. Khalifa, N. Dunlap, B. Wang, M. El-Ghar, and A. El-Baz, "An iso-surfaces based local deformation handling framework of lung tissues," in *Biomedical Imaging (ISBI), 2016 IEEE 13th International Symposium on*. IEEE, 2016, pp. 1253–1259.
182. A. Soliman, F. Khalifa, A. Shaffie, N. Dunlap, B. Wang, A. Elmaghraby, and A. El-Baz, "Detection of lung injury using 4D-CT chest images," in *Biomedical Imaging (ISBI), 2016 IEEE 13th International Symposium on*. IEEE, 2016, pp. 1274–1277.
183. A. Soliman, F. Khalifa, A. Shaffie, N. Dunlap, B. Wang, A. Elmaghraby, G. Gimel'farb, M. Ghazal, and A. El-Baz, "A comprehensive framework for early assessment of lung injury," in *Image Processing (ICIP), 2017 IEEE International Conference on*. IEEE, 2017, pp. 3275–3279.
184. A. Shaffie, A. Soliman, M. Ghazal, F. Taher, N. Dunlap, B. Wang, A. Elmaghraby, G. Gimel'farb, and A. El-Baz, "A new framework for incorporating appearance and shape features of lung nodules for precise diagnosis of lung cancer," in *Image Processing (ICIP), 2017 IEEE International Conference on*. IEEE, 2017, pp. 1372–1376.
185. A. Soliman, F. Khalifa, A. Shaffie, N. Liu, N. Dunlap, B. Wang, A. Elmaghraby, G. Gimel'farb, and A. El-Baz, "Image-based cad system for accurate identification of lung injury," in *Image Processing (ICIP), 2016 IEEE International Conference on*. IEEE, 2016, pp. 121–125.

186. A. Soliman, A. Shaffie, M. Ghazal, G. Gimel'farb, R. Keynton, and A. El-Baz, "A novel cnn segmentation framework based on using new shape and appearance features," in *2018 25th IEEE International Conference on Image Processing (ICIP)*. IEEE, 2018, pp. 3488–3492.
187. A. Shaffie, A. Soliman, H. A. Khalifeh, M. Ghazal, F. Taher, R. Keynton, A. Elmaghraby, and A. El-Baz, "On the integration of CT-derived features for accurate detection of lung cancer," in *2018 IEEE International Symposium on Signal Processing and Information Technology (ISSPIT)*. IEEE, 2018, pp. 435–440.
188. A. Shaffie, A. Soliman, H. A. Khalifeh, M. Ghazal, F. Taher, A. Elmaghraby, R. Keynton, and A. El-Baz, "Radiomic-based framework for early diagnosis of lung cancer," in *2019 IEEE 16th International Symposium on Biomedical Imaging (ISBI 2019)*. IEEE, 2019, pp. 1293–1297.
189. A. Shaffie, A. Soliman, M. Ghazal, F. Taher, N. Dunlap, B. Wang, V. Van Berkel, G. Gimelfarb, A. Elmaghraby, and A. El-Baz, "A novel autoencoder-based diagnostic system for early assessment of lung cancer," in *2018 25th IEEE International Conference on Image Processing (ICIP)*. IEEE, 2018, pp. 1393–1397.
190. A. Shaffie, A. Soliman, L. Fraiwan, M. Ghazal, F. Taher, N. Dunlap, B. Wang, V. van Berkel, R. Keynton, A. Elmaghraby et al., "A generalized deep learning-based diagnostic system for early diagnosis of various types of pulmonary nodules," *Technology in Cancer Research & Treatment*, vol. 17, 2018, p. 1533033818798800.
191. A. Mahmoud, A. El-Barkouky, H. Farag, J. Graham, and A. Farag, "A non-invasive method for measuring blood flow rate in superficial veins from a single thermal image," in *Proceedings of the IEEE Conference on Computer Vision and Pattern Recognition Workshops*, 2013, pp. 354–359.
192. N. Elsaid, A. Saied, H. Kandil, A. Soliman, F. Taher, M. Hadi, G. Giridharan, R. Jennings, M. Casanova, R. Keynton et al., "Impact of stress and hypertension on the cerebrovasculature," *Frontiers in Bioscience-Landmark*, vol. 26, no. 12, 2021, p. 1643.
193. F. Taher, H. Kandil, Y. Gebru, A. Mahmoud, A. Shalaby, S. El-Mashad, and A. El-Baz, "A novel MRA-based framework for segmenting the cerebrovascular system and correlating cerebral vascular changes to mean arterial pressure," *Applied Sciences*, vol. 11, no. 9, 2021, p. 4022.
194. H. Kandil, A. Soliman, F. Taher, M. Ghazal, A. Khalil, G. Giridharan, R. Keynton, J. R. Jennings, and A. El-Baz, "A novel computer- aided diagnosis system for the early detection of hypertension based on cerebrovascular alterations," *NeuroImage: Clinical*, vol. 25, 2020, p. 102107.
195. H. Kandil, A. Soliman, M. Ghazal, A. Mahmoud, A. Shalaby, R. Keynton, A. Elmaghraby, G. Giridharan, and A. El-Baz, "A novel framework for early detection of hypertension using magnetic resonance angiography," *Scientific Reports*, vol. 9, no. 1, 2019, pp. 1–12.
196. Y. Gebru, G. Giridharan, M. Ghazal, A. Mahmoud, A. Shalaby, and A. El-Baz, "Detection of cerebrovascular changes using magnetic resonance angiography," in *Cardiovascular imaging and image analysis*. CRC Press, 2018, pp. 1–22.
197. A. Mahmoud, A. Shalaby, F. Taher, M. El-Baz, J. S. Suri, and A. El-Baz, "Vascular tree segmentation from different image modalities," in *Cardiovascular imaging and image analysis*. CRC Press, 2018, pp. 43–70.
198. F. Taher, A. Mahmoud, A. Shalaby, and A. El-Baz, "A review on the cerebrovascular segmentation methods," in *2018 IEEE International Symposium on Signal Processing and Information Technology (ISSPIT)*. IEEE, 2018, pp. 359–364.

199. H. Kandil, A. Soliman, L. Fraiwan, A. Shalaby, A. Mahmoud, A. ElTanboly, A. Elmaghraby, G. Giridharan, and A. El-Baz, "A novel MRA framework based on integrated global and local analysis for accurate segmentation of the cerebral vascular system," in *2018 IEEE 15th International Symposium on Biomedical Imaging (ISBI 2018)*. IEEE, 2018, pp. 1365–1368.
200. F. Taher, A. Soliman, H. Kandil, A. Mahmoud, A. Shalaby, G. Gimel'farb, and A. El-Baz, "Accurate segmentation of cerebrovasculature from TOF-MRA images using appearance descriptors," *IEEE Access*, vol. 8, 2020, pp. 96139–96149.
201. F. Taher, A. Soliman, H. Kandil, A. Mahmoud, A. Shalaby, G. Gimel'farb, and A. El-Baz, "Precise cerebrovascular segmentation," in *2020 IEEE International Conference on Image Processing (ICIP)*. IEEE, 2020, pp. 394–397.
202. M. Elsharkawy, A. Sharafeldeen, A. Soliman, F. Khalifa, M. Ghazal, E. El-Daydamony, A. Atwan, H. S. Sandhu, and A. El-Baz, "A novel computer-aided diagnostic system for early detection of diabetic retinopathy using 3D-OCT higher-order spatial appearance model," *Diagnostics*, vol. 12, no. 2, 2022, p. 461.
203. M. Elsharkawy, M. Elrazzaz, M. Ghazal, M. Alhalabi, A. Soliman, A. Mahmoud, E. El-Daydamony, A. Atwan, A. Thanos, H. S. Sandhu et al., "Role of optical coherence tomography imaging in predicting progression of age-related macular disease: a survey," *Diagnostics*, vol. 11, no. 12, 2021, p. 2313.
204. H. S. Sandhu, M. Elmogy, A. T. Sharafeldeen, M. Elsharkawy, N. El-Adawy, A. Eltanboly, A. Shalaby, R. Keynton, and A. El- Baz, "Automated diagnosis of diabetic retinopathy using clinical biomarkers, optical coherence tomography, and optical coherence tomography angiography," *American Journal of Ophthalmology*, vol. 216, 2020, pp. 201–206.
205. A. Sharafeldeen, M. Elsharkawy, F. Khalifa, A. Soliman, M. Ghazal, M. AlHalabi, M. Yaghi, M. Alrahmawy, S. Elmougy, H. Sandhu et al., "Precise higher-order reflectivity and morphology models for early diagnosis of diabetic retinopathy using oct images," *Scientific Reports*, vol. 11, no. 1, 2021, pp. 1–16.
206. A. A. Sleman, A. Soliman, M. Elsharkawy, G. Giridharan, M. Ghazal, H. Sandhu, S. Schaal, R. Keynton, A. Elmaghraby, and A. El- Baz, "A novel 3D segmentation approach for extracting retinal layers from optical coherence tomography images," *Medical Physics*, vol. 48, no. 4, 2021, pp. 1584–1595.
207. A. A. Sleman, A. Soliman, M. Ghazal, H. Sandhu, S. Schaal, A. Elmaghraby, and A. El-Baz, "Retinal layers oct scans 3-d segmentation," in *2019 IEEE International Conference on Imaging Systems and Techniques (IST)*. IEEE, 2019, pp. 1–6.
208. N. Eladawi, M. Elmogy, M. Ghazal, O. Helmy, A. Aboelfetouh, A. Riad, S. Schaal, and A. El-Baz, "Classification of retinal diseases based on oct images," *Frontiers in Bioscience* (Landmark Edition), vol. 23, 2018, pp. 247–264.
209. A. ElTanboly, M. Ismail, A. Shalaby, A. Switala, A. El-Baz, S. Schaal, G. Gimel'farb, and M. El-Azab, "A computer-aided diagnostic system for detecting diabetic retinopathy in optical coherence tomography images," *Medical Physics*, vol. 44, no. 3, 2017, pp. 914–923.
210. H. S. Sandhu, A. El-Baz, and J. M. Seddon, "Progress in automated deep learning for macular degeneration," *JAMA Ophthalmology*, vol. 136, no. 12, 2018, pp. 1366–1367.
211. M. Ghazal, S. S. Ali, A. H. Mahmoud, A. M. Shalaby, and A. El-Baz, "Accurate detection of non-proliferative diabetic retinopathy in optical coherence tomography images using convolutional neural networks," *IEEE Access*, vol. 8, 2020, pp. 34387–34397.

212. K. Hammouda, F. Khalifa, A. Soliman, M. Ghazal, M. Abou El-Ghar, A. Haddad, M. Elmogy, H. Darwish, A. Khalil, A. Elmaghraby et al., "A cnn-based framework for bladder wall segmentation using MRI," in *2019 Fifth International Conference on Advances in Biomedical Engineering (ICABME)*. IEEE, 2019, pp. 1–4.
213. K. Hammouda, F. Khalifa, A. Soliman, M. Ghazal, M. Abou El-Ghar, A. Haddad, M. Elmogy, H. Darwish, R. Keynton, and A. El-Baz, "A deep learning-based approach for accurate segmentation of bladder wall using MR images," in *2019 IEEE International Conference on Imaging Systems and Techniques (IST)*. IEEE, 2019, pp. 1–6.
214. K. Hammouda, F. Khalifa, A. Soliman, H. Abdeltawab, M. Ghazal, M. Abou El-Ghar, A. Haddad, H. E. Darwish, R. Keynton, and A. El-Baz, "A 3D CNN with a learnable adaptive shape prior for accurate segmentation of bladder wall using MR images," in *2020 IEEE 17th International Symposium on Biomedical Imaging (ISBI)*. IEEE, 2020, pp. 935–938.
215. K. Hammouda, F. Khalifa, A. Soliman, M. Ghazal, M. Abou El-Ghar, M. Badawy, H. Darwish, A. Khelifi, and A. El-Baz, "A multiparametric mri-based cad system for accurate diagnosis of bladder cancer staging," *Computerized Medical Imaging and Graphics*, vol. 90, 2021, p. 101911.
216. K. Hammouda, F. Khalifa, A. Soliman, M. Ghazal, M. Abou El-Ghar, M. Badawy, H. Darwish, and A. El-Baz, "A CAD system for accurate diagnosis of bladder cancer staging using a multiparametric MRI," in *2021 IEEE 18th International Symposium on Biomedical Imaging (ISBI)*. IEEE, 2021, pp. 1718–1721.
217. A. Alksas, M. Shehata, G. A. Saleh, A. Shaffie, A. Soliman, M. Ghazal, H. A. Khalifeh, A. A. Razek, and A. El-Baz, "A novel computer-aided diagnostic system for early assessment of hepatocellular carcinoma," in *2020 25th International Conference on Pattern Recognition (ICPR)*. IEEE, 2021, pp. 10375–10382.
218. A. Alksas, M. Shehata, G. A. Saleh, A. Shaffie, A. Soliman, M. Ghazal, A. Khelifi, H. A. Khalifeh, A. A. Razek, G. A. Giridharan et al., "A novel computer-aided diagnostic system for accurate detection and grading of liver tumors," *Scientific Reports*, vol. 11, no. 1, 2021, pp. 1–18.
219. A. A. K. A. Razek, R. Khaled, E. Helmy, A. Naglah, A. AbdelKhalek, and A. El-Baz, "Artificial intelligence and deep learning of head and neck cancer," *Magnetic Resonance Imaging Clinics*, vol. 30, no. 1, 2022, pp. 81–94.
220. A. Sharafeldeen, M. Elsharkawy, R. Khaled, A. Shaffie, F. Khalifa, A. Soliman, A. A. K. Abdel Razek, M. M. Hussein, S. Taman, A. Naglah et al., "Texture and shape analysis of diffusion-weighted imaging for thyroid nodules classification using machine learning," *Medical Physics*, vol. 49, no. 2, 2022, pp. 988–999.
221. A. Naglah, F. Khalifa, R. Khaled, A. A. K. Abdel Razek, M. Ghazal, G. Giridharan, and A. El-Baz, "Novel MRI-based cad system for early detection of thyroid cancer using multi-input CNN," *Sensors*, vol. 21, no. 11, 2021, p. 3878.
222. A. Naglah, F. Khalifa, A. Mahmoud, M. Ghazal, P. Jones, T. Murray, A. S. Elmaghraby, and A. El-Baz, "Athlete-customized injury prediction using training load statistical records and machine learning," in *2018 IEEE International Symposium on Signal Processing and Information Technology (ISSPIT)*. IEEE, 2018, pp. 459–464.
223. A. H. Mahmoud, "Utilizing radiation for smart robotic applications using visible, thermal, and polarization images", Ph.D. dissertation, University of Louisville, 2014.
224. A. Mahmoud, A. El-Barkouky, J. Graham, and A. Farag, "Pedestrian detection using mixed partial derivative based his togram of oriented gradients," in *2014 IEEE International Conference on Image Processing (ICIP)*. IEEE, 2014, pp. 2334–2337.
225. A. El-Barkouky, A. Mahmoud, J. Graham, and A. Farag, "An interactive educational drawing system using a humanoid robot and light polarization," in *2013 IEEE International Conference on Image Processing*. IEEE, 2013, pp. 3407–3411.

226. A. H. Mahmoud, M. T. El-Melegy, and A. A. Farag, "Direct method for shape recovery from polarization and shading," in *2012 19th IEEE International Conference on Image Processing*. IEEE, 2012, pp. 1769–1772.
227. M. A. Ghazal, A. Mahmoud, A. Aslantas, A. Soliman, A. Shalaby, J. A. Benediktsson, and A. El-Baz, "Vegetation cover estimation using convolutional neural networks," *IEEE Access*, vol. 7, 2019, pp. 132563–132576.
228. M. Ghazal, A. Mahmoud, A. Shalaby, and A. El-Baz, "Automated framework for accurate segmentation of leaf images for plant health assessment," *Environmental Monitoring and Assessment*, vol. 191, no. 8, 2019, p. 491.
229. M. Ghazal, A. Mahmoud, A. Shalaby, S. Shaker, A. Khelifi, and A. El-Baz, "Precise statistical approach for leaf segmentation," in *2020 IEEE International Conference on Image Processing (ICIP)*. IEEE, 2020, pp. 2985–2989.

3 Enhancing Accuracy in Liver Tumor Detection and Grading

A Computer-Aided Diagnostic System

Asmaa El-Sayed Hassan, Mohamed Shehata, Hossam Magdy Balaha, Gehad A. Saleh, Ahmed Alksas, Ali H. Mahmoud, Ahmed Shaffie, Ahmed Soliman, Homam Khattab, Yassin Mohamed-Hassan, Mohammed Ghazal, Adel Khelifi, Hadil Abu Khalifeh, and Ayman El-Baz

3.1 INTRODUCTION

Liver tumors, which include a variety of benign and malignant growths that affect the liver, are a serious medical problem on a global scale [1]. The liver, a crucial organ responsible for many crucial bodily processes, is prone to diverse tumor types. Primary liver cancer, which develops inside the liver cells themselves, or secondary liver cancer, which happens when cancer cells from other organs disseminate and move to the liver, can both cause liver tumors [2]. With a high incidence and fatality rate, liver cancer is a serious health problem on a worldwide scale. Liver cancer's global prevalence is a significant concern, with approximately 800,000 new cases being reported each year, making it the sixth most known cancer worldwide. Moreover, liver cancer represents a serious public health challenge, ranking as the third leading cause of cancer-related deaths globally, resulting in around 700,000 deaths annually [3].

The two most common primary liver cancers are bile duct cancer (cholangiocarcinoma) and hepatocellular carcinoma (HCC). HCC is the most discovered form, responsible for over 90% of primary liver cancer cases [4 ,5]. Globally, one in every 5,000 persons is at risk of developing HCC. Chronic viral hepatitis infections, particularly hepatitis B and C; non-alcoholic fatty liver disease; excessive alcohol use; and exposure to certain environmental toxins are risk factors for HCC. Additionally, those older than 60 are most likely to develop adult primary liver cancer. The HCC cases frequency is greater in nations with low healthcare and social services access.

DOI: 10.1201/9780367486082-3

With 42,030 and 31,780 new cases and fatalities, respectively, per year, HCC has the fastest rate of increase in the USA, and its management is still challenging [6–8].

Treatment options include surgery and liver transplant [9]. A partial hepatectomy (removal of the cancerous portion of the liver) may be performed. It may be necessary to remove a wedge of tissue, a complete lobe, or a more significant part of the liver, along with some of the surrounding healthy tissue. During a liver transplant, the complete liver is removed and replaced with a healthy donor liver. Although a liver transplant offers patients with HCC the best results, not enough organ donors are available. It is worth noting that the severity of the HCC (i.e., stage of the HCC) affects the prognosis. Early detection combined with good medical therapy can reduce the demand for donor organs, preserving the local liver [10–11]. Currently, imaging techniques (such as ultrasound, magnetic resonance imaging [MRI], computed tomography [CT], angiography, etc.) are often used to identify and grade HCC in addition to physical (e.g., biopsy) and blood tests (e.g., alpha-fetoprotein tumor marker) [7, 9, 12, 13].

To ensure consistency in interpreting and reporting imaging findings for patients with HCC risk factors (e.g., cirrhosis), the American College of Radiology (ACR) supports the implementation of the Liver Imaging Reporting and Data System (LI-RADS) [14]. Liver tumors are classified into various categories by LI-RADS, which include LRM (malignant but not necessarily HCC), LR1 (definitively benign), LR2 (presumably benign), LR3 (indeterminate), LR4 (probably HCC), or LR5 (definitively HCC). The LI-RADS system was originally introduced in 2011, and since then, four modifications have been made, the most recent of which was in 2018 [15]. Although the LI-RADS criteria have a high specificity [12], they may have poor sensitivity since they are subjective, particularly if the liver tumor is categorized as LRM, intermediate (LR3), or probably HCC (LR4). Moreover, recent studies have shown that the occurrence of HCC varies across the different LI-RADS grades, with rates of 0.00, 0.13, 0.38, 0.74, and 0.94 observed for LR5, LR4, LR3, LR2, and LR1 grades, respectively [15, 16], which could decrease the ultimate diagnostic decision's impartiality and degree of confidence. The only method left to reliably identify liver cancer tumors is a biopsy. However, biopsies can have unfavorable outcomes, including bleeding or infection, and are invasive, costly, and risky [16, 17]. Given these circumstances, there is an urgent demand for a computer-aided diagnostic (CAD) system that is objective, non-invasive, and capable of accurately identifying and classifying liver cancers.

Machine learning (ML) has made significant advances in recent years, which have created new opportunities for the early identification and grading of liver tumors. The accuracy of cancer detection can be greatly improved by using ML algorithms, which have proven to be capable of handling enormous volumes of data and extracting useful information [18–23]. The capacity of ML algorithms to process and evaluate massive datasets quickly is one of its main benefits. With the growing availability of CT and MRI scan data, machine learning algorithms can discover patterns and traits indicating liver cancers. ML systems can detect subtle and complicated connections between imaging characteristics and tumor existence or grade by training on large datasets [24]. In the investigation of liver tumors, the incorporation of ML algorithms into computer-aided diagnostic systems has produced encouraging results. These systems are efficient in processing medical images, segmenting liver tumors, and categorizing them according to their degree or malignancy. Using ML algorithms, these systems may automate the identification and grading procedure, providing doctors with an accurate and objective assessment of liver tumors [25].

The following is a summary of the book chapter's contributions:

1. Create a two-stage CAD system designed to:
   a. Accurately differentiate between malignant, benign, and intermediate HCC liver cancers.
   b. Effectively classify and distinguish between different grades of malignant and benign tumor subtypes.
2. Outlining the most recent performance measures and comparing them to relevant research.

The rest of this chapter is organized as follows: Section 3.2 reviews and summarizes the relevant literature and the research gap. Section 3.3 addresses the utilized materials. It talks about the MR data-collecting methodology and outlines the research design and patient population. Section 3.4 covers the features/markers extraction, selection, markers integration, and diagnosis. The findings of the experiments are described in detail in Section 3.5. In Section 3.6, the presented work is discussed, the chapter is concluded, and future work is discussed.

3.2 RELATED STUDIES

HCC detection using ML algorithms encompasses a wide range of research efforts to enhance the early detection and diagnosis of HCC. The previous work involves the application of various algorithms (e.g., SVM, RFs, NNs, and deep learning models) to different types of imaging data (e.g., CT, MRI, and ultrasound). In the study presented by Sato et al. [26], real-world data gathered from clinical practice was used to create a novel machine learning model for predicting the HCC. The created framework was trained and evaluated using 539 and 1,043 patients with and without HCC, respectively. The reference standard (i.e., ground truth) for their study was established using contrast-enhanced CT images, which served as the basis for an accurate and reliable assessment of HCC. To create the best diagnostic classifier, they examined the effectiveness of several ML methods. The implementation of gradient boosting resulted in an accuracy of 0.87 in the detection of malignant tumors. However, it is important to note that their method did not utilize any imaging markers to differentiate between malignant and benign HCC tumors. Furthermore, the assessment of these identified HCC tumors in terms of appropriate strategies for their treatment was not examined. To distinguish between liver tumors at different stages, Yang et al. [27] used CE-MR images. They examined 51 liver tumors, of which 9, 35, and 7 were malignant, intermediate, and benign, respectively. Initially, they used image rendering tools to manually draw 2D ROIs on the liver tumors. They then fed these ROIs into a three-dimensional multi-channel fusion convolutional neural network (CNN). A detection accuracy of 0.91 ± 0.03 was obtained for the malignant group alone; however, an accuracy of only 0.74 ± 0.01 was obtained for distinguishing the three groups from one another.

The ability of texture analysis to distinguish between different HCC tumors was investigated by Stocker et al. [28]. In their study, 108 patients with preoperative 2D CE-MRI (55 with malignant HCC and 53 with benign HCC) were included. They chose 13 textural markers after setting 2D manual ROIs. The retrieved markers were then subjected to a binary logistic regression analysis, which was followed by statistical testing. Using the arterial phase images, their model distinguished between benign and malignant HCC tumors with an accuracy of 0.85. Nevertheless, the diagnostic performance of logistic regression for intermediate-grade HCC tumors has yet to be examined. Yamashita et al. [29] conducted a study to classify liver tumors using multi-phase CT and MRI images. The 314 patients included in the study were divided into four groups (LR1–2, LR3, LR4, and LR5) based on the LI-RADS system, with a total of 89, 62, 65, and 98 cases, respectively. From each phase, one 2D image per subject was selected, resulting in four images per subject. The images were first cropped and scaled using manual regions of interest (ROIs). Subsequently, these modified images were fed into two CNN models: a customized CNN and a transfer learning CNN. The transfer learning model demonstrated superior performance, achieving an accuracy of 0.60 in accurately distinguishing between LI-RADS categories (LR1–2, LR3, LR4, and LR5). However, when the CNN was tested on two external datasets (comprising 68 CT cases and 44 MRI cases), the accuracy dropped to 47.7% for the MRI dataset and 41.2% for the CT dataset. It should be noted that their investigation was limited to using only one image from each phase to represent the entire subject, which may have resulted in the exclusion of important morphological and anatomical characteristics of the tumors.

A study conducted by Kim et al. [30] aimed to develop and evaluate a threshold-based CAD system for assessing the risk of hepatocellular carcinoma (HCC). The study involved liver tumors from 41 patients who underwent CE-MRI scans. The tumors were classified into three categories: LR5 (26 cases), LR3 (12 cases), and LR4 (3 cases). Three characteristics were determined following semi-automated segmentation: the size of the tumor (particularly maximum diameter), capsule appearance, and the presence of a wash-out. The optimal threshold for each feature was then established using receiver operating characteristic (ROC) curves. To compare the computed markers with the markers generated by the radiologists, the researchers proceeded to conduct a statistical t-test along with evaluating intraclass correlation coefficients (ICC). This analysis allowed them to assess the agreement between the two sets of markers and determine any significant differences. Their CAD method correctly classified 76% to 83% of the tumors. However, the inclusion of LR1 and LR2 patients restricted their study. Further, they did not look into ML techniques to improve classification performance. Also, there was only a 78% agreement between the two reference radiologists. Multiphase CE-MRI data from 89 individuals were used in research presented by Wu et al. [31] to distinguish between malignant (LR4 = 14 and LR5 = 40) and intermediate (LR3 = 35) HCC tumors. By drawing 2D manual ROIs on the center image for each phase, the researchers fed this information into a pre-trained AlexNet model. Their model classified the two groups with an accuracy of 90%. However, eliminating benign tumors restricted their presented study. Furthermore, grading tumors into LR4 and LR5 was not considered.

3.3.1 Research Gap

The majority of the previously mentioned studies focused solely on examining the effectiveness of individual clinical records or textural markers in combination with machine learning techniques to distinguish between various types of liver tumors. While these investigations provided valuable insights, only a few studies explored the crucial aspect of grading the liver tumor itself. The accurate grading of liver tumors is essential for initiating appropriate treatment measures at an early stage. However, these studies were restricted by their limited diagnostic performance, which indicates the need for further improvements in this area. Furthermore, it is worth noting that none of these studies integrated a comprehensive approach by combining different types of markers, that is, morphological markers, textural markers (e.g., first- and second-order features), and functional markers, to achieve a more precise and accurate diagnosis. The integration of morphological markers, which assess the visual structure of the tumor, along with textural markers that analyze its texture patterns and functional markers that evaluate its physiological properties, holds great potential in enhancing the diagnostic capabilities for liver tumors.

3.3 MATERIALS

Population of Patients and Study Design. This research included individuals with liver tumors who had a high risk of developing HCC and had never received loco-regional therapy. Patients with chronic hepatitis, cirrhosis, and those who had previously experienced HCC were also included. Separate analyses were done for each hepatic tumor that existed in a patient who had more than one tumor. The application of techniques followed all applicable rules and regulations. Both Mansoura University in Egypt and the University of Louisville in the United States have approved all experimental procedures. Between November 2018 and January 2021, contrast-enhanced MR pictures of 97 subjects were collected. All participants gave their informed consent after being fully aware of the study's objectives. However, due to the withdrawal of permission, two patients were excluded from the study. The remaining 95 patients (male = 65 and female = 30) with liver cancers varied in age from 34 to 82 (average 56 ± 10 years). All CE-MR images of the participants were independently examined by three experienced radiologists, each with over ten years of practical experience in liver imaging. These radiologists were blinded to each other's assessments and used a secondary workstation (Phillips Advantage Windows workstation with functional tool software) following the LI-RADS v2018 guidelines [10]. The image analysis focused on four main markers: liver tumor size, enhancing capsule appearance, non-peripheral wash-out appearance, and non-rim arterial-phase hyperenhancement (APHE). Three options were presented for each case, and the choice was made based on the agreement of at least two of them. A total of 95 liver tumors in the participating patients was classified as benign (i.e., LR1 and LR2), malignant (LR4 and LR5), and intermediate (LR3), in which each grade contains 19 cases.

Protocols for MR Data Acquisition. The patient population's ($N=95$) CE-MR images were acquired utilizing a 1.5T Philips Ingenia scanner with a phased-array torso surface coil. Using an automated MR injector, 1×10^{-4} mmol/kg of gadolinium chelates,

TABLE 3.1
The Employed Settings for CE-MRI Sequence Acquisition

Acquisition Parameters of the CE-MRI Scans

TE (s)	TR (s)	Slice Size (pixels)	FOV (cm)	Slice Thickness (cm)	Slice Gap (cm)	Flip Angle
3.1×10^{-3}	7.3×10^{-3}	256×128	500×10^{-3}	3×10^{-3}	1×10^{-3}	40°

Note: FOV: field of view; TE: echo time; and TR: repetition time.

an extracellular contrast agent was administered at a rate of 2 ml/s, followed by a 20 ml saline flush. Pre-contrast (at t = 0 s), late arterial (at t = 35 s), portal venous (at t = 50 s), and delayed-contrast phase (at t = 180 s) are the four phases of the abdomen MR scan. To reduce any potential respiratory consequences, all patients were instructed to hold their breath while images were being taken. Table 3.1 lists the acquisition settings for MRIs.

3.4 METHODOLOGY

The proposed CAD system employs various processes in the detection and classification of liver cancer tumors. The procedures involved in this study encompass the extraction of morphological markers from segmented liver tumors. To quantify the enhancement characteristics across different CE-MR phases, a parametric spherical harmonic model was employed. Additionally, textural markers are estimated using a rotation-invariant model. The evaluation of functional markers involves analyzing the wash-in/wash-out slopes to assess enhanced features during different CE-MR phases. Ultimately, the CAD system utilizes a two-stage RF-based classifier that integrates the detected markers to categorize liver tumors as malignant, intermediate, or benign, along with their corresponding grade (LR5, LR4, LR3, LR2, or LR1).

3.4.1 Features/Markers Extraction

One essential phase in the ML process is the extraction of features or markers [32, 33]. In ML, *markers* refer to measurable properties or attributes of observations that can be independently evaluated. The predictive power of an ML model is increased by using appropriate markers that discriminate between object classes with clarity. To help the learning algorithm solve the primary classification issue, the extraction step was employed. During this phase, the primary objective is to convert the raw data into markers that are distinctive, standardized, and machine-understandable. In discussion with several medical collaborators, several markers' categories that are appropriate for the nature of the issue were chosen. To enable quantitative differentiation between various grades of HCC liver tumors, three distinct types of markers (namely, morphological, functional, and textural markers) are retrieved from the segmented liver tumors. The morphological markers, which are based on spherical harmonics (SH), are capable of describing the intricacy of the morphology of liver tumors. In contrast, functional markers measure the properties of augmentation across several phases based on the computation of wash-in/wash-out slopes. In this study, textural markers were extracted by the CAD system to capture the variations in texture among different grades of HCC tumors. These markers included first-order histogram markers and rotation-invariant second-order markers based on GLCM and GLRLM. By utilizing these markers, the system effectively captured and analyzed the unique textural characteristics of liver tumors, enabling the differentiation and classification of tumors based on their texture profiles.

3.4.1.1 Imaging Markers

Accurate manual segmentation of liver tumors was performed by experienced radiologists using specialized software. This meticulous segmentation aimed to enhance the extraction and estimation of textural, morphological, and functional markers. As a result, three categories of image markers (i.e., morphological, textural, and functional) were utilized to enable precise differentiation between different grades of HCC. The segmentation process yielded 3D representations of the liver tumor objects. These markers were utilized to characterize the liver tumor objects and enable precise discrimination.

3.4.1.2 Morphological Markers

The parametric morphological markers that can represent the nature of the discovered HCC have been found to increase the specificity and sensitivity of an early tumor diagnosis. The justification for applying these markers is based on the idea that malignant tumors grow more quickly and have more intricate shapes than benign ones. Liver tumors exhibit variations in morphology and surface complexity, which are influenced by their malignancy status and grade [34]. By incorporating morphology description, the automated diagnosis capabilities can be enhanced, enabling improved accuracy in distinguishing different types and grades of liver tumors. However, attaining such improvement requires precise modeling. In the suggested framework, the morphological markers for liver tumor diagnosis were extracted using state-of-the-art spectrum analysis using spherical harmonics (SH) [35]. By selecting a specific point within the tumor as the reference, a spherical coordinate system can be established to describe the tumor's surface. This surface can be represented as a function of polar and azimuthal angles, which can be expressed as a linear combination of basis functions known as $Y_{\tau\beta}$, defined on the unit sphere. The spherical harmonics (SH) modeling technique entails the construction of a triangulated mesh that closely represents the surface of the tumor. This mesh is then accurately mapped onto the unit sphere using an attraction–repulsion technique [36]. By maintaining a uniform unit distance between the origin and each remapped node while also preserving the distances between adjacent nodes, this approach ensures accurate modeling.

In the attraction–repulsion algorithm, let $C_{\alpha,i}$ with $\|C_{\alpha,i}\| = 1$ be the coordinates of node i on iteration α, in which $i \in \{1, \ldots, I\}$. Let the displacement from node i to node j be represented by $\boldsymbol{d}_{\alpha,ji} = C_{\alpha,j} - C_{\alpha,i}$; in such case, the distance (i.e., Euclidean distance) between the two nodes i and j is $d_{\alpha,i} = \left\|\boldsymbol{d}_{\alpha,ij}\right\|$. Finally, the set of neighboring nodes of node i in the triangulated mesh is represented by J_i. Following, the attraction phase, adjusts each node's location to maintain it central in relation to its neighbors:

$$\boldsymbol{C}'_{\alpha+1,i} = \boldsymbol{C}_{\alpha,i} + \boldsymbol{C}_{A,1} \sum_{j \in J_i} \left(\boldsymbol{d}_{\alpha,ji} \boldsymbol{d}^2_{\alpha,ji} + \boldsymbol{C}_{A,2} \frac{\boldsymbol{d}_{\alpha,ji}}{\boldsymbol{d}_{\alpha,ji}} \right) \tag{1}$$

Where $C_{A,1}$ and $C_{A,2}$ are parameters (i.e., attraction factors) of the algorithm. Following the attraction step, the repulsion step inflates the whole mesh to prevent degeneration, because the attraction step alone would enable nodes to grow close together.

$$\boldsymbol{C}''_{\alpha+1,i} = \boldsymbol{C}'_{\alpha+1,\mathrm{i}} + \frac{\boldsymbol{C}_R}{2\boldsymbol{I}} \sum_{j=1, j \neq i}^{I} \left(\frac{\boldsymbol{d}_{\alpha,ji}}{\boldsymbol{d}^2_{\alpha,ji}} \right) \tag{2}$$

Where CR is the parameter (i.e., repulsion factor) of the algorithm. After that, the points are reprojected onto the unit sphere, represented by $C_{\alpha+1,i} = C''_{\alpha+1,i} / \left\|C''_{\alpha+1,i}\right\|$. At the final iteration, α_f of the attraction–repulsion algorithm, the surface of the hepatic nodule corresponds one-to-one with the unit sphere. Every node of the initial mesh, represented by its coordinates C_j: (x_j, y_j, z_j), has been transformed into a corresponding point $C_{\alpha f,j}$: $\left(\cos\phi_j \sin\theta_j, \sin\phi_j \sin\theta_j, \cos\theta_j\right)$ with azimuthal angle $\phi_j \in [0, 2\pi]$ and polar angle $\theta_j \in [0, \pi]$. The nodule can then be described using an SH series. In this illustration, higher-order harmonics show the surface's finer details, while lower-order harmonics show the nodule's general size. By considering the nodule surface as a function on the unit sphere, the spherical harmonics (SHs) are generated by solving an isotropic heat equation specific to that surface. The SH $Y_{\tau\beta}$ of order β and degree τ is defined as:

$$\boldsymbol{Y}_{\tau\beta} = \begin{cases} \boldsymbol{c}_{\tau\beta} \cos\theta \sin\left(|\beta|\varphi\right) & -\tau \leq \beta \leq -1 \\ \dfrac{\boldsymbol{c}_{\tau\beta}}{\sqrt{2}} \cos\theta & \beta = 0 \\ \boldsymbol{c}_{\tau\beta} \boldsymbol{G}_{\tau}^{|\beta|} \cos\theta \cos\left(|\beta|\varphi\right) & 1 \leq \beta \leq \tau \end{cases} \tag{3}$$

Where the SHs factor is $c_{\tau\beta}$ and the associated Legendre polynomial of order β and degree τ is $G_{\tau}^{|\beta|}$.

In the last stage, the tumor object is reconstructed by utilizing spherical harmonics (SHs) as described in equation 3. Malignant tumors, known for their complex structure, are represented using a higher-order combination of SHs, while benign tumors are represented with a lower-order combination. The total number of markers utilized to evaluate the morphological complexity of the discovered malignancies is equal to the number of SHs used for reconstructing the original tumor. In this work, any tumor may be accurately reconstructed using a sufficient number (70), beyond which there are no appreciable changes in the approximations. The shape of the tumor is evaluated by comparing the original mesh with the estimated shape for each approximation. This is done by aligning the meshes using the unit sphere mapping and calculating the Euclidean distances between corresponding nodes. These calculations are performed for each of the 70 approximations, resulting in a set of 70 numerical values called reconstruction errors. These values provide a quantitative description of the tumor's morphology. Following is an overview of the attraction–repulsion algorithm.

Initialization:

- Triangulate the nodule's exterior.
- Use Laplacian filtering to make the triangulated mesh more uniform.
- Use any arbitrary, topology-preserving map onto the unit sphere as the initial value for the spherical parameterization.
- Fix the values of the threshold T and C_R, $C_{A,1}$, $C_{A,2}$.

Attraction–Repulsion:

- For $\alpha = 0, 1, \ldots$
    - For $i = 1, \ldots, I$.
    - Use *Eq.1* to calculate $C'_{\alpha+1,i}$.
    - For $i = 1, \ldots, I$.
      Use *Eq.2* to calculate $C''_{\alpha+1,i}$.
      Let $C_{\alpha+1,i} = C''_{\alpha+1,i} / \left\| C''_{\alpha+1,i} \right\|$.
    - If $max_i \left\| C_{\alpha+1,i} - C_{\alpha,i} \right\| \leq T$, then set the value of α_f *by* $\alpha + 1$ and terminate.

3.4.1.3 Textural Markers

A thorough textural analysis was done to increase both sensitivity and specificity of early liver cancer detection. In particular, four phases (i.e., pre-contrast, portal venous, late arterial, and delayed-contrast) were used to extract first-order and second-order textural markers that can be used in defining the homogeneity of the discovered tumor. The rationale for the use of textural markers is based on the idea that malignant tumors seem more unevenly distributed than benign tumors [37–43]. Two different second-order textural markers (i.e., gray level run-length matrix [GLRLM] and gray level co-occurrence matrix [GLCM]) were utilized to detect inhomogeneity in liver tumors [44, 45].

GLCM matrix evaluates the spatial relationships between voxels within a neighborhood block, including the reference voxel and its neighboring ones. It quantifies the frequency of pairs of gray level intensity values appearing adjacent to each other within the object. To construct the GLCM, the gray level range of the object is specified, and the observed gray level values are normalized accordingly [46]. All possible pairs of gray level values are identified and used to form the columns and rows of the matrix. The rows and columns of the matrix are constructed using all conceivable pairs of gray level values. The row and column of each matrix element are represented by a pair of gray level values. The disparities between the voxel and its neighbors are examined to determine the value of each element. Since the neighborhood block is defined by a distance of $\leq \sqrt{2}$, the calculations are rotation-invariant. A GLCM with a size of 256 × 256 was produced by normalizing the gray level results during analysis to the interval (0, 255). In order to recover the differentiating textural markers, the matrix is normalized after the GLCM is constructed so that the total of all components is one [44, 47]. The extracted markers are shown in Table 3.2.

TABLE 3.2
Extracted Textural Markers and Their Definitions.

Textural Marker	Definition
First-Order	
Mean μ	The balance point of gray level values for each object is represented by calculating the average gray level value.
Variance	This describes the distribution of the gray level values in respect to the estimated mean value.
Entropy	It quantifies the degree of randomness observed in the gray level values within each structure.
Skewness	It characterizes the asymmetrical distribution of gray level values around the object's mean.
Kurtosis	It measures the level of concentration of gray level values toward the extremes (i.e., tails) of the distribution.
CDFs	It involves calculating the cumulative distribution function of the histogram density values for the entire object. This is done by summing up the gray level values at different positions, ranging from 0 to 100% of the object, with a 10% interval, and then normalizing them within the range of 0 to 1.
Percentiles	It determines the percentile values of gray levels based on the corresponding CDFs.
Second-Order	
Contrast	It quantifies the difference or disparity in gray level values between neighboring pixels or voxels.
Dissimilarity	Determines the extent to which voxels deviate or differ from their neighboring voxels in terms of gray level values.
Homogeneity	It calculates the measure of the inverse difference moment among neighboring pixels or voxels.
Angular second moment (ASM)	It assesses the local uniformity or orderliness of gray levels within a given region or neighborhood.
Energy	$\sqrt{ASM}$
Correlation	It establishes the linear relationship between the center voxel and its surrounding voxels in terms of the gray level values.

(Continued)

TABLE 3.2 (Continued)
Extracted Textural Markers and Their Definitions.

Textural Marker	Definition
Gray level non-uniformity (GLN)	It characterizes the dissimilarity or variation of gray level values within the object.
High gray level run emphasis (HGLRE)	It quantifies the concentration or prevalence of high gray level values within the structure.
Long-run emphasis (LRE)	It determines the distribution of long run lengths within the object, which indicates the coarseness or roughness of the texture.
Long-run high gray level emphasis (LRHGLE)	It measures the distribution of long runs of high gray level values within the object.
Long-run low gray level emphasis (LRLGLE)	It quantifies the distribution of long runs of low gray level values within the object.
Low gray level run emphasis (LGLRE)	It measures the prevalence of low gray level values within the structure.
Run entropy (RE)	It shows how random the gray level runs are throughout the structure.
Run-length non-uniformity (RLN)	It quantifies the inhomogeneity or variation among run lengths in the object.
Run percentage (RP)	It is calculated by dividing the total count of runs by the total number of pixels in the object.
Short-run emphasis (SRE)	It measures the number of short-run lengths inside an item, showing how fine or complex the texture is.
Short-run high gray level emphasis (SRHGLE)	It measures the concentration of short runs of high gray level values within the object.
Short-run low gray level emphasis (SRLGLE)	It measures the number of short runs of low gray level values inside an object.

GLRLM computes the occurrence rate of voxel pairings that are depicted in the GLCM and assesses the connectedness of the voxels by analyzing voxel runs. It looks at the frequency with which one gray level value occurred after another in a line of voxels [48]. The greatest run, which is normally the object's maximal dimension in the XY-plane, is used to calculate the matrix size in the GLRLM. The matrix's number of columns and rows corresponds to the size of the greatest run and the range of gray levels, respectively. The frequency of a certain gray level value within a group of related voxels is represented by each element in the matrix. This frequency is defined by the row index and the column index, respectively. The normalized gray level range matrix for each structure contained 256 rows, but the number of columns varied depending on the item. Here, the vertical runs of voxels in the Z-plane (among various layers) and the horizontal runs of voxels in the XY-plane (in the same layer) were searched. The GLRLM's differentiating measures for the structures' texture were then computed [45, 47].

3.4.1.4 Functional Markers

The functionality of the liver tumors can be measured using hyperenhancement (wash-in) and hyperenhancement (wash-out). During the delayed phase and/or portal venous phase, the wash-out of contrast material is calculated, while the estimation of wash-in occurs during the late arterial phase [49, 50]. The functional markers are derived by analyzing the fluctuations in gray level intensity during the post-contrast phases. These markers are calculated using three distinct features, which are mathematically represented by the gray level slope observed in each phase. The pace at which the gray level intensity changes throughout the course of each phase is used to determine these slopes. Typically, wash-in and wash-out are represented by positive and negative slopes, respectively. Malignant tumors exhibit faster and higher wash-in and wash-out slopes compared to benign or intermediate tumors.

3.4.2 Features/Markers Selection

Feature selection is a process of choosing the most relevant attributes from a large set of possible markers [51]. As a result, a smaller set of *m* markers is chosen from a larger pool of *n* potential candidates, with $m < n$. These selected markers represent the most relevant and significant ones for the given context. The wrapper method [52, 53] and Gini impurity–based selection [54] were both used.

3.4.1.1 Wrapper Approach

In wrapper techniques, the selection procedure is based on applying a certain ML algorithm to a given dataset, in a repeated manner. The wrapper technique chooses the set of markers that have the optimum performance by comparing the algorithm's outputs when given different marker subsets as input. However, the performance requirement relies on the issue being resolved. The wrapper technique does an aggressive search through the set of potential markers. To select the ideal collection of markers, two alternative wrapping techniques were used (i.e., forward selection and step-wise selection [bi-directional elimination]). During the forward selection process, individual feature models are sequentially fitted, starting with a null model. The marker associated with the minimum p-value is then selected as the best feature. In the case of a two-parameter model, each remaining marker is paired with the previously selected feature, and the additional marker with the minimum p-value is chosen again. The third ideal marker is then discovered by combining each of the remaining markers with the previous two, and so on. Hence, forward selection builds models with an increasing number of markers (1, 2, . . ., m), until no more candidate markers meet the pre-defined threshold for the p-value. The forward selection method is summarized in algorithm 1. The forward selection method was used with significance levels of 0.1 and 0.05. Bi-directional elimination, also known as step-wise selection, involves evaluating the significance of previously included markers before considering the addition of a new marker. If any of the previously selected markers are deemed unnecessary, they are eliminated from the model. Algorithm

2 displays the steps of this method. Additionally, the bi-directional elimination method was used, with significance levels of 0.05 and 0.1.

Algorithm 1: Forward Selection Approach

1.	Set a significance threshold (ST).
2.	Iterate through each marker individually and evaluate all possible models. • For each marker, calculate the p-value for its corresponding model. • Select the marker with the lowest p-value.
3.	Generate all possible models by including one additional marker in addition to the markers that were previously selected.
4.	Once again, choose the model with the minimum p-value. • If the p-value is less than the significance threshold, proceed to step 3. • Otherwise, terminate the process.

Algorithm 2: Bi-Directional Elimination Approach

1.	To determine which markers to include or exclude in the model, an ST is chosen.
2.	The next step is forward selection, where a newly added marker is allowed to enter the model only if its p-value is less than the ST.
3.	Elimination steps are then performed, where any previously added marker that has a p-value greater than the ST is considered for removal from the model.
4.	Steps 2 and 3 are repeated iteratively to obtain an optimal final set of markers.

3.4.1.2 Selection Based on Gini Impurity

Random forests are commonly employed in data science workflows for features or markers selection. The utilization of random forests for features or markers selection is due to the inherent nature of tree-based approaches, where the improvement of node purity plays a central role. The observed pattern is characterized by a reduction in impurity across all trees, commonly referred to as Gini impurity. The impurity drop is greatest at the nodes at the beginning of the trees, whereas the impurity decrease is least at the nodes at the end of the trees [55]. By pruning the trees below a specific node, a subset of the most significant markers can be obtained. The procedures for implementing this selection strategy are shown in algorithm 3. The selection procedure was conducted in two distinct circumstances (combined and independent markers selection) to use this method. To choose the best set of markers to employ for the combined selection, the Gini impurity–based method was applied to the entire set of markers. The textural, morphological, and functional markers were chosen individually for the distinct technique to identify the best markers for each group. Then, these constrained marker sets were joined to create the ultimate, ideal marker set.

Algorithm 3: Selection Based on Gini Impurity

1.	Prepare the initial set of markers.
2.	Train a random forest classifier.
3.	Identify the most significant markers.
4.	Construct a new, smaller set by including any additional markers.

3.4.3 Integration and Diagnosis of Liver Tumor Markers

The diagnosis of liver tumors involves a two-stage classification process that integrates three types of markers: morphological, functional, and textural markers. These markers are extracted for all liver cancers at different stages. In the first stage, the focus is on distinguishing between malignant (LR5, LR4), intermediate (LR3), and benign (LR2, LR1) tumors. In the second stage, further classification is performed for benign tumors (LR1 or LR2) and malignant tumors (LR4 or LR5). This two-stage approach enhances the accuracy and precision of liver tumor diagnosis by considering multiple markers and specific tumor characteristics. Random forests (RFs), Naive Bayes (NB), fine k-nearest neighbor (kNN_{Fine}), SVM with the quadratic kernel (SVM_{Quad}), support vector machines (SVM) with the cubic kernel (SVM_{Cub}), and linear discriminant analysis (LDA) are a few popular ML classifiers that were employed. In the initial phase, the classification performance was assessed using individual markers, including functional markers (wash-in/wash-out slopes), first-order textural markers (GLCM), and second-order textural markers (GLRLM and GLCM). Table 3.3 provides specific information on the classified numbers and descriptions of these differentiating markers. To create combined markers, all the markers were then incorporated using concatenation procedures. The final diagnosis was made using the ML classifiers discussed earlier. Further, the grid search approach was employed to identify the best collection of hyperparameters for several ML classifiers, using diagnostic accuracy as an optimization criterion. Following are the ideal settings of hyperparameters for each classifier:

- Random forests with the following hyperparameters were utilized: max_depth=30, min_samples_leaf=5, min_samples_split=2, criterion='gini', class_weight='balanced', and n_estimators=100.
- kNN_{Fine} with the following hyperparameters were utilized: n neighbors=5, weights='uniform', metric='minkowski' with power of 2, leaf size=30.
- SVM with the following hyperparameters were utilized: C=1, gamma=0.001, degree=3, cache size=200, tol=0.001, decision function shape='ovr', break ties=False.
- NB with the following hyperparameters were utilized: fit_prior=True, binarize=0.0, alpha=0.5, and class_prior=None.
- LDA with the following hyperparameters were utilized: solver='lsqr', shrinkage=0.52, tol=0.0001, priors=None, store_covariance=False, and n_components=1.

To acquire the ultimate diagnosis (LR1, LR2, LR3, LR4, or LR5) of a liver tumor using a CE-MR series, the outlined steps in algorithm 4 of the CAD system can be implemented.

3.5 EXPERIMENTS' RESULTS

To evaluate the accuracy of the presented CAD system, three different evaluation procedures were employed: a leave-one-subject-out (LOSO) test and two randomly

TABLE 3.3
Visualization of Various Categories and the Corresponding Count of Extracted Markers for Each Individual

Morphological Markers	
Spherical harmonics	70 markers
Textural Markers	
Histogram markers	A total of 104 markers, with 26 markers per phase
GLCM	A total of 24 markers, with 6 markers per phase
GLRLM	A total of 84 markers, with 12 markers per phase
Functional Markers	
Wash-in/out	3 markers
Integrated Markers	
Combined	249 markers

Algorithm 4: The Outline of the Proposed CAD System	
1.	Perform liver tumor delineation/segmentation in all MR slices at all phases, separating it from the surrounding abdominal region.
2.	Create 3D representations of the segmented liver tumor.
3.	Apply the parametric spherical harmonic model to extract morphological indicators.
4.	Generate first-order and second-order textural markers by analyzing the intensity histogram and using a rotation invariant model.
5.	Calculate functional markers by analyzing wash-in/wash-out slopes to assess enhancement characteristics across different CE-MR phases.
6.	Combine and integrate the extracted imaging markers.
7.	Employ a two-stage RF classifier trained for tumor classification.
8.	In the first stage, classify the tumor as malignant, intermediate, or benign.
9.	If the tumor is categorized as intermediate, assign the final diagnosis as LR3.
10.	If the tumor is determined to be benign or malignant, proceed to the second stage.
11.	In the second stage, classify malignant tumors as LR5 or LR4 and benign tumors as LR2 or LR1 based on the fused imaging markers.
12.	Report the final tumor grade.

stratified k-fold cross-validation procedures with k values of 10 and 5. These evaluation methods were utilized to assess the performance of the CAD system and determine its accuracy. All observations, with the exception of one subject that is set aside for testing, are used in LOSO to train the classifier. The observation that was previously excluded from the classification model is then incorporated into the training data before the next iteration, leaving the subsequent subject out for testing. The dataset utilized in this study comprises 95 participants, requiring the procedure to be executed 95 times. During each iteration, 94 samples are allocated for training purposes, while a single sample is set aside for testing. For the two used stratified k-fold cross-validation, a percentage of the data, or $\frac{1}{k}\times 100\%$, is randomly chosen and utilized for testing, while the remainder of $\frac{1-k}{k}\times 100\%$ is used as training data. In the next iteration, the classification model is reset, and the subjects from the previous iteration are included in the training set. Additionally, a portion of the subjects equal to the subsequent $\frac{1}{k}\times 100\%$ is set aside for testing. To enhance the reliability and stability of the constructed model, the procedure is repeated *k* times. A randomly stratified k-fold cross-validation strategy is utilized, with two specific values of *k* (10 and 5). This approach aims to ensure the robustness of the model and improve its overall dependability.

It is crucial to remember that stratification had been offered to assist in minimizing bias and variance when k-fold cross-validation was used. By employing the stratification approach, both the training and testing sets are ensured to have the same proportion of each class as observed in the full dataset. This not only maintains the class distribution but also allows for randomization within each set. The stratification technique ensures that the model is trained and evaluated on representative samples from each class, contributing to the reliability and generalizability of the results. According to the stratification scheme, 20% ($N = 19$), 40% ($N = 38$), and 40% ($N = 38$) of the training/testing sets will come from intermediate, malignant, and benign cases, respectively.

The ultimate diagnosis was reached after two categorization steps. Each classification procedure was carried out ten times in order to quantify the performance, and the results were presented as mean ± standard deviation. Differentiating between benign (LR1, LR2), intermediate (LR3), and malignant (LR4, LR5) tumors was the goal of the first classification step. Using the three acquired individual markers along with a number of ML classifiers, the performance of the built CAD system was initially assessed. The sensitivity, specificity, and F1-score metrics were used to compare the performance of the combined with these individual models in order to emphasize the benefit of combining these different indicators [54, 56]. Table 3.4 provides the formulae for these measures.

As demonstrated in Table 3.5, the combined model outperformed the performance of all individual models with a sensitivity, specificity, and F1-score of 0.92 ± 0.09, 0.91 ± 0.02, and 0.91 ± 0.01, respectively, when employing the RFs classifier. The improved diagnostic performance is attributed to the integration process, which allows the algorithm to consider various aspects of marker quantification, including morphology, texture, and functionality. By combining these different types of markers, the algorithm can effectively capture and analyze multiple dimensions of the data, leading to enhanced diagnostic capabilities.

Individual markers from the created CAD system, including wash-in/out slopes functional markers, second-order GLCM and GLRLM textural markers, SHs morphological markers, and first-order textural markers, were used to compare the

TABLE 3.4
Formulas to Be Used to Calculate Three Performance Metrics: Sensitivity, Specificity, and F_1-Score

Metric	Equation
Sensitivity	$TP / TP + FN$
Specificity	$TN / TN + FP$
F_1-score	$2TP / 2TP + FP + FN$

Note: In these equations, *TP* stands for the proportion of correctly classified malignant subjects, *TN* for correctly classified benign subjects, *FP* for correctly classified intermediate and malignant subjects, and *FN* for correctly classified intermediate and benign subjects.
Source: [57].

TABLE 3.5
Diagnostic Performance of Combined Model with RFs Classifier

Markers	Sensitivity	Specificity	F_1-Score
	Morphological Markers		
Spherical harmonics	0.73 ± 0.03	0.84 ± 0.02	0.78 ± 0.02
	Textural Markers		
First-order (histogram)	0.80 ± 0.04	0.86 ± 0.02	0.82 ± 0.03
Second-order (GLCM)	0.86 ± 0.02	088 ± 0.02	0.86 ± 0.01
Second-order (GLRLM)	0.82 ± 0.02	0.82 ± 0.02	0.81 ± 0.01
	Functional Markers		
Wash-in/out	0.81 ± 0.03	0.84 ± 0.02	0.82 ± 0.02
	Integrated Markers		
Combined	0.92 ± 0.01	0.91 ± 0.02	0.91 ± 0.01

performance of the first stage classification. The RFs classifier was used in the comparison to identify benign tumors (LR1, LR2), intermediate tumors (LR3), and malignant tumors (LR4, LR5).

The diagnostic findings of the combined model were analyzed using numerous ML classifiers, including RFs, KNN_{Fine}, $SVM_{Cub,Quad}$, LDA, and NB, along with several validation procedures (LOSO, five-fold, and ten-fold), to choose the best classifier to build the CAD system. The RFs establish themselves as the best ML classifier among the used ones, with sensitivity, specificity, and F1-score of 0.92 ± 0.09, 0.91 ± 0.02, and 0.91 ± 0.01, respectively, for the LOSO; 0.89 ± 0.01, 0.90 ± 0.03, and 0.89 ± 0.02, respectively, for the ten-fold; and 0.87 ± 0.02, 0.89 ± 0.03, and 0.88 ± 0.02, respectively, for the five-fold. The comparative findings between the performances of various ML classifiers and techniques are presented in Table 3.6. The successful classification performance achieved by random forests (RFs) can be attributed to

TABLE 3.6
Evaluation of Various ML Classifiers and Techniques for Classification Performance

Classifier	Approach	Sensitivity	Specificity	F_1-Score
RFs	**LOSO**	**0.92 ± 0.01**	**0.91 ± 0.02**	**0.91 ± 0.01**
	10-fold	**0.89 ± 0.01**	**0.90 ± 0.03**	**0.89 ± 0.02**
	5-fold	**0.87 ± 0.02**	**0.89 ± 0.03**	**0.88 ± 0.02**
kNNFine	LOSO	0.86 ± 0.00	0.92 ± 0.00	0.89 ± 0.00
	10-fold	0.85 ± 0.02	0.92 ± 0.01	0.88 ± 0.01
	5-fold	0.83 ± 0.02	0.91 ± 0.03	0.87 ± 0.02
$SVM_{Cub,Quad}$	LOSO	0.85 ± 0.00	0.87 ± 0.00	0.86 ± 0.00
	10-fold	0.84 ± 0.03	0.85 ± 0.02	0.84 ± 0.02
	5-fold	0.84 ± 0.03	0.85 ± 0.02	0.84 ± 0.02
NB	LOSO	0.83 ± 0.00	0.87 ± 0.00	0.84 ± 0.00
	10-fold	0.80 ± 0.04	0.89 ± 0.03	0.83 ± 0.02
	5-fold	0.79 ± 0.04	0.89 ± 0.03	0.83 ± 0.03
LDA	LOSO	0.86 ± 0.00	0.88 ± 0.00	0.86 ± 0.00
	10-fold	0.83 ± 0.05	0.86 ± 0.02	0.84 ± 0.03
	5-fold	0.81 ± 0.05	0.82 ± 0.03	0.80 ± 0.03

their well-established reputation as robust machine learning techniques [58, 59]. RFs have been extensively utilized in the medical field to address various classification challenges, demonstrating their effectiveness in solving medical classification problems [60]. A decision tree collection and the random subspace approach are the foundation of the ensemble classifier known as RFs. In a typical bootstrap sample, this bagging method aids in identifying all potential correlations between the decision trees. When some markers are discovered to be effective predictors of the intended outcome, these markers will be chosen in several decision trees and develop correlation. The final results are often acquired via a majority vote or a model averaging technique after the training phase [58, 59, 61]. Since it performed better than all other classifiers evaluated, the RFs classifier was chosen to be used in the proposed CAD system.

The classification performance of the first stage was evaluated using the integrated markers of the developed CAD system. The comparison was made between benign tumors (LR1, LR2), intermediate tumors (LR3), and malignant tumors (LR4, LR5) using various ML classifiers (e.g., KNN, RFs, SVM, NB, and LDA). Three different validation approaches were employed for each classifier: leave-one-subject-out (LOSO), five-fold, and ten-fold cross-validation.

For the second classification stage, grading was performed for each class: the malignant (LR4 vs. LR5) and the benign (LR1 vs. LR2) classes. To obtain the final diagnosis, all markers were combined and inputted into the RFs classifier, utilizing the leave-one-subject-out (LOSO), five-fold, and ten-fold cross-validation techniques. According to Table 3.7 (using the LOSO technique), the overall accuracy for grading

TABLE 3.7
Diagnostic Performance for Grading Benign and Malignant Tumors Using LOSO Technique

Approach	Accuracy		
Benign			
	LR1	LR2	Overall
LOSO	0.89 ± 0.04	0.90 ± 0.02	0.89 ± 0.02
10-fold	0.86 ± 0.05	0.88 ± 0.03	0.87 ± 0.03
5-fold	0.85 ± 0.07	0.86 ± 0.05	0.85 ± 0.05
Malignant			
	LR4	LR5	Overall
LOSO	0.93 ± 0.03	0.85 ± 0.02	0.89 ± 0.02
10-fold	0.90 ± 0.04	0.80 ± 0.02	0.85 ± 0.02
5-fold	0.87 ± 0.07	0.75 ± 0.02	0.81 ± 0.03

benign tumors was 0.89 0.02, whereas the overall accuracy for grading malignant tumors was 0.89 0.02. After combining the data from the two stages, the tumors were graded into five categories: LR1, LR2, LR3, LR4, and LR5. It is important to note that the created CAD system's diagnostic performance was improved by adopting a two-stage RFs classifier when compared with a single-stage model.

Using the combined markers, the developed CAD system's diagnostic performance was evaluated in the second stage classification: LR1 vs. LR2 and LR4 vs. LR5, employing the RFs classifier.

The created CAD system's diagnostic performance was assessed by comparing it to six distinct feature reduction scenarios. These scenarios included forward selection (ST and m of 0.05 and 19 markers), forward selection (ST and m of 0.10 and 196 markers), bi-directional elimination (ST and m of 0.05 and 13 markers), bi-directional elimination (ST and m of 0.10 and 16 markers), Gini impurity–based selection on separate marker groups (morphological and textural markers with m = 109 markers), and Gini impurity–based selection (combined markers with m = 134 markers). The final diagnostic performance of the presented system was compared to the results obtained from these different reduction scenarios. The output-reduced markers from each scenario are used to compare the outcomes of the proposed two-stage model. The results of this comparison are presented in Table 3.8. The final diagnosis was made and classified as LR1, LR2, LR3, LR4, or LR5. The table includes accuracy values for each LI-RAD category as well as the overall accuracy.

The performance of the developed CAD system, which utilized combined markers, was evaluated by comparing it to the performance of six different scenarios involving the selection of markers. These scenarios involved varying numbers of markers, denoted by m, and significance thresholds, denoted by ST.

TABLE 3.8
The Diagnostic Performance for LI-RAD Categories

Markers Selection Approach		m	Accuracy					
			LR1	**LR2**	**LR3**	**LR4**	**LR5**	**Overall**
Proposed CAD system (combined)		249	0.88 ± 0.05	0.85 ± 0.02	0.78 ± 0.03	0.83 ± 0.04	0.79 ± 0.03	0.83 ± 0.02
Wrapper approach	Forward (ST = 0.05)	19	0.81 ± 0.04	0.84 ± 0.03	0.67 ± 0.05	0.86 ± 0.03	0.67 ± 0.04	0.77 ± 0.02
	Forward (ST = 0.10)	196	0.75 ± 0.03	0.88 ± 0.0	0.65 ± 0.05	0.85 ± 0.03	0.64 ± 0.04	0.75 ± 0.02
	Bi-directional (ST = 0.05)	13	0.76 ± 0.05	0.86 ± 0.03	0.72 ± 0.07	0.85 ± 0.03	0.66 ± 0.05	0.77 ± 0.02
	Bi-directional (ST = 0.10)	16	0.76 ± 0.04	0.91 ± 0.04	0.73 ± 0.03	0.84 ± 0.04	0.64 ± 0.06	0.78 ± 0.02
Gini impurity–based	Combined selection	134	0.75 ± 0.03	0.76 ± 0.0	0.73 ± 0.04	0.87 ± 0.02	0.65 ± 0.05	0.75 ± 0.02
	Separate selection	109	0.74 ± 0.03	0.82 ± 0.0	0.70 ± 0.03	0.88 ± 0.02	0.71 ± 0.03	0.77 ± 0.01

TABLE 3.9
Comparative Diagnostic Performance for Liver Tumor Grading

Markers Selection Approach	Accuracy					
	LR1	**LR2**	**LR3**	**LR4**	**LR5**	**Overall**
Proposed CAD system	0.88 ± 0.05	**0.85 ± 0.02**	0.78 ± 0.03	**0.83 ± 0.04**	0.79 ± 0.03	**0.83 ± 0.02**
Stocker et al. [28]	0.71 ± 0.00	0.82 ± 0.00	0.71 ± 0.00	0.53 ± 0.00	0.70 ± 0.00	0.69 ± 0.00
Wu et al. [31]	**0.91 ± 0.02**	0.58 ± 0.09	**0.81 ± 0.04**	0.60 ± 0.08	**0.88 ± 0.02**	0.76 ± 0.04

The diagnostic performance of the developed CAD system was evaluated by comparing it with two existing approaches from the literature, Stocker et al. [28] and Wu et al. [31]. The assessment was conducted using the same dataset ($N = 95$) utilized in this study, with a focus on HCC grading. This approach ensured an equitable assessment between the CAD system and the other methods. The final diagnostic results obtained from the CAD system were then compared with the results achieved by the two alternative techniques. The comparison, as shown in Table 3.9, revealed that the CAD system outperformed all the previously discussed methods in terms of diagnostic performance for grading liver tumors.

The performance of grading tumors into the five grades was assessed using three approaches: (a) the proposed CAD system, (b) the approach by Stocker et al. [28], and (c) the approach by Wu et al. [31]. The final diagnostic performance for each approach was evaluated and compared to determine their effectiveness in accurately grading the tumors.

3.6 CONCLUSIONS AND FUTURE WORK

At advanced stages, HCC has a significant death rate. Early thorough screening system identification is crucial, and it must be adapted to a more general management methodology. Professional research organizations have promoted recommendations to help radiologists and doctors manage the HCC. To improve the uniformity of data collecting and image reporting, LI-RADS intends to standardize the HCC-related terminology and develop an image algorithm. The established clinical gold standard for diagnosing hepatocellular carcinoma (HCC) involves image analysis conducted by expert radiologists who are blinded to the patient information. This analysis encompasses several key factors, including arterial phase hyperenhancement, washout appearance, enhancing capsule appearance, and size of the tumor [7–15, 62]. These criteria are widely recognized and accepted as reliable indicators for HCC diagnosis in clinical practice. In contrast, the field of radiogenomics and emerging imaging advancements aim to explore the heterogeneity of the HCC using imaging techniques. The objective is to gain a deeper understanding of the unique characteristics and genomic features associated with individual tumors. By harnessing this information, personalized care can be tailored based on the specific signature of each tumor, leading to more targeted and effective treatment strategies. These advancements in imaging technology hold promise for advancing our knowledge of HCC and improving patient care, the capacity of innovative algorithms and trends to enable more accuracy in diagnosis and grading, as well as prospective recommendations on customized healthcare [18–20, 63–66].

In this study, 3D objects were created from mixed tumor lesions extracted from CE-MR images captured at various stages. These 3D models represent the participants at different phases, utilizing multiple voxels within the lesions and surrounding liver parenchyma. Each voxel exhibits a grayscale value based on its signal intensity, influenced by various histopathological variables. Although visually indistinguishable, complex geometric patterns specific to tumor types can be observed when analyzing 3D arrays of grayscale values. To capture this information, texture analysis was employed, which explains how voxel values depend on the grayness of neighboring voxels in a given area. Previous research [37–43] has shown that texture information plays a crucial role in the performance of classification approaches. In this study, textural markers were extracted using first- and second-order texture analysis techniques and various algorithms. First-order texture analysis allocates voxel intensities across tumor lesions at each phase, resulting in descriptors that are largely dependent on the individual voxel values. Calculated first-order markers include mean, standard deviation, variance, kurtosis, skewness, cumulative distribution functions, entropy, and gray level percentiles [47]. In contrast, second-order texture analysis techniques rely on the spatial relationships between neighboring voxels. These algorithms are sensitive to voxel arrangements and their associations with neighboring voxels. GLRLM and GLCM were utilized in previous studies to perform second-order texture analysis [44, 45].

Because it is sensitive to spatial interrelationships, second-order texture analysis using the GLCM and GLRLM has demonstrated the capacity to distinguish between benign, malignant, and intermediate liver cancers. Malignant tumors can have complicated internal structures because of their aggressive growth patterns, high levels of neovascularity, and advanced neoangiogenesis. Due to this, liver lesions with various levels of malignancy show great variability in their microenvironment. By considering the spatial interrelationships and analyzing voxel attenuation, it becomes possible to detect subtle variations in tumor heterogeneity. In contrast to intermediate and benign lesions, malignant tumor lesions have more textural heterogeneity. The GLCM can identify whether the voxels are randomly dispersed (benign) or clustered (malignant), and the GLRLM demonstrates if these voxels are linked in long runs (homogeneous) or short runs (heterogeneous) over the whole lesion. These retrieved second-order textural indicators allowed for the observation, interpretation, and quantification of all these differences.

Additionally, it was shown that functional indicators had the ability to determine if a particular liver tumor was malignant. As a result, three markers (i.e., late arterial wash-in, portal venous wash-out, and delayed wash-out) extractable from the post-contrast phases were investigated. These markers capture the differences in enhancement markers by mathematically calculating the gray level slope in each phase. The results of this investigation, which were obtained using measuring curves of functioning, are fair and show how well these markers may distinguish between various types of liver cancers.

The form of a liver tumor depends on its level of aggressiveness. Compared to benign tumors, malignant tumors often have a more complicated appearance. To distinguish between benign, intermediate, and malignant HCC, morphological markers were employed. Grades of liver tumors were determined by morphological, textural, and functional indicators applied to 3D objects constructed from CE-MR images. In the categorization procedure, ml models were used to assess each marker. There is still significant overlap between these indicators, even if several of them demonstrated significant differences across the various stages of liver cancers. The utilization of various CE-MR sequences introduces variability, posing challenges in accurately detecting liver cancers using a single class of markers. To overcome this limitation, a combination of markers was employed, leading to a more precise approach to distinguish malignant tumors from intermediate and benign ones. Initially, all markers were integrated to classify liver tumors into the categories of benign, intermediate, and malignant HCC. Subsequently, the LR1 benign tumors were distinguished from LR2, and LR4 malignant tumors were differentiated from LR5 using the same classification and validation techniques. The achieved results confirm the effectiveness of the proposed methodology and emphasize its potential clinical applicability when combined with CE-MR imaging for computer-aided liver tumor detection. Detailed information regarding these results can be found in Tables 3.5 and 3.6.

In summary, the developed CAD system demonstrated excellent diagnostic performance, as evidenced by its high sensitivity, specificity, and F1-score values. The sensitivity, representing the system's ability to correctly identify malignant tumors, was found to be 0.92 ± 0.01. The specificity, which measures the system's accuracy in identifying benign tumors, achieved a value of 0.91 ± 0.02. Additionally, the F1-score, which combines precision and recall, yielded a value of 0.91 ± 0.01. This remarkable performance can be attributed to the integration of multiple types of markers, including morphological, textural, and functional markers. By combining these markers, the CAD system was able to enhance its diagnostic capabilities and surpass the diagnostic performance achieved by each marker alone. This integration of multiple markers allows for a more comprehensive and accurate assessment of liver tumors, leading to improved diagnostic accuracy and confidence. Furthermore, overall accuracies of 0.88 ± 0.05, 0.85 ± 0.02, 0.78 ± 0.03, 0.83 ± 0.04, and 0.79 ± 0.03 in grading liver tumors across the five phases were obtained by the proposed system. These results highlight the effectiveness of integrating multiple markers that capture various elements of liver tumor features.

In future studies, one of the primary objectives is to expand the dataset used for training and evaluation. This will involve collecting a larger and more diverse set of subjects with liver tumors to ensure a comprehensive representation of different tumor types and characteristics. By incorporating a wider range of subjects, the performance of the CAD system can be further improved in accurately distinguishing and grading various types of liver tumors. Moreover, to enhance the diagnostic capabilities of the CAD system, the dataset will include additional liver tumors with diverse LI-RADS categories. By incorporating a broader range of LI-RADS categories, the CAD system will be better equipped to handle a wider variety of liver tumor cases and provide more reliable diagnostic results.

In addition to the liver [67, 68], this work could also be applied to various other applications in medical imaging, such as the prostate [69–73], the kidney [74–92], the heart [93–110], the lung [111–160], the brain [161–183], the vascular system [184–194], the retina [195–204], the bladder [205–209], the head and neck [210–212], and injury prediction [213], as well as several non-medical applications [214–220].

REFERENCES

1. A. Alksas, M. Shehata, G. A. Saleh, A. Shaffie, A. Soliman, M. Ghazal,. . .and A. El-Baz, "A novel computer-aided diagnostic system for accurate detection and grading of liver tumors," *Scientific Reports*, vol. 11, no. 1, p. 13148, 2021.
2. National Cancer Institute, *Liver and bile duct cancer*, 2021. Available on line: https://www.cancer.gov/types/liver.
3. A. Alksas, M. Shehata, G. A. Saleh, A. Shaffie, A. Soliman, M. Ghazal,. . .and A. El-Baz, "A novel computer-aided diagnostic system for accurate detection and grading of liver tumors," *Scientific Reports*, vol. 11, no. 1, pp. 1–18, 2021.
4. K. Schütte, C. Schulz, and P. Malfertheiner, "Hepatocellular carcinoma: current concepts in diagnosis, staging and treatment," *Gastrointestinal Tumors*, vol. 1, no. 2, pp. 84–92, 2014.
5. D. Fahmy, A. Alksas, A. Elnakib, A. Mahmoud, H. Kandil, A. Khalil,. . .and A. El-Baz, "The role of radiomics and AI technologies in the segmentation, detection, and management of hepatocellular carcinoma," *Cancers*, vol. 14, no. 24, p. 6123, 2022.
6. Blue Faery: The Adrienne Wilson Liver Cancer Association, *Statistics, HCC in the United States*, 2022. Available on line: https://www.bluefaery.org/statistics/.
7. P. J. Navin, and S. K. Venkatesh, "Hepatocellular carcinoma: state of the art imaging and recent advances," *Journal of Clinical and Translational Hepatology*, vol. 7, no. 1, p. 72, 2019.
8. K. M. Elsayes, J. C. Hooker, M. M. Agrons, A. Z. Kielar, A. Tang, K. J. Fowler,. . .and C. B. Sirlin, "2017 version of LI-RADS for CT and MR imaging: an update," *Radiographics*, vol. 37, no. 7, pp. 1994–2017, 2017.

9. Cancer.Net, *Liver cancer*, 2023. Available on line: https://www.cancer.net/cancer-types/liver-cancer.
10. A. A. K. A. Razek, L. G. El-Serougy, G. A. Saleh, R. Abd El-Wahab, and W. Shabana, "Interobserver agreement of magnetic resonance imaging of liver imaging reporting and data system version 2018," *Journal of Computer Assisted Tomography*, vol. 44, no. 1, pp. 118–123, 2020.
11. F. Patella, F. Pesapane, E. M. Fumarola, I. Emili, R. Spairani, S. A. Angileri,. . .and G. Carrafiello, "CT-MRI LI-RADS v2017: a comprehensive guide for beginners," *Journal of Clinical and Translational Hepatology*, vol. 6, no. 2, p. 222, 2018.
12. S. Bota, F. Piscaglia, S. Marinelli, A. Pecorelli, E. Terzi, and L. Bolondi, "Comparison of international guidelines for noninvasive diagnosis of hepatocellular carcinoma," *Liver Cancer*, vol. 1, no. 3–4, pp. 190–200, 2012.
13. M. Badawy, A. M. Almars, H. M. Balaha, M. Shehata, M. Qaraad, and M. Elhosseini, "A two-stage renal disease classification based on transfer learning with hyperparameters optimization," *Frontiers in Medicine*, vol. 10, p. 1106717, 2023.
14. M. Tanabe, A. Kanki, T. Wolfson, E. A. Costa, A. Mamidipalli, M. P. Ferreira,. . .and C. B. Sirlin, "Imaging outcomes of liver imaging reporting and data system version 2014 category 2, 3, and 4 observations detected at CT and MR imaging," *Radiology*, vol. 281, no. 1, pp. 129–139, 2016.
15. K. M. Elsayes, A. Z. Kielar, M. M. Elmohr, V. Chernyak, W. R. Masch, A. Furlan,. . .and C. B. Sirlin, "White paper of the society of abdominal radiology hepatocellular carcinoma diagnosis disease-focused panel on LI-RADS v2018 for CT and MRI," *Abdominal Radiology*, vol. 43, pp. 2625–2642, 2018.
16. M. Alvarez, and F. G. Romeiro, "What should be done for patients with liver lesions in the LI-RADS 2 and 3 categories," *JAME Medical Journal*, vol. 4, 2019.
17. X. H. Li, Q. Liang, T. W. Chen, J. Wang, and X. M. Zhang, "Diagnostic value of imaging examinations in patients with primary hepatocellular carcinoma," *World Journal of Clinical Cases*, vol. 6, no. 9, p. 242, 2018.
18. A. Alksas, M. Shehata, G. A. Saleh, A. Shaffie, A. Soliman, M. Ghazal,. . .and A. El-Baz, "A novel computer-aided diagnostic system for early assessment of hepatocellular carcinoma," in *2020 25th International Conference on Pattern Recognition (ICPR)*. IEEE, 2021, January, pp. 10375–10382.
19. B. E. Bejnordi, M. Veta, P. J. Van Diest, B. Van Ginneken, N. Karssemeijer, G. Litjens,. . .and CAMELYON16 Consortium, "Diagnostic assessment of deep learning algorithms for detection of lymph node metastases in women with breast cancer," *Jama*, vol. 318, no. 22, pp. 2199–2210, 2017.
20. S. Chilamkurthy, R. Ghosh, S. Tanamala, M. Biviji, N. G. Campeau, V. K. Venugopal,. . .and P. Warier, "Deep learning algorithms for detection of critical findings in head CT scans: a retrospective study," *The Lancet*, vol. 392, no. 10162, pp. 2388–2396, 2018.
21. H. M. Balaha, and A. E. S. Hassan, "A variate brain tumor segmentation, optimization, and recognition framework," *Artificial Intelligence Review*, vol. 56, no. 7, pp. 7403–7456, 2023.
22. H. M. Balaha, and A. E. S. Hassan, "Comprehensive machine and deep learning analysis of sensor-based human activity recognition," *Neural Computing and Applications*, vol. 35, no. 17, pp. 12793–12831, 2023.
23. A. T. Shalata, M. Shehata, E. Van Bogaert, K. M. Ali, A. Alksas, A. Mahmoud,. . .and A. El-Baz, "Predicting recurrence of non-muscle-invasive bladder cancer: current techniques and future trends," *Cancers*, vol. 14, no. 20, p. 5019, 2022.
24. M. Biswas, V. Kuppili, L. Saba, D. R. Edla, H. S. Suri, E. Cuadrado-Godia,. . .and J. S. Suri, "State-of-the-art review on deep learning in medical imaging," *Frontiers in Bioscience-Landmark*, vol. 24, no. 3, pp. 380–406, 2019.

25. K. Bera, K. A. Schalper, D. L. Rimm, V. Velcheti, and A. Madabhushi, "Artificial intelligence in digital pathology—new tools for diagnosis and precision oncology," *Nature Reviews Clinical oncology*, vol. 16, no. 11, pp. 703–715, 2019.
26. M. Sato, K. Morimoto, S. Kajihara, R. Tateishi, S. Shiina, K. Koike, and Y. Yatomi, "Machine-learning approach for the development of a novel predictive model for the diagnosis of hepatocellular carcinoma," *Scientific Reports*, vol. 9, no. 1, p. 7704, 2019.
27 D. W. Yang, X. B. Jia, Y. J. Xiao, X. P. Wang, Z. C. Wang, and Z. H. Yang, "Noninvasive evaluation of the pathologic grade of hepatocellular carcinoma using MCF-3DCNN: a pilot study," *BioMed Research International*, vol. 2019, 2019.
28. D. Stocker, H. P. Marquez, M. W. Wagner, D. A. Raptis, P. A. Clavien, A. Boss,. . .and M. C. Wurnig, "MRI texture analysis for differentiation of malignant and benign hepatocellular tumors in the non-cirrhotic liver," *Heliyon*, vol. 4, no. 11, p. e00987, 2018.
29. R. Yamashita, A. Mittendorf, Z. Zhu, K. J. Fowler, C. S. Santillan, C. B. Sirlin,. . .and R. K. Do, "Deep convolutional neural network applied to the liver imaging reporting and data system (LI-RADS) version 2014 category classification: a pilot study," *Abdominal Radiology*, vol. 45, pp. 24–35, 2020.
30. Y. Kim, A. Furlan, A. A. Borhani, and K. T. Bae, "Computer-aided diagnosis program for classifying the risk of hepatocellular carcinoma on MR images following liver imaging reporting and data system (LI-RADS)," *Journal of Magnetic Resonance Imaging*, vol. 47, no. 3, pp. 710–722, 2018.
31. Y. Wu, G. M. White, T. Cornelius, I. Gowdar, M. H. Ansari, M. P. Supanich, and J. Deng, "Deep learning LI-RADS grading system based on contrast enhanced multiphase MRI for differentiation between LR-3 and LR-4/LR-5 liver tumors," *Annals of Translational Medicine*, vol. 8, no. 11, 2020.
32. H. M. Balaha, and A. E. S. Hassan, "Skin cancer diagnosis based on deep transfer learning and sparrow search algorithm," *Neural Computing and Applications*, vol. 35, no. 1, pp. 815–853, 2023.
33. M. Shehata, A. Alksas, R. T. Abouelkheir, A. Elmahdy, A. Shaffie, A. Soliman,. . .and A. El-Baz, "A comprehensive computer-assisted diagnosis system for early assessment of renal cancer tumors," *Sensors*, vol. 21, no. 14, p. 4928, 2021.
34. I. Sharaby, A. Alksas, A. Nashat, H. M. Balaha, M. Shehata, M. Gayhart,. . .and A. El-Baz, "Prediction of wilms' tumor susceptibility to preoperative chemotherapy using a novel computer-aided prediction system," *Diagnostics*, vol. 13, no. 3, p. 486, 2023.
35. A. Shaffie, A. Soliman, M. Ghazal, F. Taher, N. Dunlap, B. Wang,. . .and A. El-Baz, "A novel autoencoder-based diagnostic system for early assessment of lung cancer," in *2018 25th IEEE international conference on image processing (ICIP)*. IEEE, 2018, October, pp. 1393–1397.
36. M. J. Nitzken, *Shape analysis of the human brain*. 2015.
37. L. Moya, H. Zakeri, F. Yamazaki, W. Liu, E. Mas, and S. Koshimura, "3D gray level co-occurrence matrix and its application to identifying collapsed buildings," *ISPRS Journal of Photogrammetry and Remote Sensing*, vol. 149, pp. 14–28, 2019.
38. R. C. Gonzales, and P. Wintz, *Digital image processing*. Addison-Wesley Longman Publishing Co., Inc., 1987.
39. A. S. Kurani, D. H. Xu, J. Furst, and D. S. Raicu, "Co-occurrence matrices for volumetric data," *Heart*, vol. 27, p. 25, 2004.
40. N. Tustison, and J. Gee, "Run-length matrices for texture analysis," *Insight Journal*, vol. 1, pp. 1–6, 2008.
41. B. Barry, K. Buch, J. A. Soto, H. Jara, A. Nakhmani, and S. W. Anderson, "Quantifying liver fibrosis through the application of texture analysis to diffusion weighted imaging," *Magnetic Resonance Imaging*, vol. 32, no. 1, pp. 84–90, 2014.
42. G. Castellano, L. Bonilha, L. M. Li, and F. Cendes, "Texture analysis of medical images," *Clinical Radiology*, vol. 59, no. 12, pp. 1061–1069, 2004.

43. S. W. Anderson, J. A. Soto, H. N. Milch, A. Ozonoff, O'Brien, M., J. A. Hamilton, and H. J. Jara, "Effect of disease progression on liver apparent diffusion coefficient values in a murine model of NASH at 11.7 Tesla MRI," *Journal of Magnetic Resonance Imaging*, vol. 33, no. 4, pp. 882–888, 2011.
44. R. M. Haralick, "Statistical and structural approaches to texture," *Proceedings of the IEEE*, vol. 67, no. 5, pp. 786–804, 1979.
45. M. M. Galloway, "Texture analysis using gray level run lengths," *Computer Graphics and Image Processing*, vol. 4, no. 2, pp. 172–179, 1975.
46. A. Alksas, M. Shehata, H. Atef, F. Sherif, M. Yaghi, M. Alhalabi,. . .and A. El-Baz, "A Comprehensive non-invasive system for early grading of gliomas," in *2022 26th International Conference on Pattern Recognition (ICPR)*. IEEE, 2022, August, pp. 4371–4377.
47. J. J. Van Griethuysen, A. Fedorov, C. Parmar, A. Hosny, N. Aucoin, V. Narayan,. . .and H. J. Aerts, "Computational radiomics system to decode the radiographic phenotype," *Cancer Research*, vol. 77, no. 21, pp. e104–e107, 2017.
48. S. M. Ayyad, M. A. Badawy, M. Shehata, A. Alksas, A. Mahmoud, M. Abou El-Ghar,. . .and A. El-Baz, "A new framework for precise identification of prostatic adenocarcinoma," *Sensors*, vol. 22, no. 5, p. 1848, 2022.
49. E. Niendorf, B. Spilseth, X. Wang, and A. Taylor, "Contrast enhanced MRI in the diagnosis of HCC," *Diagnostics*, vol. 5, no. 3, pp. 383–398, 2015.
50. D. Yang, R. Li, X. H. Zhang, C. L. Tang, K. S. Ma, D. Y. Guo, and X. C. Yan, "Perfusion characteristics of hepatocellular carcinoma at contrast-enhanced ultrasound: influence of the cellular differentiation, the tumor size and the underlying hepatic condition," *Scientific Reports*, vol. 8, no. 1, p. 4713, 2018.
51. A. A. K. Abdel Razek, A. Alksas, M. Shehata, A. AbdelKhalek, Abdel Baky, K., A. El-Baz, and E. Helmy, "Clinical applications of artificial intelligence and radiomics in neuro-oncology imaging," *Insights into Imaging*, vol. 12, 1–17, 2021.
52. R. Kohavi, and G. H. John, *The wrapper approach*. Feature Extraction Construction and Selection, 1998, pp. 33–50.
53. C. Albon, *Machine learning with python cookbook: practical solutions from preprocessing to deep learning*. O'Reilly Media, Inc., 2018.
54. L. R. Dice, "Measures of the amount of ecologic association between species," *Ecology*, vol. 26, no. 3, pp. 297–302, 1945.
55. A. Alksas, M. Shehata, H. Atef, F. Sherif, N. S. Alghamdi, M. Ghazal,. . .and A. El-Baz, "A novel system for precise grading of glioma," *Bioengineering*, vol. 9, no. 10, p. 532, 2022.
56. A. Carass, S. Roy, A. Gherman, J. C. Reinhold, A. Jesson, T. Arbel,. . .and I. Oguz, "Evaluating white matter lesion segmentations with refined Sørensen-Dice analysis," *Scientific Reports*, vol. 10, no. 1, p. 8242, 2020.
57. H. M. Balaha, E. R. Antar, M. M. Saafan, and E. M. El-Gendy, "A comprehensive framework towards segmenting and classifying breast cancer patients using deep learning and Aquila optimizer," *Journal of Ambient Intelligence and Humanized Computing*, vol. 14, no. 6, pp. 7897–7917, 2023.
58. T. K. Ho, "The random subspace method for constructing decision forests," *IEEE Transactions on Pattern Analysis and Machine Intelligence*, vol. 20, no. 8, pp. 832–844, 1998.
59. R. Bryll, R. Gutierrez-Osuna, and F. Quek, "Attribute bagging: improving accuracy of classifier ensembles by using random feature subsets," *Pattern Recognition*, vol. 36, no. 6, pp. 1291–1302, 2003.
60. R. Suarez-Ibarrola, S. Hein, G. Reis, C. Gratzke, and A. Miernik, "Current and future applications of machine and deep learning in urology: a review of the literature on urolithiasis, renal cell carcinoma, and bladder and prostate cancer," *World Journal of Urology*, vol. 38, pp. 2329–2347, 2020.

61. M. Shehata, A. Alksas, R. T. Abouelkheir, A. Elmahdy, A. Shaffie, A. Soliman,. . .and A. El-Baz, “A new computer-aided diagnostic (cad) system for precise identification of renal tumors,” in *2021 IEEE 18th International Symposium on Biomedical Imaging (ISBI)*. IEEE, 2021, April, pp. 1378–1381.
62. A. A. K. A. Razek, L. G. El-Serougy, G. A. Saleh, W. Shabana, and R. Abd El-wahab, “Liver imaging reporting and data system version 2018: what radiologists need to know,” *Journal of Computer Assisted Tomography*, vol. 44, no. 2, pp. 168–177, 2020.
63. Z. Li, Y. Mao, W. Huang, H. Li, J. Zhu, W. Li, and B. Li, “Texture-based classification of different single liver lesion based on SPAIR T2W MRI images,” *BMC Medical Imaging*, vol. 17, pp. 1–9, 2017.
64. A. Oyama, Y. Hiraoka, I. Obayashi, Y. Saikawa, S. Furui, K. Shiraishi,. . .and J. I. Kotoku, “Hepatic tumor classification using texture and topology analysis of non-contrast-enhanced three-dimensional T1-weighted MR images with a radiomics approach,” *Scientific Reports*, vol. 9, no. 1, p. 8764, 2019.
65. K. Yasaka, H. Akai, O. Abe, and S. Kiryu, “Deep learning with convolutional neural network for differentiation of liver masses at dynamic contrast-enhanced CT: a preliminary study,” *Radiology*, vol. 286, no. 3, pp. 887–896, 2018.
66. De Fauw, J., J. R. Ledsam, B. Romera-Paredes, S. Nikolov, N. Tomasev, S. Blackwell,. . .and O. Ronneberger, “Clinically applicable deep learning for diagnosis and referral in retinal disease,” *Nature Medicine*, vol. 24, no. 9, pp. 1342–1350, 2018.
67. A. Alksas, M. Shehata, G. A. Saleh, A. Shaffie, A. Soliman, M. Ghazal, H. A. Khalifeh, A. A. Razek, and A. El-Baz, “A novel computer-aided diagnostic system for early assessment of hepatocellular carcinoma,” in *2020 25th International Conference on Pattern Recognition (ICPR)*. IEEE, 2021, pp. 10375–10382.
68. A. Alksas, M. Shehata, G. A. Saleh, A. Shaffie, A. Soliman, M. Ghazal, A. Khelifi, H. A. Khalifeh, A. A. Razek, G. A. Giridharan et al., “A novel computer-aided diagnostic system for accurate detection and grading of liver tumors,” *Scientific Reports*, vol. 11, no. 1, pp. 1–18, 2021.
69. I. Reda, M. Ghazal, A. Shalaby, M. Elmogy, A. AbouEl-Fetouh, B. O. Ayinde, M. AbouEl-Ghar, A. Elmaghraby, R. Keynton, and A. El-Baz, “A novel ADCs-based CNN classification system for precise diagnosis of prostate cancer,” in *2018 24th International Conference on Pattern Recognition (ICPR)*. IEEE, 2018, pp. 3923–3928.
70. I. Reda, A. Khalil, M. Elmogy, A. Abou El-Fetouh, A. Shalaby, M. Abou El-Ghar, A. Elmaghraby, M. Ghazal, and A. El-Baz, “Deep learning role in early diagnosis of prostate cancer,” *Technology in Cancer Research & Treatment*, vol. 17, p. 1533034618775530, 2018.
71. I. Reda, B. O. Ayinde, M. Elmogy, A. Shalaby, M. El-Melegy, M. A. El-Ghar, A. A. El-fetouh, M. Ghazal, and A. El-Baz, “A new CNN-based system for early diagnosis of prostate cancer,” in *2018 IEEE 15th International Symposium on Biomedical Imaging (ISBI 2018)*. IEEE, 2018, pp. 207–210.
72. S. M. Ayyad, M. A. Badawy, M. Shehata, A. Alksas, A. Mahmoud, M. Abou El-Ghar, M. Ghazal, M. El-Melegy, N. B. Abdel-Hamid, L. M. Labib, H. A. Ali, and A. El-Baz, “A new framework for precise identification of prostatic adenocarcinoma,” *Sensors*, vol. 22, no. 5, 2022. [Online]. Available on line: https://www.mdpi.com/1424-8220/22/5/1848
73. K. Hammouda, F. Khalifa, M. El-Melegy, M. Ghazal, H. E. Darwish, M. A. El-Ghar, and A. El-Baz, “A deep learning pipeline for grade groups classification using digitized prostate biopsy specimens,” *Sensors*, vol. 21, no. 20, p. 6708, 2021.
74. M. Shehata, A. Shalaby, A. E. Switala, M. El-Baz, M. Ghazal, L. Fraiwan, A. Khalil, M. A. El-Ghar, M. Badawy, A. M. Bakr et al., “A multimodal computer-aided diagnostic system for precise identification of renal allograft rejection: preliminary results,” *Medical Physics*, vol. 47, no. 6, pp. 2427–2440, 2020.

75. M. Shehata, F. Khalifa, A. Soliman, M. Ghazal, F. Taher, M. Abou El-Ghar, A. C. Dwyer, G. Gimel'farb, R. S. Keynton, and A. El- Baz, "Computer-aided diagnostic system for early detection of acute renal transplant rejection using diffusion-weighted MRI," *IEEE Transactions on Biomedical Engineering*, vol. 66, no. 2, pp. 539–552, 2018.
76. E. Hollis, M. Shehata, M. Abou El-Ghar, M. Ghazal, T. El-Diasty, M. Merchant, A. E. Switala, and A. El-Baz, "Statistical analysis of adcs and clinical biomarkers in detecting acute renal transplant rejection," *The British Journal of Radiology*, vol. 90, no. 1080, p. 20170125, 2017.
77. M. Shehata, A. Alksas, R. T. Abouelkheir, A. Elmahdy, A. Shaffie, A. Soliman, M. Ghazal, H. Abu Khalifeh, R. Salim, A. A. K. Abdel Razek et al., "A comprehensive computer-assisted diagnosis system for early assessment of renal cancer tumors," *Sensors*, vol. 21, no. 14, p. 4928, 2021.
78. F. Khalifa, G. M. Beache, M. A. El-Ghar, T. El-Diasty, G. Gimel'farb, M. Kong, and A. El-Baz, "Dynamic contrast-enhanced MRI- based early detection of acute renal transplant rejection," *IEEE Transactions on Medical Imaging*, vol. 32, no. 10, pp. 1910–1927, 2013.
79. F. Khalifa, M. A. El-Ghar, B. Abdollahi, H. Frieboes, T. El-Diasty, and A. El-Baz, "A comprehensive non-invasive framework for automated evaluation of acute renal transplant rejection using DCE-MRI," *NMR in Biomedicine*, vol. 26, no. 11, pp. 1460–1470, 2013.
80. F. Khalifa, A. Elnakib, G. M. Beache, G. Gimel'farb, M. A. El-Ghar, G. Sokhadze, S. Manning, P. McClure, and A. El-Baz, "3D kidney segmentation from CT images using a level set approach guided by a novel stochastic speed function," in *Proceedings of International Conference Medical Image Computing and Computer-Assisted Intervention, (MICCAI'11),* Toronto, Canada, 2011, September 18–22, pp. 587–594.
81. M. Shehata, F. Khalifa, E. Hollis, A. Soliman, E. Hosseini-Asl, M. A. El-Ghar, M. El-Baz, A. C. Dwyer, A. El-Baz, and R. Keynton, "A new non-invasive approach for early classification of renal rejection types using diffusion-weighted MRI," in *IEEE International Conference on Image Processing (ICIP)*. IEEE, 2016, pp. 136–140.
82. F. Khalifa, A. Soliman, A. Takieldeen, M. Shehata, M. Mostapha, A. Shaffie, R. Ouseph, A. Elmaghraby, and A. El-Baz, "Kidney segmentation from CT images using a 3D NMF-guided active contour model," in *IEEE 13th International Symposium on Biomedical Imaging (ISBI)*. IEEE, 2016, pp. 432–435.
83. M. Shehata, F. Khalifa, A. Soliman, A. Takieldeen, M. A. El-Ghar, A. Shaffie, A. C. Dwyer, R. Ouseph, A. El-Baz, and R. Keynton, "3D diffusion MRI-based cad system for early diagnosis of acute renal rejection," in *Biomedical Imaging (ISBI), 2016 IEEE 13th International Symposium on*. IEEE, 2016, pp. 1177–1180.
84. M. Shehata, F. Khalifa, A. Soliman, R. Alrefai, M. A. El-Ghar, A. C. Dwyer, R. Ouseph, and A. El-Baz, "A level set-based framework for 3D kidney segmentation from diffusion MR images," in *IEEE International Conference on Image Processing (ICIP)*. IEEE, 2015, pp. 4441–4445.
85. M. Shehata, F. Khalifa, A. Soliman, M. A. El-Ghar, A. C. Dwyer, G. Gimel'farb, R. Keynton, and A. El-Baz, "A promising non-invasive cad system for kidney function assessment," in *International Conference on Medical Image Computing and Computer-Assisted Intervention*. Springer, 2016, pp. 613–621.
86. F. Khalifa, A. Soliman, A. Elmaghraby, G. Gimel'farb, and A. El-Baz, "3D kidney segmentation from abdominal images using spatial-appearance models," *Computational and Mathematical Methods in Medicine*, vol. 2017, pp. 1–10, 2017.
87. E. Hollis, M. Shehata, F. Khalifa, M. A. El-Ghar, T. El-Diasty, and A. El-Baz, "Towards non-invasive diagnostic techniques for early detection of acute renal transplant rejection: a review," *The Egyptian Journal of Radiology and Nuclear Medicine*, vol. 48, no. 1, pp. 257–269, 2016.

88. M. Shehata, F. Khalifa, A. Soliman, M. A. El-Ghar, A. C. Dwyer, and A. El-Baz, "Assessment of renal transplant using image and clinical-based biomarkers," in *Proceedings °f 13th Annual Scientific Meeting of American Society for Diagnostics and Interventional Nephrology (ASDIN'17)*, New Orleans, LA, USA, 2017, February 10–12.
89. M. Shehata, F. Khalifa, A. Soliman, M. A. El-Ghar, A. C. Dwyer, and A. El-Baz, "Early assessment of acute renal rejection," in *Proceedings °f 12th Annual Scientific Meeting of American Society for Diagnostics and Interventional Nephrology (ASDIN'16)*, Pheonix, AZ, USA, 2016, February 19–21.
90. A. Eltanboly, M. Ghazal, H. Hajjdiab, A. Shalaby, A. Switala, A. Mahmoud, P. Sahoo, M. El-Azab, and A. El-Baz, "Level sets-based image segmentation approach using statistical shape priors," *Applied Mathematics and Computation*, vol. 340, pp. 164–179, 2019.
91. M. Shehata, A. Mahmoud, A. Soliman, F. Khalifa, M. Ghazal, M. A. El-Ghar, M. El-Melegy, and A. El-Baz, "3D kidney segmentation from abdominal diffusion MRI using an appearance-guided deformable boundary," *PLoS One*, vol. 13, no. 7, p. e0200082, 2018.
92. H. Abdeltawab, M. Shehata, A. Shalaby, F. Khalifa, A. Mahmoud, M. A. El-Ghar, A. C. Dwyer, M. Ghazal, H. Hajjdiab, R. Keynton et al., "A novel CNN-based cad system for early assessment of transplanted kidney dysfunction," *Scientific Reports*, vol. 9, no. 1, p. 5948, 2019.
93. K. Hammouda, F. Khalifa, H. Abdeltawab, A. Elnakib, G. Giridharan, M. Zhu, C. Ng, S. Dassanayaka, M. Kong, H. Darwish et al., "A new framework for performing cardiac strain analysis from cine MRI imaging in mice," *Scientific Reports*, vol. 10, no. 1, pp. 1–15, 2020.
94. H. Abdeltawab, F. Khalifa, K. Hammouda, J. M. Miller, M. M. Meki, Q. Ou, A. El-Baz, and T. Mohamed, "Artificial intelligence based framework to quantify the cardiomyocyte structural integrity in heart slices," *Cardiovascular Engineering and Technology*, pp. 1–11, 2021.
95. F. Khalifa, G. M. Beache, A. Elnakib, H. Sliman, G. Gimel'farb, K. C. Welch, and A. El-Baz, "A new shape-based framework for the left ventricle wall segmentation from cardiac first-pass perfusion MRI," in *Proceedings of IEEE International Symposium on Biomedical Imaging: From Nano to Macro, (ISBI'13)*, San Francisco, CA, 2013, April 7–11, pp. 41–44.
96. F. Khalifa, G. M. Beache, A. Elnakib, H. Sliman, G. Gimel'farb, K. C. Welch, and A. El-Baz, "A new nonrigid registration framework for improved visualization of transmural perfusion gradients on cardiac first–pass perfusion MRI," in *Proceedings of IEEE International Symposium on Biomedical Imaging: From Nano to Macro, (ISBI'12)*, Barcelona, Spain, 2012, May 2–5, pp. 828–831.
97. F. Khalifa, G. M. Beache, A. Firjani, K. C. Welch, G. Gimel'farb, and A. El-Baz, "A new nonrigid registration approach for motion correction of cardiac first-pass perfusion MRI," in *Proceedings of IEEE International Conference on Image Processing, (ICIP'12)*, Lake Buena Vista, Florida, 2012, September 30–October 3, pp. 1665–1668.
98. F. Khalifa, G. M. Beache, G. Gimel'farb, and A. El-Baz, "A novel CAD system for analyzing cardiac first-pass MR images," in *Proceedings of IAPR International Conference on Pattern Recognition (ICPR'12)*, Tsukuba Science City, Japan, 2012, November 11–15, pp. 77–80.
99. F. Khalifa, G. M. Beache, G. Gimel'farb, and A. El-Baz, "A novel approach for accurate estimation of left ventricle global indexes from short-axis cine MRI," in *Proceedings of IEEE International Conference on Image Processing, (ICIP'11)*, Brussels, Belgium, 2011, September 11–14, pp. 2645–2649.
100. F. Khalifa, G. M. Beache, G. Gimel'farb, G. A. Giridharan, and A. El-Baz, "A new image-based framework for analyzing cine images," in *Handbook of multi modality state-of-the-art medical image segmentation and registration methodologies*, A. El-Baz, U. R. Acharya, M. Mirmedhdi, and J. S. Suri, Eds. Springer, 2011, vol. 2, ch. 3, pp. 69–98.

101. F. Khalifa, G. M. Beache, G. Gimel'farb, G. A. Giridharan, and A. El-Baz, "Accurate automatic analysis of cardiac cine images," *IEEE Transactions on Biomedical Engineering*, vol. 59, no. 2, pp. 445–455, 2012.
102. F. Khalifa, G. M. Beache, M. Nitzken, G. Gimel'farb, G. A. Giridharan, and A. El-Baz, "Automatic analysis of left ventricle wall thickness using short-axis cine CMR images," in *Proceedings of IEEE International Symposium on Biomedical Imaging: From Nano to Macro, (ISBI'11)*, Chicago, Illinois, 2011, March 30–April 2, pp. 1306–1309.
103. M. Nitzken, G. Beache, A. Elnakib, F. Khalifa, G. Gimel'farb, and A. El-Baz, "Accurate modeling of tagged CMR 3D image appearance characteristics to improve cardiac cycle strain estimation," in *Image Processing (ICIP), 2012 19th IEEE International Conference on*, Orlando, Florida, USA. IEEE, 2012, September, pp. 521–524.
104. M. Nitzken, G. Beache, A. Elnakib, F. Khalifa, G. Gimel'farb, and A. El-Baz, "Improving full-cardiac cycle strain estimation from tagged CMR by accurate modeling of 3D image appearance characteristics," in *Biomedical Imaging (ISBI), 2012 9th IEEE International Symposium on*, Barcelona, Spain. IEEE, 2012, May, pp. 462–465. (Selected for oral presentation).
105. M. J. Nitzken, A. S. El-Baz, and G. M. Beache, "Markov-Gibbs random field model for improved full-cardiac cycle strain estimation from tagged CMR," *Journal of Cardiovascular Magnetic Resonance*, vol. 14, no. 1, pp. 1–2, 2012.
106. H. Sliman, A. Elnakib, G. Beache, A. Elmaghraby, and A. El-Baz, "Assessment of myocardial function from cine cardiac MRI using a novel 4D tracking approach," *Journal of Computer Science and Systems Biology*, vol. 7, pp. 169–173, 2014.
107. H. Sliman, A. Elnakib, G. M. Beache, A. Soliman, F. Khalifa, G. Gimel'farb, A. Elmaghraby, and A. El-Baz, "A novel 4D PDE-based approach for accurate assessment of myocardium function using cine cardiac magnetic resonance images," in *Proceedings of IEEE International Conference on Image Processing (ICIP'14)*, Paris, France, 2014, October 27–30, pp. 3537–3541.
108. H. Sliman, F. Khalifa, A. Elnakib, G. M. Beache, A. Elmaghraby, and A. El-Baz, "A new segmentation-based tracking framework for extracting the left ventricle cavity from cine cardiac MRI," in *Proceedings of IEEE International Conference on Image Processing, (ICIP'13)*, Melbourne, Australia, 2013, September 15–18, pp. 685–689.
109. H. Sliman, F. Khalifa, A. Elnakib, A. Soliman, G. M. Beache, A. Elmaghraby, G. Gimel'farb, and A. El-Baz, "Myocardial borders segmentation from cine MR images using bi-directional coupled parametric deformable models," *Medical Physics*, vol. 40, no. 9, pp. 1–13, 2013.
110. H. Sliman, F. Khalifa, A. Elnakib, A. Soliman, G. M. Beache, G. Gimel'farb, A. Emam, A. Elmaghraby, and A. El-Baz, "Accurate segmentation framework for the left ventricle wall from cardiac cine MRI," in *Proceedings of International Symposium on Computational Models for Life Science, (CMLS'13)*, vol. 1559, Sydney, Australia, 2013, November 27–29, pp. 287–296.
111. A. Sharafeldeen, M. Elsharkawy, N. S. Alghamdi, A. Soliman, and A. El-Baz, "Precise segmentation of covid-19 infected lung from CT images based on adaptive first-order appearance model with morphological/anatomical constraints," *Sensors*, vol. 21, no. 16, p. 5482, 2021.
112. M. Elsharkawy, A. Sharafeldeen, F. Taher, A. Shalaby, A. Soliman, A. Mahmoud, M. Ghazal, A. Khalil, N. S. Alghamdi, A. A. K. A. Razek et al., "Early assessment of lung function in coronavirus patients using invariant markers from chest x-rays images," *Scientific Reports*, vol. 11, no. 1, pp. 1–11, 2021.
113. B. Abdollahi, A. C. Civelek, X.-F. Li, J. Suri, and A. El-Baz, "PET/CT nodule segmentation and diagnosis: a survey," in *Multi detector CT imaging*, L. Saba, and J. S. Suri, Eds. Taylor, Francis, 2014, ch. 30, pp. 639–651.

114. B. Abdollahi, A. El-Baz, and A. A. Amini, "A multi-scale non-linear vessel enhancement technique," in *Engineering in Medicine and Biology Society, EMBC, 2011 Annual International Conference of the IEEE*. IEEE, 2011, pp. 3925–3929.
115. B. Abdollahi, A. Soliman, A. Civelek, X.-F. Li, G. Gimel'farb, and A. El-Baz, "A novel gaussian scale space-based joint MGRF framework for precise lung segmentation," in *Proceedings of IEEE International Conference on Image Processing, (ICIP'12)*. IEEE, 2012, pp. 2029–2032.
116. B. Abdollahi, A. Soliman, A. Civelek, X.-F. Li, G. Gimel'farb, and A. El-Baz, "A novel 3D joint MGRF framework for precise lung segmentation," in *Machine learning in medical imaging*. Springer, 2012, pp. 86–93.
117. A. M. Ali, A. S. El-Baz, and A. A. Farag, "A novel framework for accurate lung segmentation using graph cuts," in *Proceedings of IEEE International Symposium on Biomedical Imaging: From Nano to Macro, (ISBI'07)*. IEEE, 2007, pp. 908–911.
118. A. El-Baz, G. M. Beache, G. Gimel'farb, K. Suzuki, and K. Okada, "Lung imaging data analysis," *International Journal of Biomedical Imaging*, vol. 2013, pp. 1–2, 2013.
119. A. El-Baz, G. M. Beache, G. Gimel'farb, K. Suzuki, K. Okada, A. Elnakib, A. Soliman, and B. Abdollahi, "Computer-aided diagnosis systems for lung cancer: challenges and methodologies," *International Journal of Biomedical Imaging*, vol. 2013, pp. 1–46, 2013.
120. A. El-Baz, A. Elnakib, M. Abou El-Ghar, G. Gimel'farb, R. Falk, and A. Farag, "Automatic detection of 2D and 3D lung nodules in chest spiral CT scans," *International Journal of Biomedical Imaging*, vol. 2013, pp. 1–11, 2013.
121. A. El-Baz, A. A. Farag, R. Falk, and R. La Rocca, "A unified approach for detection, visualization, and identification of lung abnormalities in chest spiral CT scans," in *International Congress Series*, vol. 1256. Elsevier, 2003, pp. 998–1004.
122. A. El-Baz, A. A. Farag, R. Falk, and R. La Rocca, "Detection, visualization and identification of lung abnormalities in chest spiral CT scan: phase-I," in *Proceedings of International conference on Biomedical Engineering*, Cairo, Egypt, vol. 12, no. 1, 2002.
123. A. El-Baz, A. Farag, G. Gimel'farb, R. Falk, M. A. El-Ghar, and T. Eldiasty, "A framework for automatic segmentation of lung nodules from low dose chest CT scans," in *Proceedings of International Conference on Pattern Recognition, (ICPR'06)*, vol. 3. IEEE, 2006, pp. 611–614.
124. A. El-Baz, A. Farag, G. Gimel'farb, R. Falk, and M. A. El-Ghar, "A novel level set-based computer-aided detection system for automatic detection of lung nodules in low dose chest computed tomography scans," *Lung Imaging and Computer Aided Diagnosis*, vol. 10, pp. 221–238, 2011.
125. A. El-Baz, G. Gimel'farb, M. Abou El-Ghar, and R. Falk, "Appearance-based diagnostic system for early assessment of malignant lung nodules," in *Proceedings of IEEE International Conference on Image Processing, (ICIP'12)*. IEEE, 2012, pp. 533–536.
126. A. El-Baz, G. Gimel'farb, and R. Falk, "A novel 3D framework for automatic lung segmentation from low dose CT images," in *Lung imaging and computer aided diagnosis*, A. El-Baz, and J. S. Suri, Eds. Taylor, Francis, 2011, ch. 1, pp. 1–16.
127. A. El-Baz, G. Gimel'farb, R. Falk, and M. El-Ghar, "Appearance analysis for diagnosing malignant lung nodules," in *Proceedings of IEEE International Symposium on Biomedical Imaging: From Nano to Macro (ISBI'10)*. IEEE, 2010, pp. 193–196.
128. A. El-Baz, G. Gimel'farb, R. Falk, and M. A. El-Ghar, "A novel level set-based CAD system for automatic detection of lung nodules in low dose chest CT scans," in *Lung imaging and computer aided diagnosis*, A. El-Baz, and J. S. Suri, Eds. Taylor, Francis, 2011, vol. 1, ch. 10, pp. 221–238.
129. A. El-Baz, G. Gimel'farb, R. Falk, and M. A. El-Ghar, "A new approach for automatic analysis of 3D low dose CT images for accurate monitoring the detected lung nodules," in *Proceedings of International Conference on Pattern Recognition, (ICPR'08)*. IEEE, 2008, pp. 1–4.

130. A. El-Baz, G. Gimel'farb, R. Falk, and M. A. El-Ghar, "A novel approach for automatic follow-up of detected lung nodules," in Proceedings of IEEE International Conference on Image Processing, (ICIP'07), vol. 5. IEEE, 2007, pp. V–501.
131. A. El-Baz, G. Gimel'farb, R. Falk, and M. A. El-Ghar, "A new CAD system for early diagnosis of detected lung nodules," in *Image Processing, 2007. ICIP 2007. IEEE International Conference on*, vol. 2. IEEE, 2007, pp. II–461.
132. A. El-Baz, G. Gimel'farb, R. Falk, M. A. El-Ghar, and H. Refaie, "Promising results for early diagnosis of lung cancer," in *Proceedings of IEEE International Symposium on Biomedical Imaging: From Nano to Macro, (ISBI'08)*. IEEE, 2008, pp. 1151–1154.
133. A. El-Baz, G. L. Gimel'farb, R. Falk, M. Abou El-Ghar, T. Holland, and T. Shaffer, "A new stochastic framework for accurate lung segmentation," in *Proceedings of Medical Image Computing and Computer-Assisted Intervention, (MICCAI'08)*, 2008, pp. 322–330.
134. A. El-Baz, G. L. Gimel'farb, R. Falk, D. Heredis, and M. Abou El-Ghar, "A novel approach for accurate estimation of the growth rate of the detected lung nodules," in *Proceedings of International Workshop on Pulmonary Image Analysis*, 2008, pp. 33–42.
135. A. El-Baz, G. L. Gimel'farb, R. Falk, T. Holland, and T. Shaffer, "A framework for unsupervised segmentation of lung tissues from low dose computed tomography images," in *Proceedings of British Machine Vision, (BMVC'08)*, 2008, pp. 1–10.
136. A. El-Baz, G. Gimel'farb, R. Falk, and M. A. El-Ghar, "3D MGRF-based appearance modeling for robust segmentation of pulmonary nodules in 3D LDCT chest images," in *Lung imaging and computer aided diagnosis*. CRC Press, 2011, ch. 3, pp. 51–63.
137. A. El-Baz, G. Gimel'farb, R. Falk, and M. A. El-Ghar, "Automatic analysis of 3D low dose CT images for early diagnosis of lung cancer," *Pattern Recognition*, vol. 42, no. 6, pp. 1041–1051, 2009.
138. A. El-Baz, G. Gimel'farb, R. Falk, M. A. El-Ghar, S. Rainey, D. Heredia, and T. Shaffer, "Toward early diagnosis of lung cancer," in *Proceedings of Medical Image Computing and Computer-Assisted Intervention, (MICCAI'09)*. Springer, 2009, pp. 682–689.
139. A. El-Baz, G. Gimel'farb, R. Falk, M. A. El-Ghar, and J. Suri, "Appearance analysis for the early assessment of detected lung nodules," in *Lung imaging and computer aided diagnosis*. CRC Press, 2011, ch. 17, pp. 395–404.
140. A. El-Baz, F. Khalifa, A. Elnakib, M. Nitkzen, A. Soliman, P. McClure, G. Gimel'farb, and M. A. El-Ghar, "A novel approach for global lung registration using 3D Markov Gibbs appearance model," in *Proceedings of International Conference Medical Image Computing and Computer-Assisted Intervention, (MICCAI'12)*, Nice, France, 2012, October 1–5, pp. 114–121.
141. A. El-Baz, M. Nitzken, A. Elnakib, F. Khalifa, G. Gimel'farb, R. Falk, and M. A. El-Ghar, "3D shape analysis for early diagnosis of malignant lung nodules," in *Proceedings of International Conference Medical Image Computing and Computer-Assisted Intervention, (MICCAI'11)*, Toronto, Canada, 2011, September 18–22, pp. 175–182.
142. A. El-Baz, M. Nitzken, G. Gimel'farb, E. Van Bogaert, R. Falk, M. A. El-Ghar, and J. Suri, "Three-dimensional shape analysis using spherical harmonics for early assessment of detected lung nodules," in *Lung imaging and computer aided diagnosis*. CRC Press, 2011, ch. 19, pp. 421–438.
143. A. El-Baz, M. Nitzken, F. Khalifa, A. Elnakib, G. Gimel'farb, R. Falk, and M. A. El-Ghar, "3D shape analysis for early diagnosis of malignant lung nodules," in *Proceedings of International Conference on Information Processing in Medical Imaging, (IPMI'11)*, Monastery Irsee, Germany (Bavaria), 2011, July 3–8, pp. 772–783.
144. A. El-Baz, M. Nitzken, E. Vanbogaert, G. Gimel'Farb, R. Falk, and M. Abo El-Ghar, "A novel shape-based diagnostic approach for early diagnosis of lung nodules," in *Biomedical Imaging: From Nano to Macro, 2011 IEEE International Symposium on*. IEEE, 2011, pp. 137–140.

145. A. El-Baz, P. Sethu, G. Gimel'farb, F. Khalifa, A. Elnakib, R. Falk, and M. A. El-Ghar, "Elastic phantoms generated by microfluidics technology: validation of an imaged-based approach for accurate measurement of the growth rate of lung nodules," *Biotechnology Journal*, vol. 6, no. 2, pp. 195–203, 2011.
146. A. El-Baz, P. Sethu, G. Gimel'farb, F. Khalifa, A. Elnakib, R. Falk, and M. A. El-Ghar, "A new validation approach for the growth rate measurement using elastic phantoms generated by state-of-the-art microfluidics technology," in *Proceedings of IEEE International Conference on Image Processing, (ICIP'10)*, Hong Kong, 2010, September 26–29, pp. 4381–4383.
147. A. El-Baz, P. Sethu, G. Gimel'farb, F. Khalifa, A. Elnakib, R. Falk, and M. A. E.-G. J. Suri, "Validation of a new imaged-based approach for the accurate estimating of the growth rate of detected lung nodules using real CT images and elastic phantoms generated by state-of-the-art microfluidics technology," in *Handbook of lung imaging and computer aided diagnosis*, A. El-Baz, and J. S. Suri, Eds. Taylor & Francis, 2011, vol. 1, ch. 18, pp. 405–420.
148. A. El-Baz, A. Soliman, P. McClure, G. Gimel'farb, M. A. El-Ghar, and R. Falk, "Early assessment of malignant lung nodules based on the spatial analysis of detected lung nodules," in *Proceedings of IEEE International Symposium on Biomedical Imaging: From Nano to Macro, (ISBI'12)*. IEEE, 2012, pp. 1463–1466.
149. A. El-Baz, S. E. Yuksel, S. Elshazly, and A. A. Farag, "Non-rigid registration techniques for automatic follow-up of lung nodules," in *Proceedings of Computer Assisted Radiology and Surgery, (CARS'05)*, vol. 1281. Elsevier, 2005, pp. 1115–1120.
150. A. S. El-Baz, and J. S. Suri, *Lung imaging and computer aided diagnosis*. CRC Press, 2011.
151. A. Soliman, F. Khalifa, N. Dunlap, B. Wang, M. El-Ghar, and A. El-Baz, "An ISO-surfaces based local deformation handling framework of lung tissues," in *Biomedical Imaging (ISBI), 2016 IEEE 13th International Symposium on*. IEEE, 2016, pp. 1253–1259.
152. A. Soliman, F. Khalifa, A. Shaffie, N. Dunlap, B. Wang, A. Elmaghraby, and A. El-Baz, "Detection of lung injury using 4D-CT chest images," in *Biomedical Imaging (ISBI), 2016 IEEE 13th International Symposium on*. IEEE, 2016, pp. 1274–1277.
153. A. Soliman, F. Khalifa, A. Shaffie, N. Dunlap, B. Wang, A. Elmaghraby, G. Gimel'farb, M. Ghazal, and A. El-Baz, "A comprehensive framework for early assessment of lung injury," in *Image Processing (ICIP), 2017 IEEE International Conference on*. IEEE, 2017, pp. 3275–3279.
154. A. Shaffie, A. Soliman, M. Ghazal, F. Taher, N. Dunlap, B. Wang, A. Elmaghraby, G. Gimel'farb, and A. El-Baz, "A new framework for incorporating appearance and shape features of lung nodules for precise diagnosis of lung cancer," in *Image Processing (ICIP), 2017 IEEE International Conference on*. IEEE, 2017, pp. 1372–1376.
155. A. Soliman, F. Khalifa, A. Shaffie, N. Liu, N. Dunlap, B. Wang, A. Elmaghraby, G. Gimel'farb, and A. El-Baz, "Image-based cad system for accurate identification of lung injury," in *Image Processing (ICIP), 2016 IEEE International Conference on*. IEEE, 2016, pp. 121–125.
156. A. Soliman, A. Shaffie, M. Ghazal, G. Gimel'farb, R. Keynton, and A. El-Baz, "A novel CNN segmentation framework based on using new shape and appearance features," in *2018 25th IEEE International Conference on Image Processing (ICIP)*. IEEE, 2018, pp. 3488–3492.
157. A. Shaffie, A. Soliman, H. A. Khalifeh, M. Ghazal, F. Taher, R. Keynton, A. Elmaghraby, and A. El-Baz, "On the integration of CT-derived features for accurate detection of lung cancer," in *2018 IEEE International Symposium on Signal Processing and Information Technology (ISSPIT)*. IEEE, 2018, pp. 435–440.
158. A. Shaffie, A. Soliman, H. A. Khalifeh, M. Ghazal, F. Taher, A. Elmaghraby, R. Keynton, and A. El-Baz, "Radiomic-based framework for early diagnosis of lung cancer," in *2019 IEEE 16th International Symposium on Biomedical Imaging (ISBI 2019)*. IEEE, 2019, pp. 1293–1297.

159. A. Shaffie, A. Soliman, M. Ghazal, F. Taher, N. Dunlap, B. Wang, V. Van Berkel, G. Gimelfarb, A. Elmaghraby, and A. El-Baz, "A novel autoencoder-based diagnostic system for early assessment of lung cancer," in *2018 25th IEEE International Conference on Image Processing (ICIP)*. IEEE, 2018, pp. 1393–1397.
160. A. Shaffie, A. Soliman, L. Fraiwan, M. Ghazal, F. Taher, N. Dunlap, B. Wang, V. van Berkel, R. Keynton, A. Elmaghraby et al., "A generalized deep learning-based diagnostic system for early diagnosis of various types of pulmonary nodules," *Technology in Cancer Research & Treatment*, vol. 17, p. 1533033818798800, 2018.
161. A. A. K. Abdel Razek, A. Alksas, M. Shehata, A. AbdelKhalek, K. Abdel Baky, A. El-Baz, and E. Helmy, "Clinical applications of artificial intelligence and radiomics in neuro-oncology imaging," *Insights into Imaging*, vol. 12, no. 1, pp. 1–17, 2021.
162. Y. ElNakieb, M. T. Ali, O. Dekhil, M. E. Khalefa, A. Soliman, A. Shalaby, A. Mahmoud, M. Ghazal, H. Hajjdiab, A. Elmaghraby et al., "Towards accurate personalized autism diagnosis using different imaging modalities: sMRI, fMRI, and DTI," in *2018 IEEE International Symposium on Signal Processing and Information Technology (ISSPIT)*. IEEE, 2018, pp. 447–452.
163. Y. ElNakieb, A. Soliman, A. Mahmoud, O. Dekhil, A. Shalaby, M. Ghazal, A. Khalil, A. Switala, R. S. Keynton, G. N. Barnes et al., "Autism spectrum disorder diagnosis framework using diffusion tensor imaging," in *2019 IEEE International Conference on Imaging Systems and Techniques (IST)*. IEEE, 2019, pp. 1–5.
164. R. Haweel, O. Dekhil, A. Shalaby, A. Mahmoud, M. Ghazal, R. Keynton, G. Barnes, and A. El-Baz, "A machine learning approach for grading autism severity levels using task-based functional MRI," in *2019 IEEE International Conference on Imaging Systems and Techniques (IST)*. IEEE, 2019, pp. 1–5.
165. O. Dekhil, M. Ali, R. Haweel, Y. Elnakib, M. Ghazal, H. Hajjdiab, L. Fraiwan, A. Shalaby, A. Soliman, A. Mahmoud et al., "A comprehensive framework for differentiating autism spectrum disorder from neurotypicals by fusing structural MRI and resting state functional MRI," in *Seminars in pediatric neurology*. Elsevier, 2020, p. 100805.
166. R. Haweel, O. Dekhil, A. Shalaby, A. Mahmoud, M. Ghazal, A. Khalil, R. Keynton, G. Barnes, and A. El-Baz, "A novel framework for grading autism severity using task-based fMRI," in *2020 IEEE 17th International Symposium on Biomedical Imaging (ISBI)*. IEEE, 2020, pp. 1404–1407.
167. A. El-Baz, A. Elnakib, F. Khalifa, M. A. El-Ghar, P. McClure, A. Soliman, and G. Gimel'farb, "Precise segmentation of 3-D magnetic resonance angiography," *IEEE Transactions on Biomedical Engineering*, vol. 59, no. 7, pp. 2019–2029, 2012.
168. A. El-Baz, A. Farag, A. Elnakib, M. F. Casanova, G. Gimel'farb, A. E. Switala, D. Jordan, and S. Rainey, "Accurate automated detection of autism related corpus callosum abnormalities," Journal of Medical Systems, vol. 35, no. 5, pp. 929–939, 2011.
169. A. El-Baz, G. Gimel'farb, R. Falk, M. A. El-Ghar, V. Kumar, and D. Heredia, "A novel 3D joint Markov-Gibbs model for extracting blood vessels from PC–MRA images," in *Medical Image Computing and Computer-Assisted Intervention–MICCAI 2009*, vol. 5762. Springer, 2009, pp. 943–950.
170. A. Elnakib, A. El-Baz, M. F. Casanova, G. Gimel'farb, and A. E. Switala, "Image-based detection of corpus callosum variability for more accurate discrimination between dyslexic and normal brains," in *Proceedings of IEEE International Symposium on Biomedical Imaging: From Nano to Macro (ISBI'2010)*. IEEE, 2010, pp. 109–112.
171. A. Elnakib, M. F. Casanova, G. Gimel'farb, A. E. Switala, and A. El-Baz, "Autism diagnostics by centerline-based shape analysis of the corpus callosum," in *Proceedings of IEEE International Symposium on Biomedical Imaging: From Nano to Macro (ISBI'2011)*. IEEE, 2011, pp. 1843–1846.
172. A. Elnakib, M. Nitzken, M. Casanova, H. Park, G. Gimel'farb, and A. El-Baz, "Quantification of age-related brain cortex change using 3D shape analysis," in *Pattern Recognition (ICPR), 2012 21st International Conference on*. IEEE, 2012, pp. 41–44.

173. M. Nitzken, M. Casanova, G. Gimel'farb, A. Elnakib, F. Khalifa, A. Switala, and A. El-Baz, "3D shape analysis of the brain cortex with application to dyslexia," in *Image Processing (ICIP), 2011 18th IEEE International Conference on*, Brussels, Belgium. IEEE, 2011, September, pp. 2657–2660. (Selected for oral presentation. Oral acceptance rate is 10 percent and the overall acceptance rate is 35 percent).
174. F. E.-Z. A. El-Gamal, M. M. Elmogy, M. Ghazal, A. Atwan, G. N. Barnes, M. F. Casanova, R. Keynton, and A. S. El-Baz, "A novel cad system for local and global early diagnosis of Alzheimer's disease based on PIB-PET scans," in *2017 IEEE International Conference on Image Processing (ICIP)*. IEEE, 2017, pp. 3270–3274.
175. M. M. Ismail, R. S. Keynton, M. M. Mostapha, A. H. ElTanboly, M. F. Casanova, G. L. Gimel'farb, and A. El-Baz, "Studying autism spectrum disorder with structural and diffusion magnetic resonance imaging: a survey," *Frontiers in Human Neuroscience*, vol. 10, p. 211, 2016.
176. A. Alansary, M. Ismail, A. Soliman, F. Khalifa, M. Nitzken, A. Elnakib, M. Mostapha, A. Black, K. Stinebruner, M. F. Casanova et al., "Infant brain extraction in t1-weighted MR images using bet and refinement using LCDG and MGRF models," *IEEE Journal of Biomedical and Health Informatics*, vol. 20, no. 3, pp. 925–935, 2016.
177. E. H. Asl, M. Ghazal, A. Mahmoud, A. Aslantas, A. Shalaby, M. Casanova, G. Barnes, G. Gimel'farb, R. Keynton, and A. El-Baz, "Alzheimer's disease diagnostics by a 3D deeply supervised adaptable convolutional network," *Frontiers in Bioscience* (Landmark Edition), vol. 23, pp. 584–596, 2018.
178. O. Dekhil, M. Ali, Y. El-Nakieb, A. Shalaby, A. Soliman, A. Switala, A. Mahmoud, M. Ghazal, H. Hajjdiab, M. F. Casanova, A. Elmaghraby, R. Keynton, A. El-Baz, and G. Barnes, "A personalized autism diagnosis cad system using a fusion of structural MRI and resting-state functional MRI data," *Frontiers in Psychiatry*, vol. 10, p. 392, 2019. [Online]. Available on line: https://www.frontiersin.org/article/10.3389/fpsyt.2019.00392
179. O. Dekhil, A. Shalaby, A. Soliman, A. Mahmoud, M. Kong, G. Barnes, A. Elmaghraby, and A. El-Baz, "Identifying brain areas correlated with ADOS raw scores by studying altered dynamic functional connectivity patterns," *Medical Image Analysis*, vol. 68, p. 101899, 2021.
180. Y. A. Elnakieb, M. T. Ali, A. Soliman, A. H. Mahmoud, A. M. Shalaby, N. S. Alghamdi, M. Ghazal, A. Khalil, A. Switala, R. S. Keynton et al., "Computer aided autism diagnosis using diffusion tensor imaging," *IEEE Access*, vol. 8, pp. 191298–191308, 2020.
181. M. T. Ali, Y. A. Elnakieb, A. Shalaby, A. Mahmoud, A. Switala, M. Ghazal, A. Khelifi, L. Fraiwan, G. Barnes, and A. El-Baz, "Autism classification using sMRI: a recursive features selection based on sampling from multi-level high dimensional spaces," in *2021 IEEE 18th International Symposium on Biomedical Imaging (ISBI)*. IEEE, 2021, pp. 267–270.
182. M. T. Ali, Y. ElNakieb, A. Elnakib, A. Shalaby, A. Mahmoud, M. Ghazal, J. Yousaf, H. Abu Khalifeh, M. Casanova, G. Barnes et al., "The role of structure MRI in diagnosing autism," *Diagnostics*, vol. 12, no. 1, p. 165, 2022.
183. Y. ElNakieb, M. T. Ali, A. Elnakib, A. Shalaby, A. Soliman, A. Mahmoud, M. Ghazal, G. N. Barnes, and A. El-Baz, "The role of diffusion tensor MR imaging (DTI) of the brain in diagnosing autism spectrum disorder: promising results," *Sensors*, vol. 21, no. 24, p. 8171, 2021.
184. A. Mahmoud, A. El-Barkouky, H. Farag, J. Graham, and A. Farag, "A non-invasive method for measuring blood flow rate in superficial veins from a single thermal image," in *Proceedings of the IEEE Conference on Computer Vision and Pattern Recognition Workshops*, 2013, pp. 354–359.

185. N. Elsaid, A. Saied, H. Kandil, A. Soliman, F. Taher, M. Hadi, G. Giridharan, R. Jennings, M. Casanova, R. Keynton et al., "Impact of stress and hypertension on the cerebrovasculature," *Frontiers in Bioscience-Landmark*, vol. 26, no. 12, p. 1643, 2021.
186. F. Taher, H. Kandil, Y. Gebru, A. Mahmoud, A. Shalaby, S. El-Mashad, and A. El-Baz, "A novel MRA-based framework for segmenting the cerebrovascular system and correlating cerebral vascular changes to mean arterial pressure," *Applied Sciences*, vol. 11, no. 9, p. 4022, 2021.
187. H. Kandil, A. Soliman, F. Taher, M. Ghazal, A. Khalil, G. Giridharan, R. Keynton, J. R. Jennings, and A. El-Baz, "A novel computer- aided diagnosis system for the early detection of hypertension based on cerebrovascular alterations," *NeuroImage: Clinical*, vol. 25, p. 102107, 2020.
188. H. Kandil, A. Soliman, M. Ghazal, A. Mahmoud, A. Shalaby, R. Keynton, A. Elmaghraby, G. Giridharan, and A. El-Baz, "A novel framework for early detection of hypertension using magnetic resonance angiography," *Scientific Reports*, vol. 9, no. 1, pp. 1–12, 2019.
189. Y. Gebru, G. Giridharan, M. Ghazal, A. Mahmoud, A. Shalaby, and A. El-Baz, "Detection of cerebrovascular changes using magnetic resonance angiography," in *Cardiovascular imaging and image analysis*. CRC Press, 2018, pp. 1–22.
190. A. Mahmoud, A. Shalaby, F. Taher, M. El-Baz, J. S. Suri, and A. El-Baz, "Vascular tree segmentation from different image modalities," in *Cardiovascular imaging and image analysis*. CRC Press, 2018, pp. 43–70.
191. F. Taher, A. Mahmoud, A. Shalaby, and A. El-Baz, "A review on the cerebrovascular segmentation methods," in *2018 IEEE International Symposium on Signal Processing and Information Technology (ISSPIT)*. IEEE, 2018, pp. 359–364.
192. H. Kandil, A. Soliman, L. Fraiwan, A. Shalaby, A. Mahmoud, A. ElTanboly, A. Elmaghraby, G. Giridharan, and A. El-Baz, "A novel MRA framework based on integrated global and local analysis for accurate segmentation of the cerebral vascular system," in *2018 IEEE 15th International Symposium on Biomedical Imaging (ISBI 2018)*. IEEE, 2018, pp. 1365–1368.
193. F. Taher, A. Soliman, H. Kandil, A. Mahmoud, A. Shalaby, G. Gimel'farb, and A. El-Baz, "Accurate segmentation of cerebrovasculature from TOF-MRA images using appearance descriptors," *IEEE Access*, vol. 8, p. 96139, 96149, 2020.
194. F. Taher, A. Soliman, H. Kandil, A. Mahmoud, A. Shalaby, G. Gimel'farb, and A. El-Baz, "Precise cerebrovascular segmentation," in *2020 IEEE International Conference on Image Processing (ICIP)*. IEEE, 2020, pp. 394–397.
195. M. Elsharkawy, A. Sharafeldeen, A. Soliman, F. Khalifa, M. Ghazal, E. El-Daydamony, A. Atwan, H. S. Sandhu, and A. El-Baz, "A novel computer-aided diagnostic system for early detection of diabetic retinopathy using 3D-OCT higher-order spatial appearance model," *Diagnostics*, vol. 12, no. 2, p. 461, 2022.
196. M. Elsharkawy, M. Elrazzaz, M. Ghazal, M. Alhalabi, A. Soliman, A. Mahmoud, E. El-Daydamony, A. Atwan, A. Thanos, H. S. Sandhu et al., "Role of optical coherence tomography imaging in predicting progression of age-related macular disease: a survey," *Diagnostics*, vol. 11, no. 12, p. 2313, 2021.
197. H. S. Sandhu, M. Elmogy, A. T. Sharafeldeen, M. Elsharkawy, N. El-Adawy, A. Eltanboly, A. Shalaby, R. Keynton, and A. El- Baz, "Automated diagnosis of diabetic retinopathy using clinical biomarkers, optical coherence tomography, and optical coherence tomography angiography," *American Journal of Ophthalmology*, vol. 216, pp. 201–206, 2020.
198. A. Sharafeldeen, M. Elsharkawy, F. Khalifa, A. Soliman, M. Ghazal, M. AlHalabi, M. Yaghi, M. Alrahmawy, S. Elmougy, H. Sandhu et al., "Precise higher-order reflectivity and morphology models for early diagnosis of diabetic retinopathy using OCT images," *Scientific Reports*, vol. 11, no. 1, pp. 1–16, 2021.

199. A. A. Sleman, A. Soliman, M. Elsharkawy, G. Giridharan, M. Ghazal, H. Sandhu, S. Schaal, R. Keynton, A. Elmaghraby, and A. El-Baz, "A novel 3D segmentation approach for extracting retinal layers from optical coherence tomography images," *Medical Physics*, vol. 48, no. 4, pp. 1584–1595, 2021.
200. A. A. Sleman, A. Soliman, M. Ghazal, H. Sandhu, S. Schaal, A. Elmaghraby, and A. El-Baz, "Retinal layers OCT scans 3-D segmentation," in *2019 IEEE International Conference on Imaging Systems and Techniques (IST)*. IEEE, 2019, pp. 1–6.
201. N. Eladawi, M. Elmogy, M. Ghazal, O. Helmy, A. Aboelfetouh, A. Riad, S. Schaal, and A. El-Baz, "Classification of retinal diseases based on OCT images," *Frontiers in Bioscience* (Landmark Edition), vol. 23, pp. 247–264, 2018.
202. A. ElTanboly, M. Ismail, A. Shalaby, A. Switala, A. El-Baz, S. Schaal, G. Gimel'farb, and M. El-Azab, "A computer-aided diagnostic system for detecting diabetic retinopathy in optical coherence tomography images," *Medical Physics*, vol. 44, no. 3, pp. 914–923, 2017.
203. H. S. Sandhu, A. El-Baz, and J. M. Seddon, "Progress in automated deep learning for macular degeneration," *JAMA Ophthalmology*, vol. 136, no. 12, pp. 1366–1376, 2018.
204. M. Ghazal, S. S. Ali, A. H. Mahmoud, A. M. Shalaby, and A. El-Baz, "Accurate detection of non-proliferative diabetic retinopathy in optical coherence tomography images using convolutional neural networks," *IEEE Access*, vol. 8, pp. 34387–34397, 2020.
205. K. Hammouda, F. Khalifa, A. Soliman, M. Ghazal, M. Abou El-Ghar, A. Haddad, M. Elmogy, H. Darwish, A. Khalil, A. Elmaghraby et al., "A CNN-based framework for bladder wall segmentation using MRI," in *2019 Fifth International Conference on Advances in Biomedical Engineering (ICABME)*. IEEE, 2019, pp. 1–4.
206. K. Hammouda, F. Khalifa, A. Soliman, M. Ghazal, M. Abou El-Ghar, A. Haddad, M. Elmogy, H. Darwish, R. Keynton, and A. El-Baz, "A deep learning-based approach for accurate segmentation of bladder wall using MR images," in *2019 IEEE International Conference on Imaging Systems and Techniques (IST)*. IEEE, 2019, pp. 1–6.
207. K. Hammouda, F. Khalifa, A. Soliman, H. Abdeltawab, M. Ghazal, M. Abou El-Ghar, A. Haddad, H. E. Darwish, R. Keynton, and A. El-Baz, "A 3D CNN with a learnable adaptive shape prior for accurate segmentation of bladder wall using MR images," in *2020 IEEE 17th International Symposium on Biomedical Imaging (ISBI)*. IEEE, 2020, pp. 935–938.
208. K. Hammouda, F. Khalifa, A. Soliman, M. Ghazal, M. Abou El-Ghar, M. Badawy, H. Darwish, A. Khelifi, and A. El-Baz, "A multiparametric MRI-based cad system for accurate diagnosis of bladder cancer staging," *Computerized Medical Imaging and Graphics*, vol. 90, p. 101911, 2021.
209. K. Hammouda, F. Khalifa, A. Soliman, M. Ghazal, M. Abou El-Ghar, M. Badawy, H. Darwish, and A. El-Baz, "A CAD system for accurate diagnosis of bladder cancer staging using a multiparametric MRI," in *2021 IEEE 18th International Symposium on Biomedical Imaging (ISBI)*. IEEE, 2021, pp. 1718–1721.
210. A. A. K. A. Razek, R. Khaled, E. Helmy, A. Naglah, A. AbdelKhalek, and A. El-Baz, "Artificial intelligence and deep learning of head and neck cancer," *Magnetic Resonance Imaging Clinics*, vol. 30, no. 1, pp. 81–94, 2022.
211. A. Sharafeldeen, M. Elsharkawy, R. Khaled, A. Shaffie, F. Khalifa, A. Soliman, A. A. K. Abdel Razek, M. M. Hussein, S. Taman, A. Naglah et al., "Texture and shape analysis of diffusion-weighted imaging for thyroid nodules classification using machine learning," *Medical Physics*, vol. 49, no. 2, pp. 988–999, 2021.
212. A. Naglah, F. Khalifa, R. Khaled, A. A. K. Abdel Razek, M. Ghazal, G. Giridharan, and A. El-Baz, "Novel MRI-based CAD system for early detection of thyroid cancer using multi-input CNN," *Sensors*, vol. 21, no. 11, p. 3878, 2021.

213. A. Naglah, F. Khalifa, A. Mahmoud, M. Ghazal, P. Jones, T. Murray, A. S. Elmaghraby, and A. El-Baz, "Athlete-customized injury prediction using training load statistical records and machine learning," in *2018 IEEE International Symposium on Signal Processing and Information Technology (ISSPIT)*. IEEE, 2018, pp. 459–464.
214. A. H. Mahmoud, "Utilizing radiation for smart robotic applications using visible, thermal, and polarization images," Ph.D. dissertation, University of Louisville, 2014.
215. A. Mahmoud, A. El-Barkouky, J. Graham, and A. Farag, "Pedestrian detection using mixed partial derivative based his togram of oriented gradients," in *2014 IEEE International Conference on Image Processing (ICIP)*. IEEE, 2014, pp. 2334–2337.
216. A. El-Barkouky, A. Mahmoud, J. Graham, and A. Farag, "An interactive educational drawing system using a humanoid robot and light polarization," in *2013 IEEE International Conference on Image Processing*. IEEE, 2013, pp. 3407–3411.
217. A. H. Mahmoud, M. T. El-Melegy, and A. A. Farag, "Direct method for shape recovery from polarization and shading," in *2^{01}2 19th IEEE International Conference on Image Processing*. IEEE, 2012, pp. 1769–1772.
218. M. A. Ghazal, A. Mahmoud, A. Aslantas, A. Soliman, A. Shalaby, J. A. Benediktsson, and A. El-Baz, "Vegetation cover estimation using convolutional neural networks," *IEEE Access*, vol. 7, pp. 132563–132576, 2019.
219. M. Ghazal, A. Mahmoud, A. Shalaby, and A. El-Baz, "Automated framework for accurate segmentation of leaf images for plant health assessment," *Environmental Monitoring and Assessment*, vol. 191, no. 8, p. 491, 2019.
220. M. Ghazal, A. Mahmoud, A. Shalaby, S. Shaker, A. Khelifi, and A. El-Baz, "Precise statistical approach for leaf segmentation," in *2020 IEEE International Conference on Image Processing (ICIP)*. IEEE, 2020, pp. 2985–2989.

4 Texture Analysis in Radiology

Charles Pierce and Daniel Thomas Ginat

4.1 INTRODUCTION

Radiologists primarily use qualitative information from medical images in order to interpret findings and make conclusions. Frequently, the qualitative information found in medical imaging is sufficient to identify key findings. However, there are often cases of findings that are difficult to definitely identify or characterize. As a result of potentially equivocal findings, the interpretation of medical imaging can vary significantly between different radiologists. These equivocal findings are not just a matter of training and exposure but, in some cases, represent the limits of interpreting medical imaging using conventional qualitative techniques. Radiomics may represent a solution to these problems.

Although conventional medical imaging assessment relies on visual inspection to produce qualitative data, radiomics relies on quantitative data extracted from an image. Quantitative data within an image includes things like grayscale intensity.[1] While radiologists are capable of detecting patterns, software is more suitable for analyzing statistical features of the pixel values and other quantitative data.[2] In turn, quantitative data can provide parameters to classify a lesion, such as enhancing or non-enhancing.[2] Features like this, coupled with information about the heterogeneity of a finding, can be useful in identifying the malignancy risk of the lesion in some cases.[2]

Different features of radiomics can be useful in assessing the characteristics of a finding. For instance, features within tumors that may be difficult to detect on visual inspection could be obvious from a radiomics-based approach. A widespread example of this involves risk-stratifying cystic appearing renal lesions based on the attenuation of the cystic contents and enhancement patterns. Low risk benign cystic lesions have thin walls without septations and fluid attenuating contents. Although radiologists tend to visually assess the presence or absence of septations, they frequently will measure the cystic contents in Houndsfield units before classifying renal cysts as benign.[2] In the case of a "clearly malignant" renal cystic lesions, radiologist will qualitatively assess for prominent soft tissue components/nodularity and quantitively interrogate the cysts for "measurable enhancement."[2] This objective quantitative data is essential for helping providers determine when no further imaging workup is indicated or when surgery may be required.

More advanced analytic techniques can be used in radiomics to go beyond describing the attenuation or enhancement pattern of a finding. Texture analysis is an approach within radiomics that utilizes the quantitative data to describe textual

DOI: 10.1201/9780367486082-4

features of an image.[3] Textual features can be used to determine the underlying relationships between grayscale values of an image.[3] Grayscale values in an image depict patterns and visually represent relationships between different portions of an image.[3] The relationships between these grayscale patterns can represent pathological findings, the underlying genetic features, as well as provide information about possible underlying progress of pathology or treatment.[4]

Texture analysis continues to be a rapidly developing field. Although there is a lot of potential for quantitative analysis through radiomics to provide insights that qualitative analysis through visual inspection cannot capture, there are considerable challenges that remain. Texture analysis within radiomics lacks standardized guidelines.[5] Texture analysis has been used to further imaging based characterization of pathological findings, treatment responses, prognostic ability, and other areas of physiologic and medical interest. Further studies need to be conducted on large scales with high-quality data in order to continue to advance texture analysis.

To further the advancement of texture analysis, we sought to write this chapter to provide a foundation in the rapidly developing methods and advancements. This chapter will explore a background of texture analysis and its uses in radiological assessment. Clinical imaging has the potential to be very powerful with texture analysis because it can provide insights into the aspects of medical imaging that escape visual inspection. To appreciate texture analysis within radiomics, we will discuss the background of texture analysis, methodology and analysis, and medical applications.

4.1.1 Texture Analysis Background

Texture analysis was originally developed to describe the variation in the surface of an object. Originally, texture analysis was built on the notion that the coarseness of a texture could be appreciated by the high degree of non-uniformity within the material itself.[6] This translates into radiology to differences in intensity of pixels and voxels.[5] When considering a heterogeneous area of interest in a medical image, it is apparent that the intensity of the pixels and voxels of the region of interest will have significant variation.[5] In comparison, homogenous findings within medical imaging will have pixels or voxels of similar intensity that appear to be uniform throughout.[5] Borrowing on the principles of texture analysis, radiomics can investigate the differences in pixel and voxels intensity of grayscale images in order to determine the textual features of the image.[5]

4.2 TEXTURAL ANALYSIS WORKFLOW

Texture analysis workflow can mirror the workflow for conventional visual assessment. Conventional radiology depends on image acquisition, the identification of an important finding, description of that finding, and analysis of the likely significance or pathological correlation of the findings. In texture analysis, identifying an important finding or area of interest requires limiting focus to a particular area within the image. This is called image segmentation.[5,7] Once a region of interest has been identified and possibly segmented, further characterization could be performed using quantifiable data. This characterization in texture analysis often relies on statistical analysis of pixel values along with other methods.[1,3–5] Finally, once the region of interest identification, segmentation, and characterization have been performed, the significance of a finding could be determined.[1,4,5,7]

Additionally, while a radiologist can manually tune signal to noise ratios through "window" and "level" settings, computational textural analysis requires mathematical denoising and data formatting.[5] The denoising and formatting are part of data preparation for texture analysis.[5] While radiologists can perform similar functions automatically, computational texture analysis requires some upfront work to prepare the data before texture analysis can be applied to extract features and provide an analysis of the data.

4.3 IMAGE ACQUISITION

Texture analysis can be applied to a multitude of images from any number of sources. Texture analysis can be utilized in CT, US, MRI, PET, and a variety of $n >= 2$ dimensional visual data generators.[5] Although there are no limitations on the protocol or equipment used to obtain an image for texture analysis, the reproducibility of the data will depend significantly on the reproducibility of the acquisition process.[1,5] Image acquisition requires manual effort and can be a significant source of variation in data. Acquisition is dependent on a multitude of variables. When a patient undergoes medical imaging, differences will arise depending on the protocol, how faithfully and precisely the technician follows the protocol, the equipment, and the patient's own efforts to help obtain a quality image.[5] However, the multitude of processes for image acquisition needs to account for variation before continuing with analysis. Differences within parameters between different image acquisitions will lead to inconsistent data.[5]

4.4 IMAGE SEGMENTATION

Radiologists rely on conventional visual inspection to identify the salient findings on an image. For computational analysis, this area of interest must be rigorously defined and distinguished. This region of interest can be demarcated in analysis through the process of image segmentation.[1,5] It can be performed manually for computational analysis using graphical user interfaces that delineate the area of an image to undergo texture analysis.[1,5] Regions of interest can also be segmented automatically using machine learning algorithms.[1,5] Regions of interest can also be identified using a combination of manual segmentation along with automatic segmentation.[1,5]

Manual segmentation is time-consuming for several reasons. First, it requires going through large sets of medical images that may have multiple slices to view. Secondly, it frequently requires referencing comparison imaging to confirm the location of the finding. Lastly, manual segmentation frequently requires multiple radiologists segmenting multiple times to increase internal validity. Automated segmentation has the potential to become a more rapid and robust way of highlighting regions of interest, but currently, automatic image segmentation is limited by requiring significant computational power along with time. Machine learning algorithms continue to become more sophisticated. However, machine learning algorithms require significant computational time to develop. Automated segmentation is also very sensitive to noise and artifacts.[1,5] The benefit of automated segmentation is that it is an objective approach to segmentation.[1,5] Automated segmentation with machine learning continues to improve as deep learning algorithms become more widely utilized. Texture analysis itself can be used for segmentation if the segmented region of interest contains discerning textual features. Ultimately, automated segmentation relies on classification strategies, and texture analysis can be a suitable approach for classification problems.

4.5 FEATURE EXTRACTION

There are multiple approaches that can be used for feature extraction. The most commonly used method for feature extraction is statistical analysis of the image.[1,3–5,7] Other approaches include transform-based and model-based feature extraction.[1,3–5] Statistics allow textual features to be captured in a multitude of ways.[1,3–5,7] Statistical extraction of texture features relies on describing the region of interest using various statistical methods. Statistical analysis of a region of interest uses the grayscale values within a specific area as inputs.[1,3–5,7] Statistical extraction of texture features ranges in sophistication from a breakdown of normal distribution-based methods to matrix-based methods.[4,7] First-order methods for statistical extraction of texture features use a histogram of the frequency distribution of grayscale intensity in the region of interest.[4,7] Once the frequency distribution for a region of interest is known, basic first-order statistics can be used to further characterize the region of interest. This includes features like mean intensity of the region and standard deviation.[1,4,7] Even these basic features can help give an idea of the homogeneity or heterogeneity of a region of interest. For instance, a highly heterogeneous region of interest is likely to have a much greater standard deviation than a homogeneous region of interest. Other first-order features include kurtosis and skewness.[4,7] Second-order features are more mathematically sophisticated and include gray level run-length matrices and gray level co-occurrence matrices. Non-statistical methods for extracting textual features include transforming methods.[4,7] Transform-based analysis extracts textural features by comparing textural matrices to wave spectrums and describing and identifying peaks within a periodic spectrum.[4,5] By capturing the periodicity and the variation in peaks, patterns emerge that can be used for classification. Fractal analysis is an example of structural methods that rely on decomposing an image and then identifying basic rules that can be used to reconstruct the image from basic units.[4]

4.6 FEATURE REDUCTION

Once features have been successfully extracted from an imaging study, the next step is to determine how these features can be used to either classify or predict pertinent findings. For many images, the theoretical limit of texture of features that could be extracted is infinite. To build useful models or a useful understanding of how textural features can help us understand imaging, we need to be able to focus on or reduce the number of features in our analysis. Without feature reduction, texture of feature analysis runs the risk of overfitting any model produced.[1,5]

Feature reduction is a multi-step process that is focused on finding features that are reproducible, relevant, and unique. The first step in feature reduction is to identify robust and reproducible features. Reproducible features in studies using manual or semi-automated feature extraction require intra- and interobserver variability analysis.[5] Test/retest robustness should be assessed with a phantom.[5] Non-reproducible features should be excluded at the end of analysis when the investigation can proceed to the next step of feature reduction. The next step of feature reduction is to identify the most relevant features for the investigation. There are a multitude of approaches that could be used to identify the most relevant features. These often rely on machine learning algorithms and can involve recursive analysis.[5] When the most relevant features are identified, it is important to understand the relationships between relevant features. This can be part of the second feature reduction step or can be its own additional step.[5] As a separate step, this involves the generation of correlation clusters. These correlation clusters can help identify the most correlated features in the analysis.[5] Conventional statistics or machine learning can help with the selection of highly correlated features.[5] The last step of feature reduction is to build a model and test the features that have been extracted.

4.7 STATISTICAL EXTRACTION OF TEXTURAL FEATURES

When extracting texture features using statistical analysis, it is important to understand what the goal of the project is. If the goal of the project is simply to report on the textual features of a specific finding, for example normal vs. steatotic liver parenchyma, many familiar and commonly used statistical tools, including the t-test can be utizilized.[7] In this example, analysis could be as simple as determining the average Houndsfield units associated with pathology confirmed normal vs. steatotic liver parenchyma. If the goal of the investigation is to use textual features to predict a treatment outcome or prognosis, then a model should be developed using recursive analysis or similar approach. An example of this could be automating a process to identify lung nodules and predict their risk for malignancy based on texture features.

4.8 FIRST-ORDER METHODS

First-order analysis is based on the representation of an image as a pixel intensity histogram and has the highest reproducibility.[1,3–5] With this model of a pixel intensity histogram for the region of interest, fundamental frequency relationships can be observed. These include simple and familiar statistics, such as mean, standard deviation, and other familiar statistics that can be used to describe a normal distribution.[1,3–5,7] Other textural features that are typically described in the literature include kurtosis. Kurtosis is the peakedness versus the flatness of a pixel histogram.[1,3,4] This can be useful in describing the heterogeneity of an image. Very homogeneous images may be highly kurtotic, while heterogeneous images are more likely to have a very flat distribution. Skewness represents asymmetry within the histogram. A skewed distribution can be representative of several things. Entropy is the first-order textual feature that represents irregularity of the histogram.[1,3,4] This can be particularly useful when trying to categorize homogeneous versus heterogeneous regions of interest. Threshold represents the percentage of pixels within a defined area having a defined range of intensity.[4]

4.9 SECOND-ORDER METHODS

Second-order statistics can employ several different methods. This approach uses pairs of pixels and focuses on the spatial relationships to each other using second-order statistics. Two commonly reported methods are gray level co-occurrence matrices and run-length mattresses[4,7] (Figure 4.1). Gray level co-occurrence matrices are used as a framework to model spatial relationships of pixel intensity.[4,7] The set-up for this approach breaks an image down and focuses on two pixels within the image, and asks what is the probability that they will have a certain grayscale

TABLE 4.1
First-Order Statistics Frequently Employed in Texture Analysis

First-Order Statistics		
Description	**Statistics**	**Description**
Provide descriptions of gray level frequency within a particular area of interest. A histogram of the image is generated and used to determine a number of characteristics.	Mean gray level intensity	The average gray level intensity determined by the mean.
	Entropy	Measures the homogeneity of an image.
	Variance	The square of the standard deviation of the distribution of gray level intensity.
	Standard deviation	Measures the dispersion from the mean. High SD indicates values with a wide distribution from the mean, and low SD indicates values with a close cluster around the mean.
	Skewness	The measure of the asymmetry of the gray level intensity histogram. Negative skew indicates a high number of dark objects, and positive skew indicates a high number of light objects.
	Kurtosis	The measure of how peaked a histogram is compared to a normal distribution. Histograms that are more peaked than a normal distribution have positive kurtotic values, and histograms that are comparatively flatter have negative kurtotic values.
	Mean positive pixels	Determines the mean value of pixels that have positive values in order to reduce the impact of dark areas.

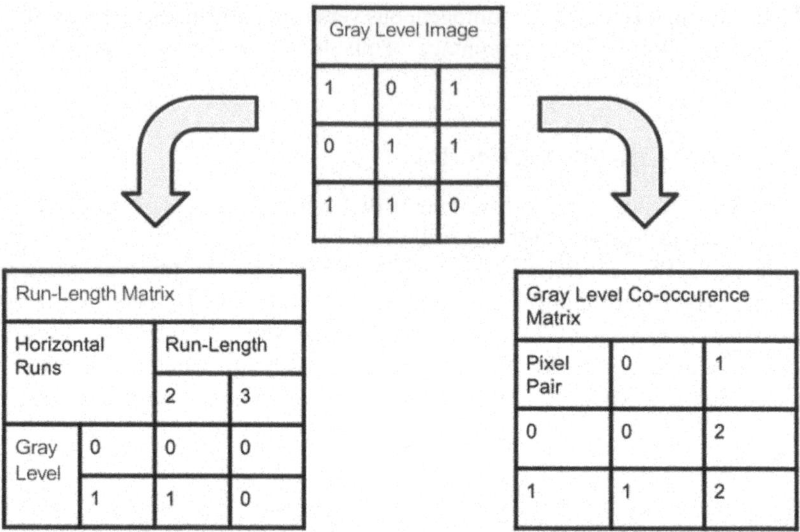

FIGURE 4.1 An example of run-length and co-occurrence matrices.

intensity.[8] Run-length matrices capture the frequency of adjacent or consecutive pixels having the same grade level in a specific direction.[6] The statistics obtained through second-order analysis include entropy, energy, sum of variance, sum of averages, and uniformity.[1,3,4] Even higher-order features can be explored by analyzing the relationship between multiple pixels and voxels. Some higher-order methods include advanced matrix methods, wavelet energy methods, and autoregressive models.[1,3,4]

4.10 APPLICATIONS

Although there are many potential applications for texture feature analysis, texture features are typically investigated for their uses as biomarkers and for classification tasks in medical imaging. Classification can be used to characterize unknown lesions, to assess treatment progress, and to help capture information within medical imaging that is not obvious with qualitative analysis. Feature analysis is frequently employed in instances where a difficult-to-define finding with an indeterminate characterization could be either benign or malignant.[4] The goal being that robust texture analysis could one day serve as a virtual biopsy.

4.11 TUMOR CLASSIFICATION

Recent investigations into this focused on indeterminate tumors, where the risk is that biopsy or surgical excision will result in the removal of an ultimately benign tumor. Within the kidney, multiple benign masses can resemble the malignant renal cell carcinoma.[2,4] Textual analysis has been investigated as a way to differentiate renal cyst and renal cell carcinoma (RCC).[4] Ramen et al. investigation utilizing texture features found that mean, standard deviation, and entropy are able to classify cysts, oncocytoma, clear-cell RCC, and papillary RCC with a random forest model.[9] Another study by Hodgdon et al. was able to show that homogeneity and entropy could be used to distinguish angiomyolipoma (AML) from RCC using a non-enhanced CT image.[10] The area under the receiver operating curve in the study was 0.89 and performed better than subjective assessment.[10] Not all studies showed that texture analysis was superior to subjective analysis. Although Leng et al. were able to successfully classify clear cell RCC versus AML by using textual features such as standard deviation, entropy, and uniformity, the subjective assessment of heterogeneity that was the basis for classification performed better than the objective assessment with textual analysis.[11]

4.12 TUMOR GRADING

Texture analysis can be used to help distinguish low-grade lesions from high-grade lesions when malignancy is part of the differential. For instance, texture analysis was used in a recent study to classify high-grade intraductal papillary mucinous neoplasms with high-grade dysplasia from intraductal papillary mucinous neoplasm with low-grade dysplasia.[12] In the study, texture analysis was able to perform better than the Fukuoka criteria.[12] In another study, linear discriminant analysis was performed on a group of texture features with successful discrimination between astrocytoma and oligodendroglioma.[13]

Classification of colorectal masses has also been investigated with texture analysis. In a recent study, textual features were able to classify advanced rectal cancer versus advanced adenoma with an AUC of 0.94.[14] Texture features, including gradient and curvature, were able to improve the AUC for classifying neoplastic versus non-neoplastic colonic lesions.[15]

4.13 RADIOGENOMICS

Genetic features can sometimes be correlated with texture features. Classifying genetic status based on texture features can provide important information that can change both prognosis and treatment options. For example, multiple studies have been able to use textual analysis to predict the IDH status of patients with a high degree of accuracy utilizing first-order and grade-level co-occurrence matrices.[16,17] IDH status has a documented impact on prognosis and response to treatment in glioblastoma.[18] Patients who have mutated IDH have a better response to treatment with temozolomide and an improved prognosis over patients with non-mutated IDH.[18] The utility of predicting IDH status with imaging could lead to an earlier and more informed understanding of a patient's prognosis and options.

4.14 LIMITATION OF TEXTURE FEATURE EXTRACTION

As discussed earlier, there is significant variation in image acquisition within medical imaging. There are a multitude of different protocols across CT, MRI, and other image acquisition studies. For texture analysis to prove useful, actual features should be reversed across different image acquisition devices. If operators employ the same protocols, textual features should be able to be extracted with ideally little variation. Analyzing the variation in texture features derived from different CT images has recently been investigated.[5] Although many studies have shown that consistent application of imaging protocols can provide robust textual feature extraction with satisfactory inter-scanner agreement, there are several artifacts that can confound feature extraction.[5] One important confounder is motion artifacts and must be minimized where possible. Additionally, intra-scanner agreement can be variable. Textual features can be highly dependent on protocol parameters, such as slice thickness.[5] Although some features can be reproducible even with changes in protocol, many features are not reproducible.

The current limitations of textural features include, perhaps most importantly, the variability in imaging acquisition and imaging processing. Different protocols can have significant effects on textural feature extraction. This can greatly reduce reproducibility. At this time, there are no guidelines or image acquisition with the goal of texture feature extraction.[5] Additionally, a significant confounder for texture of future reproducibility is the variability that is introduced with region of interest segmentation.[5] Manual segmentation continues to cause increased noise.[5] Lastly, although first-order statistical analysis software such as PyRadiomics, ImageJ, and others have respectable reproducibility, second-order analysis only has moderate to poor agreement.[5] Classification using textual features runs the risk of overfitting, especially when there are more textual features than there are samples in the training dataset.[1,5]

REFERENCES

1. Rogers W, Thulasi Seetha S, Refaee TAG, et al. Radiomics: from qualitative to quantitative imaging. *Br J Radiol*. 2020;93(1108):20190948. doi:10.1259/bjr.20190948
2. Silverman SG, Pedrosa I, Ellis JH, et al. Bosniak classification of cystic renal masses, version 2019: an update proposal and needs assessment. *Radiology*. 2019;292(2):475–488. doi:10.1148/radiol.2019182646
3. Rizzo S, Botta F, Raimondi S, et al. Radiomics: the facts and the challenges of image analysis. *Eur Radiol Exp*. 2018;2(1):36. doi:10.1186/s41747-018-0068-z
4. Lubner MG, Smith AD, Sandrasegaran K, Sahani DV, Pickhardt PJ. CT texture analysis: definitions, applications, biologic correlates, and challenges. *Radiographics*. 2017;37(5):1483–1503. doi:10.1148/rg.2017170056
5. van Timmeren JE, Cester D, Tanadini-Lang S, Alkadhi H, Baessler B. Radiomics in medical imaging-"how-to" guide and critical reflection. *Insights Imaging*. 2020;11(1):91. doi:10.1186/s13244-020-00887-2
6. Conners RW, Harlow CA. A theoretical comparison of texture algorithms. *IEEE Trans Pattern Anal Mach Intell*. 1980;2(3):204–222. doi:10.1109/tpami.1980.4767008
7. Varghese BA, Cen SY, Hwang DH, Duddalwar VA. Texture analysis of imaging: what radiologists need to know. *AJR Am J Roentgenol*. 2019;212(3):520–528. doi:10.2214/AJR.18.20624
8. Haralick RM, Shanmugam K, Dinstein I. Textural features for image classification. *IEEE Trans Syst Man Cybern*. 1973;SmC-3(6):610–621. doi:10.1109/TSMC.1973.4309314
9. Raman SP, Chen Y, Schroeder JL, Huang P, Fishman EK. CT texture analysis of renal masses: pilot study using random forest classification for prediction of pathology. *Acad Radiol*. 2014;21(12):1587–1596. doi:10.1016/j.acra.2014.07.023

10. Hodgdon T, McInnes MDF, Schieda N, Flood TA, Lamb L, Thornhill RE. Can quantitative CT texture analysis be used to differentiate fat-poor renal angiomyolipoma from renal cell carcinoma on unenhanced CT images? *Radiology*. 2015;276(3):787–796. doi:10.1148/radiol.2015142215
11. Leng S, Takahashi N, Gomez Cardona D, et al. Subjective and objective heterogeneity scores for differentiating small renal masses using contrast-enhanced CT. *Abdom Radiol (NY)*. 2017;42(5):1485–1492. doi:10.1007/s00261-016-1014-2
12. Hanania AN, Bantis LE, Feng Z, et al. Quantitative imaging to evaluate malignant potential of IPMNs. *Oncotarget*. 2016;7(52):85776–85784. doi:10.18632/oncotarget.11769
13. Zhang S, Chiang GCY, Magge RS, et al. MRI based texture analysis to classify low grade gliomas into astrocytoma and 1p/19q codeleted oligodendroglioma. *Magn Reson Imaging*. 2019;57:254–258. doi:10.1016/j.mri.2018.11.008
14. Pooler BD, Lubner MG, Theis JR, Halberg RB, Liang Z, Pickhardt PJ. Volumetric textural analysis of colorectal masses at CT colonography: differentiating benign versus malignant pathology and comparison with human reader performance. *Acad Radiol*. 2019;26(1):30–37. doi:10.1016/j.acra.2018.03.002
15. Song B, Zhang G, Lu H, et al. Volumetric texture features from higher-order images for diagnosis of colon lesions via CT colonography. *Int J Comput Assist Radiol Surg*. 2014;9(6):1021–1031. doi:10.1007/s11548-014-0991-2
16. Hsieh KLC, Chen CY, Lo CM. Radiomic model for predicting mutations in the isocitrate dehydrogenase gene in glioblastomas. *Oncotarget*. 2017;8(28):45888–45897. doi:10.18632/oncotarget.17585
17. Zhang X, Tian Q, Wu YX, et al. *IDH mutation assessment of glioma using texture features of multimodal MR images*. Armato SG, Petrick NA, eds. 2017:101341S. doi:10.1117/12.2254212
18. SongTao Q, Lei Y, Si G, et al. IDH mutations predict longer survival and response to temozolomide in secondary glioblastoma. *Cancer Sci*. 2012;103(2):269–273. doi:10.1111/j.1349-7006.2011.02134.x

5 Texture Analysis Using a Self-Organizing Feature Map

Emad Alsyed, Rhodri Smith, Christopher Marshall, and Emiliano Spezi

5.1 INTRODUCTION AND BACKGROUND

Medical imaging plays a critical role in all phases of cancer management. Different imaging modalities contribute to the staging, diagnosis, and treatment of many types of cancer (Griffeth 2005; Higgins and Pomper 2015). Radiographic technologies such as magnetic resonance imaging (MRI), computed tomography (CT), and positron emission tomography (PET) form the core medical imaging modalities within oncology. Whilst MRI provides substantial diagnostic advantages over CT, functional molecular imaging with fluorodeoxyglucose (18F-FDG) PET provides metabolic information of tumors and the characterization of tumor biology. Aggressive tumors tend to have higher levels of FDG uptake, while less-aggressive tumors tend to have lower levels of FDG uptake, which has been corroborated histologically. This additional dimension of diagnostic information that is provided by FDG-PET can be used to improve determination of disease prognosis and effectiveness of treatment planning (Dhingra, Mahajan, and Basu 2015) whilst also providing a personalized blueprint of the varied (intra- and inter-lesionally) cellular biology and metabolic and receptor phenotypic expression of the imaged tumor. This makes PET imaging an ideal candidate for biomarker exploration via radiomics and the quest to achieve personalized cancer management with the associated promised progress in survival (Naqa et al. 2009; Gillies, Kinahan, and Hricak 2016; Cook et al. 2018).

Several technical challenges, however, currently exist in the path toward radiomics-driven precision medicine and its incorporation into multi-center clinical trials and clinical practice; these barriers are summarized in Table 5.1. Standardization forms a crucial aspect of overcoming many such barriers. Standardization would assist in alleviating the effect of confounding variables on the calculation of radiomic features. Work toward standardizing the image processing workflow (Zwanenburg et al. 2020) along with harmonizing image acquisition is ongoing (Mulshine et al. 2015). Standardization for PET imaging is, however, still in the early stages of refinement (Cook et al. 2018) in comparison to other imaging modalities such as MR and CT (Cattell, Chen, and Huang 2019; Berenguer et al. 2018). PET images with larger voxels, poorer resolution, and more complex patient preparation (i.e., post-injection imaging time, fasting status, activity injected) result in this image modality being

 DOI: 10.1201/9780367486082-5

TABLE 5.1
Barriers Affecting Standardization of Radiomics for Widespread Incorporation into the Clinic

Radiomics Barriers (Standardization Of):
Image segmentation
Image pre/post-processing
Image reconstruction
Image acquisition
Feature calculation
Image protocol
Statistical power (number of patients)
Respiratory motion (correction)
Feature selection
Availability of standard datasets
Replication of analysis

the most under-developed w.r.t radiomics standardization. Multiple investigations have evaluated the effect of image respiratory motion (Oliver et al. 2015), segmentation (Velden et al. 2016; Alsyed et al. 2019b), reconstruction type (Galavis et al. 2010), contour size, and post-injection imaging time (Alsyed et al. 2019a) on texture features. To date, only a limited number of research studies have used advanced machine learning techniques to assess the importance of texture analysis in medical imaging (Hatt et al. 2017; Ackerley et al. 2019), and to the authors' knowledge, no research has explored the use of machine learning in an attempt to understand the relationship between the aforementioned confounding variables and radiomic signatures (Alsyed et al. 2021).

5.2 THE SELF-ORGANIZING MAP

A self-organizing feature map (SOM) is a type of artificial neural network (ANN) that is trained using unsupervised learning to produce a lower-dimensional representation of the input data on an underlying manifold. Manifold learning has demonstrated to be effective in other areas of image processing, such as image segmentation (Kadoury 2018; Smith et al. 2018) and respiratory motion correction (Smith et al. 2019). No study has applied SOMs to examine, detect, and capture the statistical variability of medical image texture parameters. The Kohonen SOM takes a set of input data, for example, L texture parameters, and maps them onto a two-dimensional grid of neurons (Figure 5.1) (Kohonen 1982; Köküer et al. 2007). Each neuron in the grid is assigned an initial weight vector $M = \left(m_{g1}, m_{g2}, .., m_{gD}\right)$ with the same dimensionality as that of the input data $g = \left(1, 2, \ldots L\right)$. The training uses competitive learning (nodes compete to respond to input data). Training examples are fed into the network at random, and the Euclidean distance to all weight vectors is calculated

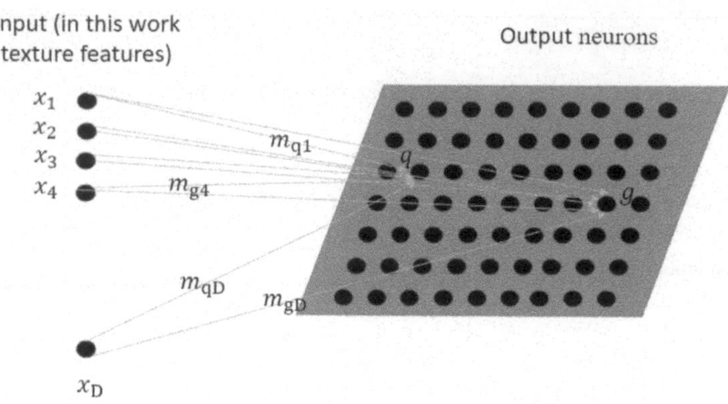

FIGURE 5.1 An illustrative example of an SOM. The learned neuron matrix maps the input data values onto a regular two-dimensional grid.

(equation 1) and is used to update the neuron weights; the neuron whose vector is similar to the input is defined as the best matching unit (BMU).

$$m_{t+1}^{gd} = m_t^{gd} + \eta h(g,q)(x_d - m_{jd}^t), \quad for \quad 1 \le d \le D \tag{5.1}$$

Where $h(g,q)$ is the neighborhood function which has the value 1 at the winning neuron q and decreases as the distance between g and q increases, and η is the learning rate parameter that controls the size of weight vector. Each high-dimensional data point is thus embedded onto a single neuron which reproduces its structure. Neuronal weights act as pointers to the input space and form a discrete approximation of the distribution of the training samples. More neurons point to regions with high training sample concentration, and fewer where the samples are scarce. The SOM thereby compresses information while preserving the most important topological and/or metric relationships of the primary data elements on the display (Kohonen 2001). Following training, the SOM's neurons represent a low-dimensional representation of high-dimensional input data without disturbing the shape of the data distribution and relationship between each input data element. Each unit in the grid contains a different "codebook" vector that represents the typical objects for that area in the map. For a simple example, a 150 cm tall 25-year-old woman may be included on node A on the SOM, one would expect a 160 cm tall 27-year-old woman to be included either on node A or close to it, whilst a 220 cm tall 70-year-old man would be located on a node with a large distance from A. Counts or distribution plots can also be created in the form of a "heatmap" representing the number of data samples with a chosen criteria falling into a particular codebook (pie representations of the representative vectors for the grid). This criterion is independent of the learning of the SOM and in this work represents the confounding variables explored (e.g., contour size, image reconstruction setting). Referring to the simple example, the heatmap could represent the voting preference of an individual, and information around what type of people vote for a given party can be inferred from the heatmap plot.

Segmentation or clustering of the heatmap will therefore reveal any relationships between the input data and the potentially confounding variables.

5.3 METHOD FOR APPLYING SOM TO ANALYZE RADIOMIC CONFOUNDING VARIABLES

The SOM can be utilized to explore the effect of confounding variables on extracted texture features. Applying the SOM to the radiomic pipeline involves the steps listed in the following and in Figure 5.2.

1. Obtain radiomic features for the underlying data at numerous plausible acquisition/image processing settings that encompass the natural variability that would be present during routine processing. For example, if the confounding variable of interest is segmentation volume, collating texture features at numerous contour sizes allows this variability to be probed using the SOM. Similarly, if the confounding variable of interest is a particular acquisition/reconstruction parameter setting, such as an image reconstruction variable (e.g., number of iterations), the image must be processed at numerous reconstruction settings to produce a set *S*. The range of variables used in constructing the set should reflect those that may be observed during routine image acquisition/processing.
2. Segment the region of interest extracting texture features from each volume using standardized texture analysis methodology.
3. Standardize each texture feature using Z-score normalization across the distribution of texture features acquired for inter-comparison of different textures.
4. Using the mean Z-omic for each group of texture features, learn the self-organizing map.

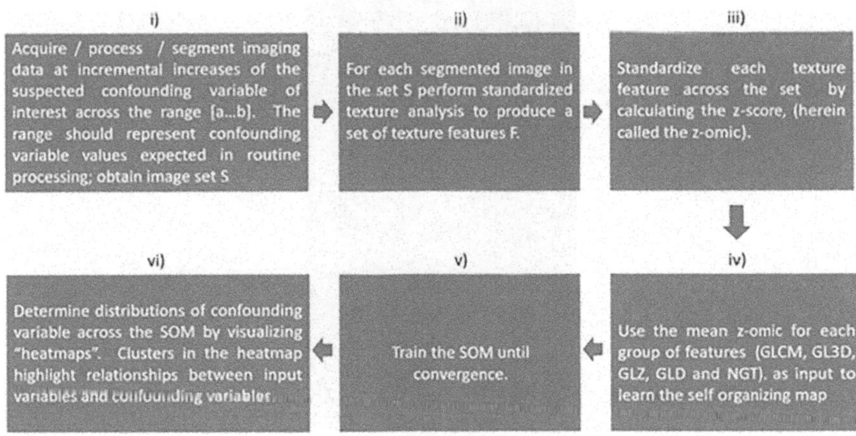

FIGURE 5.2 Workflow for applying SOM to explore the effect of a confounding variable on tree analysis.

5. Train the SOM until convergence.
6. Finally, visualize the "heatmaps" of the confounding variable across the codebook of the SOM.

5.3.1 Evaluation of the Effectiveness of the SOM in Assessing Radiomic Confounding Variables

Our previous work has demonstrated the effectiveness of the SOM in identifying confounding variables present during radiomics image process of pre-clinical subjects (Alsyed et al. 2021). Radiomic features were extracted from PET images of eight mice with 4T1 tumors (mammography carcinoma xenografts) utilizing an in-house developed tool kit. The mice were injected with 10.0 ± 2.0 MBq of 18F-FDG and imaged 50 minutes post-injection, dynamically for 20 minutes, with a Mediso Nanoscan PET/CT. Images were re-binned into 4 × 5 minutes PET scans (50–55, 55–60, 60–65, and 65–70 minutes post-injection). Four different systematic 3D-Contour sizes (4, 4.5, 5, 5.5 mm) were segmented on the first time point (55 minutes) using Velocity 3.2.1 software (Varian Medical Systems, Palo Alto, CA). Contours obtained on the first time point were overlaid on all other images, which were re-binned into subsequent time points. Figure 5.3 shows coronal slice of lower right flank with four different contours for the first mouse. Texture features were extracted for each volume at each time point; thus, each feature has 128 observable values that resulted from eight mice, four different time points, and four different contour sizes.

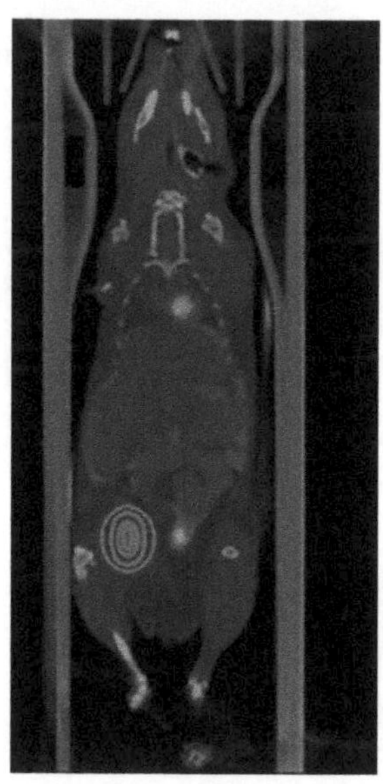

FIGURE 5.3 Coronal slice of lower right flank (left of image) with four different contours for the first mouse.

The feature set was standardized to give the Z-omic. The mean Z-omic for eight pre-defined groups of texture was calculated (GLCM, GL3D, GLZ, GLD, and NGT). The R software was used to learn the self-organizing map of the averaged Z-omic using 16 organizing neural networks, a learning rate of 0.05, and a Gaussian neighborhood function with standard deviation 1. From the input dataset, each instance of measured texture (Z-omic) is assigned to a single node that best represents its distribution of variability.

The standard Kohonen SOM plot creates pie representations of the representative vectors for the grid. These are commonly referred to as codebooks (Figure 5.4), where the radius of a wedge corresponds to the magnitude in a particular dimension. To assess the impact of pre-defined confounding variables, a heatmap can be visualized. The heatmap demonstrates the distribution or frequency of the confounding variables over the nodes of the SOM. The SOM is blind to the suspected confounding variables, so any visible clusters in the heatmap demonstrate an association between the confounding variable and the input data. An example of the distribution of the confounding variable, contour size, with respect to the self-organized features, is shown in Figure 5.5, examining the distribution of extremes of contour size across the codebook. Two distinct clusters can be visualized in the heatmap, allowing us to

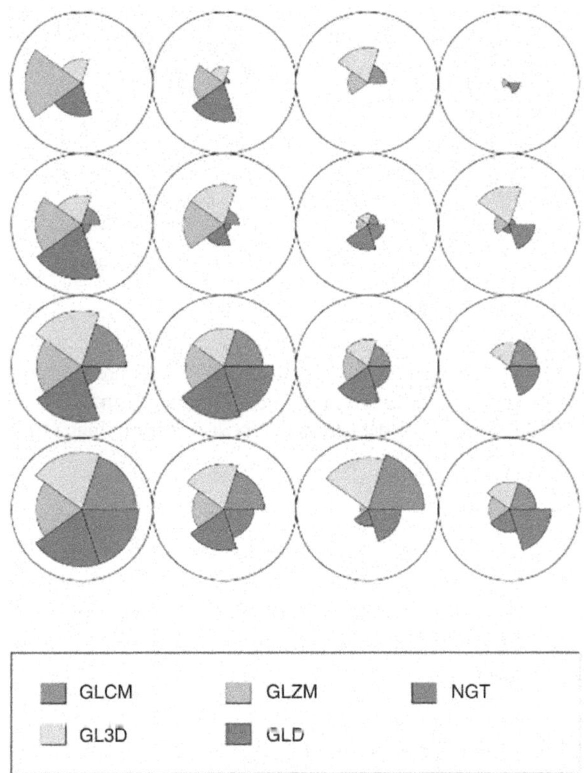

FIGURE 5.4 Codebook representing the distribution of texture features onto the grid.

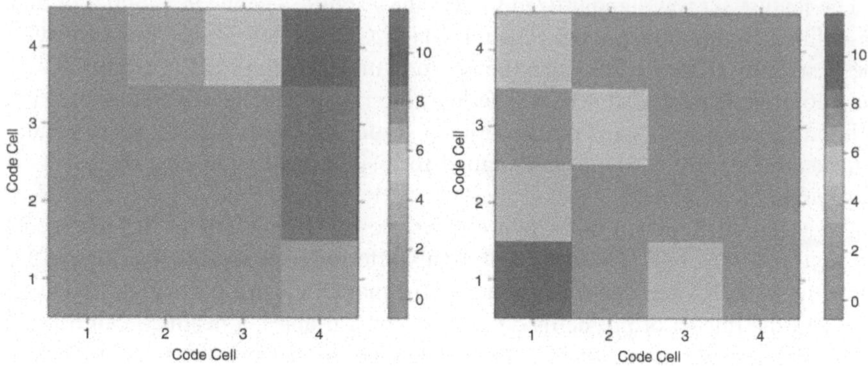

FIGURE 5.5 Codebook representing the distribution of texture features onto the grid. Leftmost, the distribution of contour size 4 mm. Rightmost, the distribution of contour size 5.5 mm.

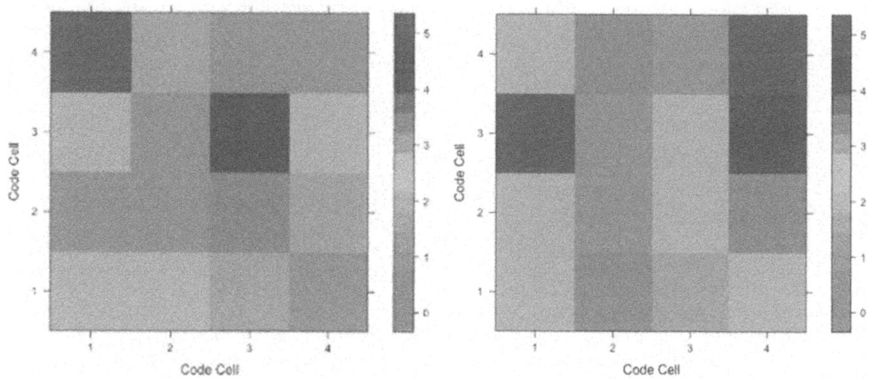

FIGURE 5.6 Codebook representing the distribution of texture features onto the grid. Leftmost, the heatmap or occurrence of post-injection image time (50–55 min) over the codebook. Rightmost, the heatmap or occurrence of post-injection image time (65–70 min).

interpret that contour size acts as a confounding variable when performing texture analysis, and changing contour size alters the distribution of standardized (and normalized [z-score]) texture features. No such relationship was observed when probing post-injection imaging time as the confounding variable (Figure 5.6), as no clusters were observed in the heatmap.

5.4 APPLICATION OF THE SOM FOR PREDICTIVE ANALYSIS

Section 1.3 introduced the unsupervised SOM for visualization and clustering of high-dimensional data onto a lower-dimensional representation whilst preserving the topological structure of the data. Whilst this allows us to gain insights into the radiomic dataset and drivers behind radiomic distribution, clustering is necessary for interpretation and classification. The SOM can, however, be used in a supervised learning framework, whereby independent variables or class information can be trained with the SOM to make classifications/predictions of testing data. The X-Y fused SOM (Melssen, Wehrens, and Buydens 2006) has been proposed for these purposes, using an additional grid of nodes (Ymap) to map the class information. During the training, a "fused" similarity measure is used which is based on a weighted sum of similarities between an input vector and all units in the Xmap, and similarities between the corresponding output vector and the units in the Ymap. The common winning unit is determined by the location of the minimum in the fused similarity measure. Both maps are updated simultaneously according to the standard SOM formalism: for the update of the Xmap, the input vector is used, while for the update of the Ymap, the corresponding output vector is used. The procedure for predicting the class membership of new inputs starts with presenting a new input vector to the network. The position of the winning unit in the Xmap is used to look up the class membership of the corresponding unit in the Ymap: the maximum value of this unit's weight vector determines the actual class membership (Arnrich et al. 2010). Constructing a predictive radiomics model using this supervised SOM would allow prediction of outcome metrics to be coupled with modelling of confounding variables and may prove to be an effective method going forward.

5.5 CONCLUSION

We have introduced the novel application of a self-organizing map for radiomic analysis in PET imaging, although the generic framework has applicability in all other imaging modalities. We have demonstrated its ability in identifying confounding variables that affect radiomic features, in this case contour size. With the extension to greater suspected confounding variables such as image acquisition/reconstruction settings, image pre- and post-processing variables, and varying segmentation algorithms, we offer a flexible method for full interrogation of the radiomics pipeline. We have postulated how the SOM may also be utilized with outcome data to serve as a predictive tool for dependent variables (e.g., prognosis, therapy response). In so doing, the learnt representations of self-organized features serve as the attributes for prediction, which will take into consideration the statistical variability in the underlying dataset. This serves as a promising area for future work.

REFERENCES

Ackerley, I., R. Smith, J. Scuffham, M. Halling-Brown, E. Lewis, and E. Spezi. 2019. "Using deep learning to detect esophageal lesions in PET-CT scans." In *Biomedical Applications in Molecular, Structural, and Functional Imaging, SPIE*, pp. 138–146.

Alsyed, E., R. Smith, C. Marshall, S. Paisey, and E. Spezi. 2019a. "The statistical influence of imaging time and segmentation volume on PET radiomic features: A preclinical study." In *2019 IEEE Nuclear Science Symposium and Medical Imaging Conference, NSS/MIC 2019*. doi:10.1109/NSS/MIC42101.2019.9059863.

Alsyed, E., R. Smith, C. Marshall, S. Paisey, and E. Spezi. 2019b. "Stability of PET radiomic features: A preclinical study." In *European Journal of Nuclear Medicine and Molecular Imaging, S759*. New York, NY: Springer.

Alsyed, E., R. Smith, S. Paisey, C. Marshall, and E. Spezi. 2021. *A Self Organizing Map for Exploratory Analysis of PET Radiomic Features*. IEEE, pp. 1–3. doi:10.1109/nss/mic42677.2020.9507846.

Arnrich, B., C. Kappeler-Setz, G. Tröster, R. La Marca, and U. Ehlert. 2010. "Self organizing maps for affective state detection." *International Workshop on Machine Learning for Assistive Technologies*, 1–8.

Berenguer, R., M. D. R. Pastor-Juan, J. Canales-Vázquez, M. Castro-García, M. V. Villas, F. M. Legorburo, and S. Sabater. 2018. "Radiomics of CT features may be nonreproducible and redundant: Influence of CT acquisition parameters." *Radiology* 288 (2): 407–415. doi:10.1148/radiol.2018172361.

Cattell, R., S. Chen, and C. Huang. 2019. "Robustness of radiomic features in magnetic resonance imaging: review and a phantom study." *Visual Computing for Industry, Biomedicine, and Art* 2 (1): 19. doi:10.1186/s42492-019-0025-6.

Cook, G. J. R., G. Azad, K. Owczarczyk, M. Siddique, and V. Goh. 2018. "Challenges and promises of PET radiomics." *International Journal of Radiation Oncology Biology Physics* 102 (4): 1083–1089. doi:10.1016/j.ijrobp.2017.12.268.

Dhingra, V., A. Mahajan, and S. Basu. 2015. "Emerging clinical applications of PET based molecular imaging in oncology: The promising future potential for evolving personalized cancer care." *Indian Journal of Radiology and Imaging* 25 (4): 332–341. doi:10.4103/0971-3026.169467.

Galavis, P., C. Hollensen, N. Jallow, and B. Paliwal Al. 2010. "Variability of textural features in FDG PET images due to different acquisition modes and reconstruction parameters." *Acta Oncologica* 49 (7): 12–22. doi:10.1007/s11103-011-9767-z.

Gillies, R. J, P. E. Kinahan, and H. Hricak. 2016. "Radiomics: Images are more than pictures, they are data." *Radiology* 278 (2): 563–577.

Griffeth, L. K. 2005. "Use of Pet/Ct scanning in Cancer patients: Technical and practical considerations." *Baylor University Medical Center Proceedings* 18 (4): 321–330. doi:10.1080/08998280.2005.11928089.

Hatt, M., F. Tixier, L. Pierce, P. E. Kinahan, C. C. Le Rest, and D. Visvikis. 2017. "Characterization of PET/CT images using texture analysis: The past, the present . . . any future?" *European Journal of Nuclear Medicine and Molecular Imaging* 44 (1): 151–165. doi:10.1007/s00259-016-3427-0.

Higgins, L. J., and M. G. Pomper. 2015. "The evolution of imaging in Cancer: Current state and future challenges existing advanced imaging techniques." *Seminars in Oncology* 38 (1): 3–15. doi:10.1053/j.seminoncol.2010.11.010.

Kadoury, S. 2018. *Manifold Learning in Medical Imaging*. IntechOpen. doi:10.5772/intechopen.79989.

Kohonen, T. 1982. "Self-organized formation of topologically correct feature maps." *Biological Cybernetics* 43 (1): 59–69. doi:10.1007/BF00337288.

Kohonen, T. 2001. "An overview of SOM literature." In *Self-Organizing Maps*, 347–371.

Köküer, M., R. N. G. Naguib, P. Janc, H. B. Younghusband, and R. Green. 2007. "Towards automatic risk analysis for hereditary non-polyposis colorectal cancer based on pedigree data." 319–337. doi:10.1016/B978-0-444-52855-1.50014-3.

Melssen, W., R. Wehrens, and L. Buydens. 2006. "Supervised kohonen networks for classification problems." *Chemometrics and Intelligent Laboratory Systems* 83 (2): 99–113. doi:10.1016/j.chemolab.2006.02.003.

Mulshine, J. L., D. S. Gierada, S. G. Armato, R. S. Avila, D. F. Yankelevitz, E. A. Kazerooni, M. F. McNitt-Gray, A. J. Buckler, and D. C. Sullivan. 2015. "Role of the quantitative imaging biomarker alliance in optimizing CT for the evaluation of lung cancer screen detected nodules." *Journal of the American College of Radiology* 12 (4): 390–395. doi:10.1016/j.jacr.2014.12.003.

Naqa, I. E., P. Grigsby, A. Apte, E. Kidd, E. Donnelyy, W. Thorstad, and J. Deasy. 2009. "Exploring feature based approaches in PET images for predicting cancer treatment outcomes." *Pattern Recognition* 42 (6): 1162–1171. doi:10.1016/j.patcog.2008.08.011.

Oliver, J. A., M. Budzevich, G. G. Zhang, T. J. Dilling, K. Latifi, and E. G. Moros. 2015. "Variability of image features computed from conventional and respiratory-gated PET/CT images of lung cancer." *Translational Oncology* 8 (6) 524–534. doi:10.1016/j.tranon.2015.11.013.

Smith, R., S. Paisey, N. Evans, V. Florence, E. Fittock, F. Siebzehnrubl, and C. Marshall. 2018. "Deep learning pre-clinical medical image segmentation for automated organ-wise delineation of PET." *European Journal of Nuclear Medicine and Molecular Imaging* 45 (1): S290.

Smith, R. L., P. Dasari, C. Lindsay, M. King, and K. Wells. 2019. "Dense motion propagation from sparse samples." *Physics in Medicine and Biology* 64 (20). doi:10.1088/1361-6560/ab41a0.

Velden, F. H. P. V., G. M. Kramer, V. Frings, I. A. Nissen, E. R. Mulder, A. J. D. Langen, O. S. Hoekstra, E. F. Smit, and R. Boellaard. 2016. "Repeatability of radiomic features in non-small-cell lung cancer [18 F] FDG-PET/CT studies : Impact of reconstruction and delineation." *Molecular Imaging and Biology Molecular Imaging and Biology*: 788–795. doi:10.1007/s11307-016-0940-2.

Zwanenburg, A., M. Vallières, M. A. Abdalah, H. J. W. L. Aerts, V. Andrearczyk, A. Apte, S. Ashrafinia, et al. 2020. "The image biomarker standardization initiative: Standardized quantitative radiomics for high-throughput image-based phenotyping." *Radiology* 295 (2): 328–338. doi:10.1148/radiol.2020191145.

6 Sensor-Based Human Activity Recognition Analysis Using Machine Learning and Topological Data Analysis (TDA)

Hossam Magdy Balaha and Asmaa El-Sayed Hassan

6.1 INTRODUCTION

Human activity recognition (HAR), also known as human behavior understanding or human activity analysis, is the process of identifying and classifying human movements and behaviors from data collected by sensors or video cameras. It has a wide range of applications, including healthcare, sports, and security [1]. In healthcare, activity recognition can be used to monitor the daily activities of patients with chronic conditions, such as diabetes or heart disease. By tracking a patient's movements and behaviors, healthcare professionals can detect early signs of deterioration and take appropriate action to prevent further deterioration. For example, if a patient's activity levels suddenly decrease, it could be a sign that their condition is worsening, and healthcare professionals can take steps to address the issue [2].

Activity recognition can also be used in the sports industry to analyze an athlete's performance and identify areas for improvement. By tracking an athlete's movements and behaviors during practice or competition, coaches and trainers can identify areas where the athlete is performing well and areas where they could benefit from additional training or support. In the field of security, activity recognition can be used to identify suspicious or unusual activities in surveillance footage. By analyzing the movements and behaviors of individuals in a particular location, security professionals can quickly identify potential threats and take appropriate action to protect people and property [3, 4].

There are various methods for recognizing human activities, including machine learning algorithms, computer vision techniques, and sensor fusion. Machine learning (ML) algorithms use data from sensors or cameras to learn and classify different types of movements and behaviors [5–7]. Computer vision techniques use image recognition algorithms to identify and classify movements and behaviors from video footage. To gain a deeper knowledge of a person's movements and behaviors, sensor

DOI: 10.1201/9780367486082-6

fusion combines data from many sensors (e.g., cameras and accelerometers). HAR is a powerful tool that can be used in a variety of industries to improve safety, performance, and healthcare outcomes. As technology continues to advance, activity recognition will likely become an increasingly important tool for understanding and improving human behavior [8–10].

One of the crucial steps in the process of HAR is feature extraction. It involves extracting relevant information from raw data collected by sensors or cameras and representing it in a way that can be used by machine learning algorithms to classify different types of movements and behaviors [11–13]. There are several different approaches to feature extraction, including handcrafted features, which are designed by humans based on their expert knowledge of the data, and learned features, which are learned by the ML algorithm from the data itself. When working with small well-defined datasets, handcrafted features can be useful, but they might not be as useful when working with broader, more complex datasets. Learned features, on the other hand, can be more effective for larger, more complex datasets, but they require more computational resources and may take longer to train [14–16].

Some examples of features that are commonly extracted for HAR include acceleration, angular velocity, and orientation data from sensors such as accelerometers and gyroscopes, as well as spatial and temporal information from video footage [17]. Other features that may be extracted include statistical features, such as mean and standard deviation, and frequency domain features, such as power spectral density. In addition to extracting features from raw data, it is also important to preprocess the data to remove noise and correct errors. This can improve the accuracy and efficiency of the ML algorithm and help it classify different types of movements and behaviors more accurately [18–20].

Classification is the process of assigning a label or class to a given input, and it is a crucial step in the process of HAR. Once relevant features have been extracted and preprocessed, an ML algorithm can use them to classify different types of movements and behaviors. There are several different approaches to classification, including supervised learning, in which the ML algorithm is trained on labeled data, and unsupervised learning, in which the algorithm must learn to classify the data without the benefit of labels [21–24]. Supervised learning is generally more accurate and efficient, but it requires a large labeled dataset to train the algorithm, which can be time-consuming and resource-intensive to collect. Unsupervised learning, on the other hand, is faster and requires fewer resources, but it is generally less accurate and may require additional processing to improve the accuracy of the classification [25–27].

One common approach to classification in HAR is to use an ML algorithm such as a support vector machine (SVM) or a decision tree (DT). These algorithms can learn to classify different types of movements and behaviors based on the features that have been extracted from the data [28, 29]. Other classification algorithms that may be used include k-nearest neighbors (KNN), Naive Bayes, and random forests. It is important to carefully evaluate the performance of the classification algorithm to ensure that it is accurately recognizing different types of movements and behaviors. This may involve testing the algorithm on a separate, labeled dataset or using a cross-validation approach to evaluate its performance [30, 31].

A high-quality benchmark dataset is lacking in the field of HAR research. Most publicly accessible datasets have issues with incomplete or unbalanced data. Most of the observed activities are rudimentary and do not encompass a wide range of activities. It can be challenging to integrate different architectures for HAR using context information. Despite performing at the cutting edge on benchmark datasets, many approaches are nevertheless overconfident in their predictions. The goal of this chapter is to offer an ML algorithm analysis to handle HAR. Additionally, two challenging goals are efficiently reducing the dimensionality of the raw data and identifying the most relevant features from it.

The following is a summary of the book chapter's contributions:

1. Carrying out tasks involving the recognition of human action while thoroughly comparing a range of ML algorithms to find the best mode.
2. Examining sampling and balancing methods to deal with unbalanced data, thereby identifying the most effective courses of action.
3. Reporting cutting-edge performance indicators and contrasting them with various relevant studies and methodologies.

The remainder of this chapter is structured as follows: The associated literature is reviewed and summarized in Section 6.2. The background is covered in Section 6.3. It discusses oversampling and unbalanced data approaches, dimensionality reduction and feature engineering, topological data analysis (TDA), feature scaling, classification, and optimization. The approach, the datasets acquired, the data preprocessing phase, the features engineering and dimensionality reduction strategies, and the ML classification and optimization phase are all covered in Section 6.4. The details and discussions of the experiments' findings are presented in Section 6.5. The book chapter is concluded and future work is discussed in Section 6.6.

6.2 LITERATURE REVIEW

Although HAR is not a new idea, the unpredictable variations in the human look, attire, lighting, and background make it difficult to identify and categorize human actions. The current section concentrates on the most recent research on HAR. The mentioned prior work can be categorized as machine and deep learning–based approaches. For machine learning, Sumaira et al. [32] utilized 2D skeletal data to conduct a comparative comparison of various HAR models. Four distinct activities (i.e., sit, stand, walk, and fall) were identified by the authors using the appearance and motion characteristics of 2D landmarks of human skeletal joints. Additionally, the performance of five supervised ML approaches was compared, including linear discriminant (LD), K-nearest neighbors (KNNs), support vector machine (SVM), feed-forward backpropagation neural network, and Naive Bayes (NB). With an overall accuracy of 98%, it was found that the KNN classifier produced the best results. Additionally, applying normalization to the train and test data improved the obtained results. KNN, SVM, LDA, and Naive Bayes classifiers had 100% accuracy in recognizing falls, walks, and standing. In Nasir et al. [33], an efficient approach to categorizing human actions was proposed. Removing unnecessary frames from films, extracting segments of interest (SoIs), mining feature descriptors using geodesic distance (GD), 3D Cartesian plane features (3D-CF), joints MOCAP (JMOCAP), and n-way point trajectory generation (nPTG) were among the procedures used. Additionally, a neuro-fuzzy classifier (NFC) was applied for classification. On two publicly accessible datasets, HMDB-51 and Hollywood2, the suggested method was tested, and an accuracy of 82.55% and 91.99% were achieved, respectively. These effective outcomes demonstrated the feasibility of the suggested model. Guangming et al. [34] introduced the continuous human action recognition (CHAR) algorithm online. They used skeletal data from the Kinect depth sensor for their research. An online classification technique based on a variable-length maximum entropy Markov model and likelihood probabilities was used to recognize continuous activity (MEMM). In contrast to previously disclosed CHAR techniques, their approach does not require the earlier detection of the start and finish locations of each activity. Results from experiments utilizing the Cornell CAD-60 dataset and the MSR Daily Activity 3D dataset showed that their proposed method was quite efficient for identifying human activities. Manzi et al. [35] created a machine learning method for recognizing human activities using data retrieved from a depth camera. In contrast to earlier approaches, each activity was represented by a unique set of clusters independently fetched from activities. These models were created using a multiclass SVM that was trained on the CAD-60 and TST datasets using the SOM optimization. Based on the input sequence and activity, these numbers were subjected to vary, producing dynamically created clusters. The needed clusters to model each example varies from two to four. Their approach reached the maximum performance using only about 4 sec of input data.

Using two stacked ensemble learning models, Guo et al. [36] introduced a dual-ensemble class imbalance learning approach. Three heterogeneous sub-classifiers were created for each subset of data using an internal ensemble learning model. Then, the base classifier with the highest accuracy was chosen. Multimodal evolutionary algorithms were used to find the best combination of the base classifiers and the most accurate way to recognize human behaviors in the external ensemble learning model. The suggested learning approach offered the simplest ensemble structure with the best accuracy for imbalanced data, according to statistical experimental results based on seven datasets of imbalanced activity. On all metrics, the suggested method outperformed the other five of the most commonly applied ensemble classification approaches. Using machine learning approaches, an intelligent smart healthcare system was described by Subasi et al. [37] to perform human activity recognition (HAR) intuitively. Their suggested method was effective and accurate at modeling and identifying daily activities. Additionally, the dataset comprising body motion and vital sign recordings from individuals with various profiles while engaging in various physical activities was the main focus of this study for HAR purposes. Additionally, for a robust and accurate HAR, two separate datasets (i.e., a mobile phone and wearable body sensors) were used. Their work has demonstrated that there is exceedingly difficult-to-identify human activity using sensor data, even with the availability of many machine learning techniques. Aslan et al. [38] carried out HAR using the speeded-up robust features (SURF) to extract features from several datasets (i.e., KTH and Weizmann datasets along with their dataset). A bag of visual words (BoVW) served as reinforcement for these features, and both grayscale and binary images were utilized to obtain SURF features. In addition, the characteristics were classified using four distinct ML methods (i.e., k-NN, decision tree, SVM, and Naive Bayes). Methods were compared using the recognition performances of binary and grayscale image features. The outcomes demonstrated that for HAR, the SURF of the binary picture was superior to the SURF of the gray image. Smartwatch sensor data was used by Ahmed et al. [39] to present a hybrid feature selection algorithm that detected various human activities. The information was obtained using inertial sensors (i.e., human waist-mounted accelerometers and gyroscopes) to record different human activities. The dataset was given a total of 23 base features, and the sensors yielded a total of 138 heterogeneous features. However, the additional features result in worse performance. To choose the best features, the proposed hybrid feature selection method utilizing the filter and wrapper approaches was used. The SVM was then used to detect human activity. Using optimal features, the suggested method had an average classification performance of 96.81%, which is almost 6% better than with no feature selection.

Recently published work has already built DL-based HAR. In their analysis of current work with a focus on evaluation approaches, Julieta et al. [40] demonstrated that a straightforward baseline with no attempt to represent motion at all can achieve state-of-the-art performance. This result was investigated by taking into account the architectures, loss functions, and training procedures utilized in state-of-the-art (SOTA) approaches, and the recent RNN methods were analyzed. The traditional RNN models for human motion were proposed with three modifications. As a result, a straightforward and expandable RNN architecture with SOTA performance in predicting human activity is produced. Ferrari et al. [41] focused on the identification of human activities using data collected by a smartphone's built-in accelerometer. They created a clear validation methodology that considers personalization issues that allowed the unbiased assessment of the effectiveness of machine learning algorithms. The algorithms were also tested on three distinct open-source datasets. The personalization model took into account two factors: the similarity between individuals in terms of their physical characteristics (e.g., age, weight, and height), along with the similarity in terms of the intrinsic qualities of the signals these individuals produce when engaging in various activities. The results of the trials showed that using personalization models generally increases accuracy. Nafea et al. [42] introduced a new methodology using CNN with varying kernel dimensions along with BiLSTM to capture features at various resolutions. They suggested an effective selection of the optimal video representation and the effective extraction of spatial and temporal features from sensor data using traditional CNN and BiLSTM. Data collected from diverse methods (e.g., sensors, accelerometers, and gyroscopes) were utilized in UCI and Wireless Sensor Data Mining (WISDM) datasets. The outcomes showed that the suggested scheme was effective at enhancing HAR. Hence, the proposed method attained a better score in the WISDM dataset than the UCI dataset (i.e., 98.53% versus 97.05%). A new architecture for multi-class wearable user identification was provided in Mekruksavanich and Jitpattanakul [43], with a foundation in the deep learning models used to recognize human behavior. Sensory data from the wearable devices' tri-axial gyroscopes and accelerometers were used to gather detailed information about users while they perform various activities.

A series of experiments was also presented, and it was shown that the suggested framework worked as intended. The conducted highest accuracy by CNN and LSTM was 91.77% and 92.43%, respectively, for all users.

Chen et al. [44] proposed a deep learning model called DEBONAIR (deep learning–based multimodal complex human activity recognition). It was an end-to-end model extracting features systematically. For various sensor data, specific sub-network designs were developed, and the combined outputs of all sub-networks were used to extract fusion features. Then, to learn the sequential information of complex human activities (CHAs), an LSTM network was utilized. The model was evaluated using two multimodal CHA datasets. The experiment outcomes demonstrated that DEBONAIR was notably superior to the SOTA CHA recognition models. An attention-based multi-head model for human activity recognition (HAR) was put forth by Khan and Ahmad [45]. Three compact convolutional heads were included in this architecture, each of which used one-dimensional CNN to extract features from sensory data. The purpose of creating the lightweight multi-head model was to improve CNN's capacity for representation by enabling the automatic selection of salient features and elimination of irrelevant features. To assess the model, experiments were run on two publicly accessible benchmark datasets (i.e., WISDM and UCI HAR). The results of the experiment showed how well the suggested framework performed in activity recognition and reported improved accuracy while maintaining the computational economy. With CNN used for extracting spatial features and the LSTM network used for learning temporal information, Khan et al. [46] created a hybrid model for activity recognition using these two techniques. Additionally, a novel dataset with 12 different classes of human physical activities was gathered from 20 humans using the Kinect V2 sensor. To find the best HAR solution, thorough ablation research of many classical machine and deep learning methods was conducted. The CNN-LSTM method produced an accuracy of 90.89%, demonstrating that the suggested model was appropriate for HAR applications. A deep learning model based on residual blocks and BiLSTM was suggested by Li and Wang [47]. The residual block was utilized by the mode to extract spatial characteristics. Additionally, BiLSTM was used to derive the forward and backward dependencies of the feature sequence. Then, the Softmax layer received the obtained features to perform the HAR. A dataset with six typical human actions (i.e., sitting, standing, moving between two points, walking, and running) was created. Using three datasets (i.e., their dataset, WISDM, and PAMAP2), the proposed model was assessed. The experimental findings demonstrated that the suggested model, when applied to their datasets, WISDM, and PAMAP2, attained an accuracy of 96.95%, 97.32%, and 97.15%, respectively. The proposed model performed better and required fewer parameters than several SOTA models. A multistage deep neural network framework was created by Fan et al. [48] to interpret accelerometer, gyroscope, and magnetometer data, which can be used to learn about different human activities. Initially, critical information from the raw data of measurement devices was extracted using the stage of variational autoencoders (VAE). To create more realistic human activities, the stage of generative adversarial networks (GANs) was used. In addition, the transfer learning technique improved the performance of the target domain, which helped create a strong and reliable model to identify human activity.

6.3 BACKGROUND

6.3.1 HAR Data Acquisition

HAR is the process of automatically recognizing and classifying human actions and movements based on data collected from sensors or other sources. There are many ways to acquire data for human activity recognition, depending on the specific application and the type of data being collected. Some common methods for data acquisition in human activity recognition include [49, 50]:

1. **Wearable Sensors.** These can be used to collect data from sensors that are worn on the body, such as accelerometers, gyroscopes, and magnetometers.
2. **Video Cameras.** These can be used to collect data from visual sources, such as images or video streams.
3. **Smartphones.** These can be used to collect data from a variety of sensors, such as accelerometers, gyroscopes, and GPS.
4. **Internet of Things (IoT) Devices.** These can be used to collect data from sensors or other devices that are connected to the internet, such as smart home devices or smart city infrastructure.
5. **Manual Data Collection.** In some cases, data for human activity recognition may be collected manually by human observers or through self-report methods.

There are many other methods for data acquisition in human activity recognition, and the most appropriate one will depend on the specific application and the type of data being collected.

6.3.2 Imbalanced Datasets and Oversampling Techniques

Imbalanced data refers to a classification problem where the number of instances belonging to one class significantly outnumbers the number of instances belonging to the other class(es). This can create problems when training machine learning models, as the model may be biased toward the more numerous classes [51]. Oversampling is a technique used to address imbalanced data by generating additional synthetic data for the minority class. This can be done in several ways, including [52–55]:

1. **Random Oversampling.** This involves randomly selecting instances from the minority class and duplicating them until the class distribution is balanced.
2. **SMOTE (Synthetic Minority Oversampling Technique).** This involves generating synthetic data for the minority class by creating new instances that are interpolations between existing instances.
3. **ADASYN (Adaptive Synthetic Sampling).** This is like SMOTE, but instead of creating synthetic data for all minority class instances, it focuses on creating synthetic data for the instances that are difficult to classify.
4. **Borderline SMOTE.** This is like SMOTE, but instead of creating synthetic data for all minority class instances, it focuses on creating synthetic data for the instances that are closest to the decision boundary between the two classes.

5. **ROS (Random Oversampling of the Classes).** This is a variant of random oversampling that involves oversampling both the minority and majority classes rather than just the minority class.

Oversampling can be a useful technique for improving the performance of machine learning models on imbalanced data, as it helps balance the class distribution and prevent the model from being biased toward the majority class. However, it is important to note that oversampling can also lead to overfitting, as it increases the size of the training set and may introduce noise. Dealing with imbalanced data can be a challenging task, and choosing the right approach will depend on the specific characteristics of the data and the desired performance of the model [56, 57].

6.3.3 Features Engineering and Dimensionality Reduction Techniques

Feature engineering is the process of extracting, selecting, and constructing features from raw data that can be used in ML algorithms. It is a crucial step in the machine learning process, as the quality and quantity of the features can greatly impact the performance of the model. Effective feature engineering involves understanding the domain of the problem, as well as the characteristics of the data [58–60]. It often involves transforming or scaling the data, creating new features through feature extraction or feature construction, and selecting the most relevant features through feature selection. One common approach to feature engineering is to use domain knowledge to create new features that are likely to be relevant to the problem. For example, in the field of natural language processing, it may be useful to create features based on the parts of speech of the words in a document. Feature engineering is a time-consuming but important aspect of building successful machine learning models. It can require a combination of trial and error, domain expertise, and creativity to create features that are both useful and effective [61–64].

There is no unified taxonomy for feature engineering, as the approaches used will depend on the specific characteristics of the data and the desired performance of the model. However, some common approaches to feature engineering include [65, 66]:

1. **Feature Extraction.** This involves transforming raw data into a form that is more suitable for modeling, such as converting text into numerical vectors or extracting features from images.
2. **Feature Selection.** This involves identifying the most relevant features to use in a model, either by selecting a subset of the available features or by constructing new features from the raw data.
3. **Feature Construction.** This involves creating new features from the raw data, either by combining existing features or by using domain knowledge, to create features that are likely to be relevant to the problem.

6.3.3.1 Features Extraction

In this chapter, only feature extraction techniques are applied to perform dimensionality reduction. These techniques are operated to reduce model complexity, overfitting, and generalization error, moreover, increasing the computation efficiency of the model [67]. They are:

1. **Principal component analysis (PCA)** is a technique for extracting the key features from a dataset's extensive feature set. In high-dimensional data, it determines the direction of the maximum deviation and projects the data into a new subspace with dimensions either the same as or less than the initial ones [68].
2. **Linear discriminant analysis (LDA)** is a supervised learning feature extraction method that aims to reduce the spreading inside the class and increase the gap between the means of the classes [69].
3. **Independent component analysis (ICA)** is a linear strategy that focuses on the accurate identification of each independent component in a mixture of input data [70].
4. **Random projection (RP)** is a method used to conduct the feature reduction for a set of points that are in Euclidean space [71].
5. **Truncated singular value decomposition (T-SVD)** uses a matrix factorization technique to reduce the dimensions of the data, much to PCA. The data is not centered before determining the singular value decomposition, in contrast to PCA. It can therefore be effectively used for sparse matrices [72]. The number of columns in the t-SVD factorization of the data matrix equals the number of truncations.

6.3.3.1.1 Numerical Features Extraction

Numerical feature extraction involves extracting numerical features from raw data, typically by calculating summary statistics or applying mathematical transformations. Some common techniques for numeric feature extraction include [73, 74]:

1. **Calculation of Summary Statistics.** This involves calculating statistical measures, such as the mean, median, mode, standard deviation, and quartiles of a set of numeric values.
2. **Binning.** This involves dividing the range of numeric values into a set of bins and replacing the original values with the bin to which they belong.
3. **Normalization.** This involves scaling the numeric values to a specific range, such as 0 to 1, or –1 to 1. This can be useful when the scale of the values is different, as it allows the model to weigh the features equally.
4. **Standardization.** This involves scaling the numeric values to have zero mean and unit variance. This can be useful when the distribution of the values is not normal, as it allows the model to weigh the features equally.
5. **Log Transformation.** This involves applying the logarithmic function to the numeric values. This can be useful when the values have a skewed distribution, as it can help normalize the distribution.

The choice of numerical feature extraction technique will depend on the specific characteristics of the data and the desired performance of the model. It is important to carefully evaluate the results of different techniques to choose the one that is most effective for a given problem.

6.3.3.2 Features Selection

Feature selection is the process of identifying the most relevant features to use in a machine learning model, either by selecting a subset of the available features or by constructing new features from the raw data [75]. There are many techniques for feature selection, and the choice of technique will depend on the specific characteristics of the data and the desired performance of the model. Some common techniques for feature selection include [76]:

1. **Filter Methods.** These techniques select features based on their relevance to the target variable, using statistical measures such as correlation or ANOVA.
2. **Wrapper Methods.** These techniques use a machine learning model to evaluate the performance of different feature subsets and select the subset that performs the best.
3. **Embedded Methods.** These techniques incorporate feature selection as part of the training process of a machine learning model, such as LASSO or decision trees with pruning.
4. **Metaheuristic Algorithms.** These techniques use evolutionary search to find the optimal combination of features.

The goal of feature selection is to identify the features that are most relevant to the problem and will lead to the best performance of the machine learning model. It is important to carefully evaluate the results of different feature selection techniques to choose the one that is most effective for a given problem.

6.3.3.3 Topological Data Analysis

Topological data analysis (TDA) is a branch of mathematics that uses topological techniques to study the shape and structure of data [77]. It is a relatively new field, with applications in a wide range of areas, including machine learning, computer science, and biology. One of the key ideas of TDA is to represent data as a topological space, called a simplicial complex, which captures the relationships between the data points. This allows TDA to identify patterns and features in the data that may not be apparent using traditional techniques, such as clustering or dimensionality reduction [78].

One of the main advantages of TDA is its ability to handle high-dimensional data, as it does not rely on traditional distance measures that can break down in high dimensions. TDA can also handle data that is noisy or has missing values, as it is robust to perturbations in the data. TDA is a promising tool for data analysis and has the potential to uncover new insights and patterns in data that were previously hidden. It is an active area of research, and new applications and techniques are being developed continuously [79–81].

6.3.3.4 Features Scaling Techniques

Feature scaling is the process of transforming the values of a feature to a common range, such as 0 to 1 or –1 to 1. It is a common preprocessing step in machine learning, as it can help improve the performance of some models and make it easier to compare different features. There are several techniques for feature scaling, including [82]:

1. **Min–Max Scaling.** This technique scales the values of a feature to a specific range, such as 0 to 1 or –1 to 1.
2. **Standardization.** This technique scales the values of a feature to have zero mean and unit variance.
3. **Normalization.** This technique scales the values of a feature to a range of 0 to 1.

The choice of feature scaling technique will depend on the specific characteristics of the data and the desired performance of the model. It is important to carefully evaluate the results of different techniques to choose the one that is most effective for a given problem.

6.3.4 Machine Learning Classification and Tuning

Machine learning algorithms are a set of techniques that allow computers to learn from data, without being explicitly programmed. These algorithms can be broadly classified into two categories: supervised and unsupervised learning. Supervised learning algorithms are trained on labeled data, meaning, that the data has been labeled with the correct output. The algorithm receives input data and the corresponding correct output, and the goal is to learn a function that maps the input data to the output [83–86]. Examples of supervised learning algorithms include linear regression, logistic regression, and support vector machines. Unsupervised learning algorithms, on the other hand, do not receive labeled training data. Instead, the goal is to discover patterns or relationships in the data. Examples of unsupervised learning algorithms include k-means clustering and principal component analysis. ML algorithms have revolutionized the field of artificial intelligence and have found applications in a wide range of industries, including finance, healthcare, and marketing. They can process large amounts of data quickly and accurately, making them a valuable tool for data analysis and decision-making [87].

6.3.4.1 Machine Learning Classifiers

Many different classifiers can be used for supervised machine learning; however, the most common ones include [88–90]:

1. **Random forest (RF)** is an ensemble learning algorithm that is used for both classification and regression problems. It builds multiple decision trees and combines them to form a more accurate and stable model. The key idea behind random forest is to randomly select a subset of the features at each split of the decision tree, which helps decorrelate the trees and reduce overfitting. The final prediction is made by averaging the predictions of the individual decision trees (in case of regression) or by taking a majority vote (in case of classification).

2. **LightGBM (LGBM)** is a gradient-boosting framework that uses tree-based learning algorithms. It is designed to be efficient and scalable, particularly for large datasets or datasets with many features. One of the key features of LGBM is the use of a novel tree-building algorithm, called a histogram-based algorithm, which bins continuous feature values into discrete bins and uses the bins in the tree-building process, which can effectively deal with the categorical feature and large feature dimension. Additionally, LGBM uses a leaf-wise tree growth algorithm, which focuses on growing the leaf nodes of the tree, rather than the level-wise tree growth used by other tree-based algorithms, which leads to a faster and more efficient tree-building process.
3. **AdaBoost (adaptive boosting)** is a popular ensemble learning algorithm that is used for classification and regression problems. It combines multiple "weak" learning models, such as decision trees, to create a stronger model. The idea behind AdaBoost is to iteratively train weak models on subsets of the data, with the subsets chosen in a way that allows each model to focus on the examples that were misclassified by the previous models. The weights of the examples are adjusted in each iteration to give more importance to the misclassified examples. The final model is a weighted combination of the weak models, where the weights are determined by the performance of the models on the training data. AdaBoost is known for its good performance and versatility, as it can be used with a variety of base models and can handle both numerical and categorical features. However, it can be sensitive to outliers and noisy data. It is also sensitive to irrelevant features, so it is important to preprocess or select the feature before using this algorithm.
4. **Decision trees (DT)** are a widely used supervised learning algorithm that can be used for both classification and regression problems. The algorithm creates a tree-like model of decisions and their possible consequences. Each internal node of the tree represents a feature (or attribute), each branch represents a decision or rule, and each leaf node represents the output or prediction. The goal of the decision tree is to build a model that accurately predicts the target variable based on the input features. The tree is built by recursively splitting the data into subsets based on the values of the input features and selecting the feature and split point that results in the purest subsets. The process is repeated on each subset until the tree reaches a maximum depth or the subsets become pure.
5. **Extra trees (extremely randomized trees)** are a variation of the random forest algorithm that is used for both classification and regression problems. The algorithm creates multiple decision trees and combines them to form a more accurate and stable model, like random forest. However, extra trees differs from random forest in the way the decision trees are constructed. In extra trees, at each split of the decision tree, a random subset of features is selected, and the best feature is chosen among the random subsets to split the data. This randomness in selecting the features for each split reduces the correlation between the trees and improves the diversity of the ensemble. As a result, the extra trees algorithm tends to reduce overfitting by providing more robust results than a single decision tree.

6. **Extreme gradient boosting (XGBoost)** is an open-source implementation of the gradient boosting algorithm. It is specifically designed for large-scale datasets and is known for its high performance and efficiency. XGBoost uses a tree-based learning algorithm, like other gradient-boosting frameworks, such as LGBM and CatBoost. One of the key features of XGBoost is the use of a novel technique called "gradient-based one-side sampling," which deals with imbalanced datasets by resampling the training set to achieve a more balanced distribution of the target variable. Additionally, XGBoost uses a more regularized model formalization to control overfitting, which is called "regularized gradient boosting." It supports parallel computing for faster training speed.
7. **Histogram gradient boosting (HGB)** is a type of gradient boosting algorithm that utilizes histograms as the base models instead of traditional decision trees. The idea behind this approach is that histograms can capture non-linear relationships between the features and target variables more efficiently and compactly decision trees. In histogram gradient boosting, the algorithm starts by creating a histogram of the feature values for each input feature and the target variable. Then, it uses the gradient of the loss function with concerning histogram to update the histogram and reduce the loss. This process is repeated until a stopping criterion is met, such as a maximum number of iterations or a minimum improvement in the loss. One of the advantages of this method is that it can handle large datasets with high-dimensional features more efficiently than traditional decision tree–based methods and also often results in models with less overfitting.

6.3.4.2 Cross-Validation

k-fold cross-validation is a method used to evaluate the performance of a machine learning model. It involves randomly dividing the dataset into *k* groups, or "folds," of roughly equal size. The model is then trained on *k*-1 of the folds and tested on the last one. This process is repeated *k* times, with a different fold being used as the test set each time. The average performance across all *k* iterations is then used as an estimate of the model's performance [91].

One of the main benefits of k-fold cross-validation is that it allows you to use all the data for both training and testing, which can give you a more accurate estimate of the model's performance. It is also a good way to assess the robustness of the model, as it can give you a sense of how well it is likely to generalize to new data. The value of *k* is typically chosen to be 5 or 10, although there is no hard rule. A larger value of *k* will give you a more accurate estimate of the model's performance, but it will also take longer to run.

6.3.4.3 Grid Search

Grid search (GS) is a technique used to tune the hyperparameters of a machine learning model. It works by training the model on a range of hyperparameter values specified in a "grid" and then evaluating the model's performance for each combination of values. The goal is to find the combination of values that gives the best performance [92].

For example, if you were tuning a decision tree model, you might specify a grid of values for the maximum depth of the tree, the minimum number of samples required at a leaf node and to split a node. The grid search algorithm would then train a model for each combination of these hyperparameter values and evaluate the model's performance using cross-validation. GS is a brute-force method that can be computationally expensive, but it is often the most effective way to tune the hyperparameters of a model. It is especially useful when you have a limited understanding of which values are likely to work best, or when you have many hyperparameters to tune [93].

6.3.4.4 Performance Evaluation

Machine learning performance evaluation is the process of assessing the accuracy, precision, and other qualities of a machine learning model on a given dataset. It is an important step in the development of any machine learning system, as it allows you to determine whether the model is performing well and to identify any weaknesses or areas for improvement.

There are many different metrics that can be used to evaluate the performance of a machine learning model, depending on the type of model and the task it is being used for. Some common metrics include [94, 95]:

1. **Accuracy.** This is the proportion of correct predictions made by the model. It is often used for classification tasks but can also be used for regression tasks.
2. **Precision.** This is the proportion of true positive predictions made by the model, out of all the positive predictions it made. It is often used to evaluate the performance of a model on imbalanced datasets.
3. **Recall.** This is the proportion of true positive predictions made by the model, out of all the actual positive instances in the dataset. It is often used to evaluate the performance of a model on imbalanced datasets.
4. **F1-score.** This is the harmonic mean of precision and recall and is often used as a single metric to summarize the performance of a model on imbalanced datasets.

There are many other metrics that can be used to evaluate the performance of a machine learning model, and which one is most appropriate will depend on the specific task and goals of the model.

6.4 METHODOLOGY

6.4.1 Data Retrieval Phase

Human Activity Recognition Using Smartphones Data Set v1.0 (UCI-HAR) is the used dataset in the current chapter. It is applied to 30 volunteers with an age range of [19: 48] years. The data is retrieved using "Samsung Galaxy S II" Android mobile phone. It contains six different activities (i.e., "WALKING," "WALKING_UPSTAIRS,"

"WALKING_DOWNSTAIRS," "SITTING," "STANDING," and "LAYING"). It is split into two subsets: 70% for training inside the "train" folder and 30% for testing inside the "test" folder. The number of training samples is 7,352, while the number of testing samples is 2,947. There are 561 features, and they are defined in "features.txt." It can be retrieved and downloaded from the URL: https://archive.ics.uci.edu/ml/datasets/human+activity+recognition+using+smartphones. The training and testing subsets are combined in this book chapter and partitioned in a later process. The statistics of the UCI-HAR are presented graphically in Figure 6.1.

6.4.2 Preprocessing Phase

Since the datasets are not balanced, overfitting or classification problems may result. To choose the best technique to use in the experiments, the current book chapter made use of the strategies covered earlier in this book chapter. The SMOTE approach performs better than others since it takes less time, provides balanced datasets, and does not crash. As a result, the SMOTE technique is employed in the current work to apply synthetic oversampling to the datasets used. After the SMOTE balancing procedure, it reached 11,664 records. The categories are on the x-axis, while the counts for each category are on the y-axis.

The datasets are time series data, and a sampling technique is utilized in the classification phase as opposed to row-by-row. The records are stacked row-by-row following the sampling procedure because the sampling is applied vertically. As illustrated in Table 6.1, two configurations are used while applying it. The first column is the step size, the second column is the overlapping percentage, and the third column is the output shape after sampling of the balanced dataset.

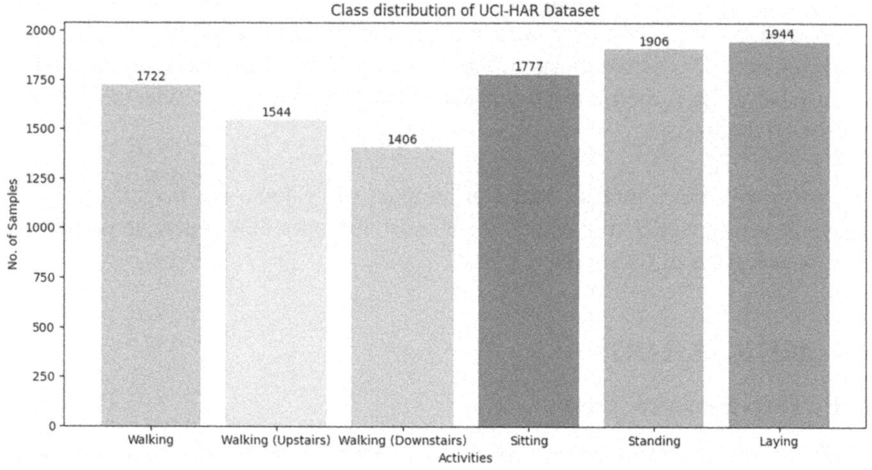

FIGURE 6.1 Statistics of the UCI-HAR dataset.

TABLE 6.1
The Data Sampling Details for the UCI-HAR Dataset

Step Size	Overlapping Percentage	Output Shape
50	50%	(232 × 100 × 561)
100	0%	(116 × 100 × 561)

6.4.3 Features Engineering

In nine different ways, TDA is utilized to extract the characteristics from the dataset. In order, they are as follows: (1) persistence entropy in two viewpoints, normalized and non-normalized; (2) number of points; (3) bottleneck; (4) Wasserstein; (5) Betti; (6) Landspace; (7) persistence image; (8) heat; and (9) silhouette. Three phases make up the feature extraction pipeline: A bottleneck scaler reduces the lifetime of the most persistent point across all diagrams and homology dimensions to two, resulting in (1) cubical persistence, (2) bottleneck scaling, and (3) a filter that eliminates points with lifetimes that are less than or equal to a cutoff value (set to 0.01 in the current book chapter). Twenty-six features are extracted from that for each row. Five strategies for feature reduction were used in the current book chapter. As previously mentioned, these are PCA, LDA, ICA, RP, and T-SVD. For the UCI-HAR dataset, the features are trimmed down to 100.

6.4.4 Classification and Tuning

The current book chapter achieves state-of-the-art (SOTA) performance metrics by utilizing several categorization techniques. They are the LGBM, XGB, AdaBoost, HGB, ETs, DT, and RF classifiers, in that order. GS is used to tune the hyperparameters of the models. It involves specifying a set of hyperparameters and their corresponding values and training the model with each combination of these values. The goal is to find the combination of hyperparameters that results in the best performance on a given validation set. This can be a time-consuming process, especially if the number of hyperparameters and their possible values are large. However, it is a simple and widely used method for hyperparameter optimization. Table 6.2 summarizes the different hyperparameters used with the classifiers utilized in this book chapter. Standardization, normalization, min–max scaling, max–absolute scaling, and resilient scaling are all employed.

The predicted performance of the classifiers is enhanced by the usage of the train-to-test splitting and k-fold cross-validation. The datasets are separated into train and test subsets using the train-to-test splitting method. The train and test subsets in the current book chapter are split 85% to 15%, and the number of folds is set to five.

TABLE 6.2
The Different Hyperparameters Used with the Classifiers

ML Classifier	Hyperparameter	Values
LGBM, XGBoost, and AdaBoost	Learning rate	0.01, 0.1, and 1.0
DT, ETs, and RF	Criterion	Gini and entropy
DT	Splitter	Best and random
HGB	N/A	N/A

6.5 EXPERIMENTS AND DISCUSSION

In order to limit the number of features to 100, the "UCI-HAR" dataset is sampled, and five feature reduction algorithms are used. The most effective combinations for the various classifiers are shown in Table 6.3, and the related performance metrics are shown in Table 6.4 for the reduced "UCI-HAR" data with dimensions 232 × 100. After the grid searching procedure, it displays the best of each classifier. It demonstrates that the best-reported accuracy, balanced accuracy, precision, specificity, recall, and F1-score by HGB classifier with the PCA feature reduction technique are, in that order, 99.43%, 98.97%, 98.28%, 99.66%, 98.28%, and 98.28%. The UCI-HAR dataset's WSM measures are summarized in Table 6.5. It demonstrates that the HGB using the PCA feature reduction technique produced the greatest WSM value, which is 98.68%.

The most effective combinations for the various classifiers discussed are shown in Table 6.6, and the related performance metrics are shown in Table 6.7 for the reduced "UCI-HAR" data with dimensions 116 × 100. After the grid searching procedure, it displays the best of each classifier. It demonstrates that the LGBM classifier with the T- SVD feature reduction technique has the best-reported accuracy, balanced accuracy, precision, specificity, recall, and F1-score values of 100%. The metrics for the WSM are summarized in Table 6.8. It demonstrates that the LGBM classifier's output using the T-SVD feature reduction technique has the greatest WSM value, which is 100%.

Following TDA and utilizing the various specified classifiers, Table 6.9 displays the optimal combinations, and Table 6.10 displays the appropriate performance metrics. After the grid searching procedure, it displays the best of each classifier. It demonstrates that the best-reported accuracy, balanced accuracy, precision, specificity, recall, and F1-score by the LGBM classifier are, respectively, 95.69%, 92.24%, 87.07%, 97.41%, 87.07%, and 87.07% for the UCI-HAR dataset with 0% overlap. The best-reported classifier in terms of elapsed time is DT, with 0.8 seconds. The best-reported accuracy, balanced accuracy, precision, specificity, recall, and F1-score using the LGBM classifier are 96.70%, 94.05%, 90.09%, 98.02%, 90.09%, and 90.09% respectively, for the UCI-HAR dataset with 50% overlap. The WSM metrics are listed in Table 6.11. It demonstrates that the highest WSM values were produced by the LGBM classifiers using the UCI-HAR + 0% overlap and UCI-HAR + 50% overlap, respectively.

TABLE 6.3
The Best Combinations on the Reduced "UCI-HAR" Dataset (232 × 100) Using Each Classifier

Feature Reduction Technique	Classifier	Scaling Technique	Criterion	Splitter	Max Depth	Estimators #	Class Weight	Learning Rate
PCA	DT	Robust	Entropy	Best	None	—	—	—
	AdaBoost	Normalization	—	—	—	300	—	10%
	RFC	Normalization	Gini	—	None	300	Balanced	—
	ETC	Max–absolute	Gini	—	None	300	—	—
	HGB	Normalization	—	—	—	—	—	—
	XGB	Normalization	—	—	None	300	—	10%
	LGBM	Standardization	—	—	None	300	—	10%
LDA	DT	Normalization	Gini	Best	None	—	—	—
	AdaBoost	Normalization	—	—	—	300	—	10%
	RFC	Normalization	Gini	—	None	300	Balanced	—
	ETC	Normalization	Gini	—	None	300	—	—
	HGB	Standardization	—	—	—	—	—	—
	XGB	Normalization	—	—	None	300	—	10%
	LGBM	Normalization	—	—	None	300	—	—
ICA	DT	Standardization	Gini	Best	None	—	—	—
	AdaBoost	Standardization	—	—	—	300	—	10%
	RFC	Normalization	Gini	—	None	300	None	—
	ETC	Standardization	Gini	—	None	300	—	—
	HGB	Normalization	—	—	—	—	—	—
	XGB	Normalization	—	—	None	300	—	10%
	LGBM	Normalization	—	—	None	300	—	10%
T-SVD	DT	Normalization	Entropy	Best	None	—	—	—
	AdaBoost	Standardization	—	—	—	300	—	10%
	RFC	Normalization	Gini	—	None	300		—
	ETC	Robust	Entropy	—	None	300	—	—
	HGB	Standardization	—	—	—	—	—	—
	XGB	Normalization	—	—	None	300	—	10%
	LGBM	Min–max	—	—	None	300	—	10%
RP	DT	Standardization	Entropy	Best	None	—	—	—
	AdaBoost	Normalization	—	—	—	300	—	1%
	RFC	Normalization	Gini	—	None	300	None	—
	ETC	Max–absolute	Gini	—	None	300	—	—
	HGB	Standardization	—	—	—	—	—	—
	XGB	Standardization	—	—	None	300	—	10%
	LGBM	Max–absolute	—	—	None	300	—	10%

TABLE 6.4
The Corresponding Performance Metrics Using Each Classifier Applied on the Reduced "UCI-HAR" Dataset (232 × 100)

Feature Reduction Technique	Classifier	Accuracy	Balanced Accuracy	Precision	Specificity	Recall	F1
PCA	DT	99.28%	98.71%	97.84%	99.57%	97.84%	97.84%
	AdaBoost	99.14%	98.45%	97.41%	99.48%	97.41%	97.41%
	RFC	99.14%	98.45%	97.41%	99.48%	97.41%	97.41%
	ETC	99.28%	98.71%	97.84%	99.57%	97.84%	97.84%
	HGB	99.43%	98.97%	98.28%	99.66%	98.28%	98.28%
	XGB	99.14%	98.45%	97.41%	99.48%	97.41%	97.41%
	LGBM	99.28%	98.71%	97.84%	99.57%	97.84%	97.84%
LDA	DT	97.70%	95.86%	93.10%	98.62%	93.10%	93.10%
	AdaBoost	86.49%	75.69%	59.48%	91.90%	59.48%	59.48%
	RFC	97.99%	96.38%	93.97%	98.79%	93.97%	93.97%
	ETC	97.99%	96.38%	93.97%	98.79%	93.97%	93.97%
	HGB	96.98%	94.57%	90.95%	98.19%	90.95%	90.95%
	XGB	97.56%	95.60%	92.67%	98.53%	92.67%	92.67%
	LGBM	97.70%	95.86%	93.10%	98.62%	93.10%	93.10%
ICA	DT	78.88%	61.98%	36.64%	87.33%	36.64%	36.64%
	AdaBoost	81.32%	66.38%	43.97%	88.79%	43.97%	43.97%
	RFC	88.22%	78.79%	64.66%	92.93%	64.66%	64.66%
	ETC	90.66%	83.19%	71.98%	94.40%	71.98%	71.98%
	HGB	86.78%	76.21%	60.34%	92.07%	60.34%	60.34%
	XGB	86.93%	76.47%	60.78%	92.16%	60.78%	60.78%
	LGBM	87.79%	78.02%	63.36%	92.67%	63.36%	63.36%
T-SVD	DT	99.28%	98.71%	97.84%	99.57%	97.84%	97.84%
	AdaBoost	97.70%	95.86%	93.10%	98.62%	93.10%	93.10%
	RFC	99.14%	98.45%	97.41%	99.48%	97.41%	97.41%
	ETC	99.28%	98.71%	97.84%	99.57%	97.84%	97.84%
	HGB	98.56%	97.41%	95.69%	99.14%	95.69%	95.69%
	XGB	98.56%	97.41%	95.69%	99.14%	95.69%	95.69%
	LGBM	99.28%	98.71%	97.84%	99.57%	97.84%	97.84%
RP	DT	91.09%	83.97%	73.28%	94.66%	73.28%	73.28%
	AdaBoost	82.47%	68.45%	47.41%	89.48%	47.41%	47.41%
	RFC	97.99%	96.38%	93.97%	98.79%	93.97%	93.97%
	ETC	98.42%	97.16%	95.26%	99.05%	95.26%	95.26%
	HGB	96.41%	93.53%	89.22%	97.84%	89.22%	89.22%
	XGB	95.26%	91.47%	85.78%	97.16%	85.78%	85.78%
	LGBM	96.84%	94.31%	90.52%	98.10%	90.52%	90.52%

TABLE 6.5
Tabular Summarization of the WSM Metrics Using the "UCI-HAR" Dataset (232 × 100)

	DT	AdaBoost	RFC	ETC	HGB	XGB	LGBM	Max Value
PCA	98.35%	98.02%	98.02%	98.35%	**98.68%**	98.02%	98.35%	**98.68%**
LDA	94.79%	71.57%	95.43%	95.43%	93.20%	94.47%	94.79%	95.43%
ICA	57.20%	61.71%	74.96%	79.86%	72.13%	72.42%	74.11%	79.86%
T-SVD	98.35%	94.79%	98.02%	98.35%	96.72%	96.72%	98.35%	98.35%
RP	80.74%	63.87%	95.43%	96.40%	91.94%	89.46%	92.89%	96.40%
Max Value	98.35%	98.02%	98.02%	98.35%	**98.68%**	98.02%	98.35%	**98.68%**

TABLE 6.6
The Best Combinations Applied on the Reduced "UCI-HAR" Dataset (116 × 100) Using Each Classifier

Feature Reduction Technique	Classifier	Scaling Technique	Criterion	Splitter	Max Depth	Estimators #	Class Weight	Learning Rate
PCA	DT	Normalization	Entropy	Best	None	—	—	—
	AdaBoost	Normalization	—	—	—	300	—	1%
	RFC	Robust	Gini	—	None	300	Balanced	—
	ETC	Standardization	Gini	—	None	300	—	—
	HGB	Standardization	—	—	—	—	—	—
	XGB	Normalization	—	—	None	300	—	10%
	LGBM	Normalization	—	—	None	300	—	10%
LDA	DT	Max–absolute	Gini	Best	None	—	—	—
	AdaBoost	Standardization	—	—	—	300	—	10%
	RFC	Normalization	Entropy	—	None	300	Balanced	—
	ETC	Min–max	Gini	—	None	300	—	—
	HGB	Normalization	—	—	—	—	—	—
	XGB	Normalization	—	—	None	300	—	1%
	LGBM	Min–max	—	—	None	300	—	1%
ICA	DT	Min–max	Gini	Random	None	—	—	—
	AdaBoost	Robust	—	—	—	300	—	100%
	RFC	Min–max	Gini	—	None	300	None	—
	ETC	Standardization	Gini	—	None	300	—	—
	HGB	Normalization	—	—	—	—	—	—
	XGB	Standardization		—	None	300	—	100%
	LGBM	Standardization	—	—	None	300	—	100%

(Continued)

TABLE 6.6 (Continued)
The Best Combinations Applied on the Reduced "UCI-HAR" Dataset (116 × 100) Using Each Classifier

Feature Reduction Technique	Classifier	Scaling Technique	Criterion	Splitter	Max Depth	Estimators #	Class Weight	Learning Rate
T-SVD	DT	Min–max	Entropy	Best	None	—	—	—
	AdaBoost	Min–max	—	—	—	300	—	1%
	RFC	Standardization	Entropy	—	None	300	None	—
	ETC	Max–absolute	Gini	—	None	300	—	—
	HGB	Standardization	—	—	—	—	—	—
	XGB	Standardization	—	—	None	300	—	1%
	LGBM	Min–max	—	—	None	300	—	100%
RP	DT	Robust	Entropy	Random	None	—	—	—
	AdaBoost	Standardization	—	—	—	300	—	1%
	RFC	Robust	Gini	—	None	300	None	—
	ETC	Min–max	Entropy	—	None	300	—	—
	HGB	Normalization	—	—	—	—	—	—
	XGB	Normalization	—	—	None	300	—	10%
	LGBM	Max–absolute	—	—	None	300	—	10%

TABLE 6.7
The Corresponding Performance Metrics Using Each Classifier Applied on the Reduced "UCI-HAR" Dataset (116 × 100)

Feature Reduction Technique	Classifier	Accuracy	Balanced Accuracy	Precision	Specificity	Recall	F1
PCA	DT	99.14%	98.45%	97.41%	99.48%	97.41%	97.41%
	AdaBoost	99.14%	98.45%	97.41%	99.48%	97.41%	97.41%
	RFC	99.14%	98.45%	97.41%	99.48%	97.41%	97.41%
	ETC	99.43%	98.97%	98.28%	99.66%	98.28%	98.28%
	HGB	99.43%	98.97%	98.28%	99.66%	98.28%	98.28%
	XGB	99.14%	98.45%	97.41%	99.48%	97.41%	97.41%
	LGBM	99.14%	98.45%	97.41%	99.48%	97.41%	97.41%
LDA	DT	93.97%	89.14%	81.90%	96.38%	81.90%	81.90%
	AdaBoost	83.91%	71.03%	51.72%	90.34%	51.72%	51.72%
	RFC	96.55%	93.79%	89.66%	97.93%	89.66%	89.66%
	ETC	95.69%	92.24%	87.07%	97.41%	87.07%	87.07%
	HGB	93.10%	87.59%	79.31%	95.86%	79.31%	79.31%
	XGB	95.69%	92.24%	87.07%	97.41%	87.07%	87.07%
	LGBM	95.98%	92.76%	87.93%	97.59%	87.93%	87.93%

(Continued)

TABLE 6.7 (Continued)
The Corresponding Performance Metrics Using Each Classifier Applied on the Reduced "UCI-HAR" Dataset (116 × 100)

Feature Reduction Technique	Classifier	Accuracy	Balanced Accuracy	Precision	Specificity	Recall	F1
ICA	DT	79.60%	63.28%	38.79%	87.76%	38.79%	38.79%
	AdaBoost	75.29%	55.52%	25.86%	85.17%	25.86%	25.86%
	RFC	77.01%	58.62%	31.03%	86.21%	31.03%	31.03%
	ETC	80.17%	64.31%	40.52%	88.10%	40.52%	40.52%
	HGB	76.15%	57.07%	28.45%	85.69%	28.45%	28.45%
	XGB	75.57%	56.03%	26.72%	85.34%	26.72%	26.72%
	LGBM	76.44%	57.59%	29.31%	85.86%	29.31%	29.31%
T-SVD	DT	98.85%	97.93%	96.55%	99.31%	96.55%	96.55%
	AdaBoost	99.14%	98.45%	97.41%	99.48%	97.41%	97.41%
	RFC	98.85%	97.93%	96.55%	99.31%	96.55%	96.55%
	ETC	99.43%	98.97%	98.28%	99.66%	98.28%	98.28%
	HGB	99.14%	98.45%	97.41%	99.48%	97.41%	97.41%
	XGB	98.56%	97.41%	95.69%	99.14%	95.69%	95.69%
	LGBM	100%	100%	100%	100%	100%	100%
RP	DT	85.34%	73.62%	56.03%	91.21%	56.03%	56.03%
	AdaBoost	77.87%	60.17%	33.62%	86.72%	33.62%	33.62%
	RFC	97.41%	95.34%	92.24%	98.45%	92.24%	92.24%
	ETC	97.99%	96.38%	93.97%	98.79%	93.97%	93.97%
	HGB	93.68%	88.62%	81.03%	96.21%	81.03%	81.03%
	XGB	93.39%	88.10%	80.17%	96.03%	80.17%	80.17%
	LGBM	95.40%	91.72%	86.21%	97.24%	86.21%	86.21%

TABLE 6.8
Tabular Summarization of the WSM Metrics Using the "UCI-HAR" Dataset (116 × 100)

	AdaBoost	DT	ETC	HGB	LGBM	RFC	XGB	Max Value
PCA	98.02%	98.02%	98.02%	98.68%	98.68%	98.02%	98.02%	98.68%
LDA	86.70%	66.59%	92.26%	90.38%	84.89%	90.38%	91.01%	92.26%
ICA	58.52%	50.73%	53.81%	59.58%	52.26%	51.24%	52.78%	59.58%
T-SVD	97.37%	98.02%	97.37%	98.68%	98.02%	96.72%	**100%**	**100%**
RP	69.34%	55.37%	94.15%	95.43%	86.10%	85.49%	89.76%	95.43%
Max Value	98.02%	98.02%	98.02%	98.68%	98.68%	98.02%	**100%**	**100%**

TABLE 6.9
The Best Combinations Applied on the Two Sampled Datasets after TDA

Dataset Keyword	Classifier	Scaling Technique	Criterion	Splitter	Max Depth	Estimators #	Class Weight	Learning Rate
UCI-HAR 0% overlap	DT	Max–absolute	Entropy	Random	None	—	—	—
	AdaBoost	Robust	—	—	—	300	—	100%
	RFC	Standardization	Gini	—	None	300	None	—
	ETC	Min–max	Entropy	—	None	300	—	—
	HGB	Standardization	—	—	—	—	—	—
	XGB	Standardization	—	—	None	300	—	10%
	LGBM	Standardization	—	—	None	300	—	1%
UCI-HAR 50% overlap	DT	Standardization	Gini	Best	None	—	—	—
	AdaBoost	Standardization	—	—	—	300	—	1%
	RFC	Min–max	Entropy	—	None	300	Balanced	—
	ETC	Standardization	Gini	—	None	300	—	—
	HGB	Standardization	—	—	—	—	—	—
	XGB	Standardization	—	—	None	300	—	1%
	LGBM	Min–max	—	—	None	300	—	100%

TABLE 6.10
The Corresponding Performance Metrics Using Each Classifier Applied on the Two Sampled Datasets after TDA

Dataset Keyword	Classifier	Accuracy	Balanced Accuracy	Precision	Specificity	Recall	F1
UCI-HAR 0% overlap	DT	92.53%	86.55%	77.59%	95.52%	77.59%	77.59%
	AdaBoost	82.47%	68.45%	47.41%	89.48%	47.41%	47.41%
	RFC	94.83%	90.69%	84.48%	96.90%	84.48%	84.48%
	ETC	94.54%	90.17%	83.62%	96.72%	83.62%	83.62%
	HGB	91.38%	84.48%	74.14%	94.83%	74.14%	74.14%
	XGB	93.10%	87.59%	79.31%	95.86%	79.31%	79.31%
	LGBM	95.69%	92.24%	87.07%	97.41%	87.07%	87.07%
UCI-HAR 50% overlap	DT	94.40%	89.91%	83.19%	96.64%	83.19%	83.19%
	AdaBoost	77.16%	58.88%	31.47%	86.29%	31.47%	31.47%
	RFC	95.55%	91.98%	86.64%	97.33%	86.64%	86.64%
	ETC	96.12%	93.02%	88.36%	97.67%	88.36%	88.36%
	HGB	95.98%	92.76%	87.93%	97.59%	87.93%	87.93%
	XGB	95.55%	91.98%	86.64%	97.33%	86.64%	86.64%
	LGBM	96.70%	94.05%	90.09%	98.02%	90.09%	90.09%

TABLE 6.11
Tabular Summarization of the WSM Metrics Using TDA and the Two Sampled Datasets

Dataset Keyword	DT	AdaBoost	RFC	ETC	HGB	XGB	LGBM	Max Value
UCI-HAR 0% Overlap	83.70%	63.87%	88.53%	87.92%	81.33%	84.89%	**90.38%**	**90.38%**
UCI-HAR 50% Overlap	87.62%	54.07%	90.07%	91.32%	91.01%	90.07%	**92.57%**	**92.57%**

6.6 CONCLUSIONS AND FUTURE WORK

HAR had recently attracted a lot of attention and had been established as an effective strategy. There are numerous potential applications for it (e.g., intelligent assistance for people suffering from cognitive disorders and elderly people). This study provides a thorough analysis for identifying human activity using feature extraction, feature reduction, and machine learning (ML) algorithms. The public dataset UCI-HAR sensor-based data was used to train and evaluate multiple algorithms to identify a variety of human activities. To address the issue of unbalanced data, five oversampling strategies were used. To take use of time series data, a sampling technique with two overlapping percentages (i.e., 50% and 0%) was also applied to the balanced dataset. Five conventional methods (i.e., independent component analysis [ICA], principal component analysis [PCA], linear discriminant analysis [LDA], random projection [RP], and truncated singular value decomposition [T-SVD]) were used for feature extraction and dimensionality reduction. Additionally, topological data analysis (TDA) was applied for feature extraction. Six of the seven different types of machine learning algorithms employed are ensemble classifiers. The decision tree (DT) classifier is the simplest, while the ensemble classifiers include the light gradient boosting machine (LGBM), XGBoost (XGB), adaptive boosting (AdaBoost), histogram-based gradient boosting (HGB), random forest (RF), and extra trees (ETs). There were two different types of experiments made. Traditional feature reduction methods and machine learning algorithms are used to create the first category, while TDA feature extraction and ML algorithms are used to create the second. For these categories of experiments, grid search was used to perform the hyperparameter optimization process. For the first category of experiments, the "UCI-HAR" data was reduced to two dimensions (i.e., 232 × 100 and 116 × 100). For the reduced "UCI-HAR" data with dimensions (232 × 100), the best-reported accuracy, balanced accuracy, precision, specificity, recall, F1-score, and WSM were 99.43%, 98.97%, 98.28%, 99.66%, 98.28%, 98.28%, and 98.68%, respectively, achieved by the HGB classifier with the PCA feature reduction technique. For the reduced "UCI-HAR" data with dimensions 116 × 100, the best-reported accuracy, balanced accuracy, precision, specificity, recall, F1-score, and WSM values of 100% were achieved by

LGBM classifier with the T- SVD feature reduction technique. For the second category experiments, the best-reported accuracy, balanced accuracy, precision, specificity, recall, F1-score, and WSM by the LGBM classifier are, respectively, 95.69%, 92.24%, 87.07%, 97.41%, 87.07%, 87.07%, and 90.38% for the UCI-HAR dataset with 0% overlap. The best-reported accuracy, balanced accuracy, precision, specificity, recall, F1-score, and WSM using the LGBM classifier are 96.70%, 94.05%, 90.09%, 98.02%, 90.09%, 90.09%, and 92.57% respectively, for the UCI-HAR dataset with 50% overlap. Metaheuristic optimizers, such as the Aquila Optimizer (AO), particle swarm optimization (PSO), and sparrow search algorithm (SpaSA), can be utilized in future analyses to tune the learning models' hyperparameters. The experiments can be done with various datasets that have more features, such as heart rate. Additionally, the undersampling techniques can be tested and compared with the oversampling ones on the datasets.

REFERENCES

1. Vrigkas, M., Nikou, C., & Kakadiaris, I. A. (2015). A review of human activity recognition methods. *Frontiers in Robotics and AI*, 2, 28.
2. Jobanputra, C., Bavishi, J., & Doshi, N. (2019). Human activity recognition: A survey. *Procedia Computer Science*, 155, 698–703.
3. Arshad, M. H., Bilal, M., & Gani, A. (2022). Human activity recognition: Review, taxonomy and open challenges. *Sensors*, 22(17), 6463.
4. Abdulazeem, Y., Balaha, H. M., Bahgat, W. M., & Badawy, M. (2021). Human action recognition based on transfer learning approach. *IEEE Access*, 9, 82058–82069.
5. Balaha, H. M., Shaban, A. O., El-Gendy, E. M., & Saafan, M. M. (2022). A multi-variate heart disease optimization and recognition framework. *Neural Computing and Applications*, 1–38.
6. Baghdadi, N. A., Malki, A., Balaha, H. M., AbdulAzeem, Y., Badawy, M., & Elhosseini, M. (2022b). Classification of breast cancer using a manta-ray foraging optimized transfer learning framework. *PeerJ Computer Science*, 8, e1054.
7. Sharafeldeen, A., Elsharkawy, M., Alghamdi, N. S., Soliman, A., & El-Baz, A. (2021). Precise segmentation of COVID-19 infected lung from CT images based on adaptive first-order appearance model with morphological/anatomical constraints. *Sensors*, 21(16), 5482.
8. Elsharkawy, M., Sharafeldeen, A., Soliman, A., Khalifa, F., Widjajahakim, R., Switala, A., . . . Seddon, J. M. (2021). Automated diagnosis and grading of dry age-related macular degeneration using optical coherence tomography imaging. *Investigative Ophthalmology & Visual Science*, 62(8), 107–107.
9. Balaha, M. M., El-Kady, S., Balaha, H. M., Salama, M., Emad, E., Hassan, M., & Saafan, M. M. (2022). A vision-based deep learning approach for independent-users Arabic sign language interpretation. *Multimedia Tools and Applications*, 1–20.
10. Alghamdi, N. S., Taher, F., Kandil, H., Sharafeldeen, A., Elnakib, A., Soliman, A., . . . El-Baz, A. (2022). Segmentation of infant brain using nonnegative matrix factorization. *Applied Sciences*, 12(11), 5377.
11. Dang, L. M., Min, K., Wang, H., Piran, M. J., Lee, C. H., & Moon, H. (2020). Sensor-based and vision-based human activity recognition: A comprehensive survey. *Pattern Recognition*, 108, 107561.
12. Sharafeldeen, A., Elsharkawy, M., Shaffie, A., Khalifa, F., Soliman, A., Naglah, A., . . . Elmougy, S. (2022). *Thyroid Cancer Diagnostic System using Magnetic Resonance Imaging*. IEEE.

13. Balaha, H. M., Balaha, M. H., & Ali, H. A. (2021). Hybrid COVID-19 segmentation and recognition framework (HMB-HCF) using deep learning and genetic algorithms. *Artificial Intelligence in Medicine*, 119, 102156.
14. Ramasamy Ramamurthy, S., & Roy, N. (2018). Recent trends in machine learning for human activity recognition—a survey. *Wiley Interdisciplinary Reviews: Data Mining and Knowledge Discovery*, 8(4), e1254.
15. Balaha, H. M., Ali, H. A., Youssef, E. K., Elsayed, A. E., Samak, R. A., Abdelhaleem, M. S., . . . Abdelhameed, M. M. (2021). Recognizing arabic handwritten characters using deep learning and genetic algorithms. *Multimedia Tools and Applications*, 80(21), 32473–32509.
16. Baghdadi, N. A., Malki, A., Balaha, H. M., Badawy, M., & Elhosseini, M. (2022). A3C-TL-GTO: Alzheimer automatic accurate classification using transfer learning and artificial gorilla troops optimizer. *Sensors*, 22(11), 4250.
17. Nguyen, B., Coelho, Y., Bastos, T., & Krishnan, S. (2021). Trends in human activity recognition with focus on machine learning and power requirements. *Machine Learning with Applications*, 5, 100072.
18. Baghdadi, N. A., Malki, A., Balaha, H. M., AbdulAzeem, Y., Badawy, M., & Elhosseini, M. (2022a). An optimized deep learning approach for suicide detection through Arabic tweets. *PeerJ Computer Science*, 8, e1070.
19. Sandhu, H. S., Elmogy, M., Sharafeldeen, A. T., Elsharkawy, M., El-Adawy, N., Eltanboly, A., . . . El-Baz, A. (2020). Automated diagnosis of diabetic retinopathy using clinical biomarkers, optical coherence tomography (OCT), and OCT angiography. *American Journal of Ophthalmology*, 216, 201–206.
20. Baghdadi, N. A., Malki, A., Abdelaliem, S. F., Balaha, H. M., Badawy, M., & Elhosseini, M. (2022). An automated diagnosis and classification of COVID-19 from chest CT images using a transfer learning-based convolutional neural network. *Computers in Biology and Medicine*, 144, 105383.
21. Yousif, N. R., Balaha, H. M., Haikal, A. Y., & El-Gendy, E. M. (2022). A generic optimization and learning framework for Parkinson disease via speech and handwritten records. *Journal of Ambient Intelligence and Humanized Computing*, 1–21.
22. Elsharkawy, M., Sharafeldeen, A., Soliman, A., Khalifa, F., Ghazal, M., El-Daydamony, E., . . . El-Baz, A. (2022a). A novel computer-aided diagnostic system for early detection of diabetic retinopathy using 3D-OCT higher-order spatial appearance model. *Diagnostics*, 12(2), 461.
23. Fahmy, D., Kandil, H., Khelifi, A., Yaghi, M., Ghazal, M., Sharafeldeen, A., . . . El-Baz, A. (2022). How AI can help in the diagnostic dilemma of pulmonary nodules. *Cancers*, 14(7), 1840.
24. Balaha, H. M., Saif, M., Tamer, A., & Abdelhay, E. H. (2022). Hybrid deep learning and genetic algorithms approach (HMB-DLGAHA) for the early ultrasound diagnoses of breast cancer. *Neural Computing and Applications*, 34(11), 8671–8695.
25. Balaha, H. M., Ali, H. A., Saraya, M., & Badawy, M. (2021). A new Arabic handwritten character recognition deep learning system (AHCR-DLS). *Neural Computing and Applications*, 33(11), 6325–6367.
26. Haggag, S., Elnakib, A., Sharafeldeen, A., Elsharkawy, M., Khalifa, F., Farag, R. K., . . . Sewelam, A. (2022). A computer-aided diagnostic system for diabetic retinopathy based on local and global extracted features. *Applied Sciences*, 12(16), 8326.
27. Elgafi, M., Sharafeldeen, A., Elnakib, A., Elgarayhi, A., Alghamdi, N. S., Sallah, M., & El-Baz, A. (2022). Detection of diabetic retinopathy using extracted 3D features from OCT images. *Sensors*, 22(20), 7833.
28. Sharafeldeen, A., Elsharkawy, M., Khaled, R., Shaffie, A., Khalifa, F., Soliman, A., . . . Naglah, A. (2022). Texture and shape analysis of diffusion-weighted imaging for thyroid nodules classification using machine learning. *Medical Physics*, 49(2), 988–999.

29. El-Baz, A. S., Shalaby, A., Elsharkawy, M., Sharafeldeen, A., Soliman, A., Mahmoud, A., . . . Giridharan, G. A. (2022). *Assessment of Pulmonary Function in Coronavirus Patients*. US Patent App. 17/685,493.
30. Balaha, H. M., & Saafan, M. M. (2021). Automatic exam correction framework (AECF) for the MCQS, essays, and equations matching. *IEEE Access*, 9, 32368–32389.
31. Sharaby, I., Alksas, A., Nashat, A., Balaha, H. M., Shehata, M., Gayhart, M., . . . El-Baz, A. (2023). Prediction of Wilms' tumor susceptibility to preoperative chemotherapy using a novel computer-aided prediction system. *Diagnostics*, 13(3), 486.
32. Ghazal, S., Khan, U. S., Mubasher Saleem, M., Rashid, N., & Iqbal, J. (2019). Human activity recognition using 2D skeleton data and supervised machine learning. *IET Image Processing*, 13(13), 2572–2578.
33. Nasir, I. M., Raza, M., Shah, J. H., Khan, M. A., & Rehman, A. (2021, April). Human action recognition using machine learning in uncontrolled environment. In *2021 1st International Conference on Artificial Intelligence and Data Analytics (CAIDA)* (pp. 182–187). IEEE.
34. Zhu, G., Zhang, L., Shen, P., & Song, J. (2016). An online continuous human action recognition algorithm based on the kinect sensor. *Sensors*, 16(2), 161.
35. Manzi, A., Dario, P., & Cavallo, F. (2017). A human activity recognition system based on dynamic clustering of skeleton data. *Sensors*, 17(5), 1100.
36. Guo, Y., Chu, Y., Jiao, B., Cheng, J., Yu, Z., Cui, N., & Ma, L. (2021). Evolutionary dual-ensemble class imbalance learning for human activity recognition. *IEEE Transactions on Emerging Topics in Computational Intelligence*, 6(4).
37. Subasi, A., Khateeb, K., Brahimi, T., & Sarirete, A. (2020). Human activity recognition using machine learning methods in a smart healthcare environment. In *Innovation in Health Informatics* (pp. 123–144). Academic Press.
38. Aslan, M. F., Durdu, A., & Sabanci, K. (2020). Human action recognition with bag of visual words using different machine learning methods and hyperparameter optimization. *Neural Computing and Applications*, 32(12), 8585–8597.
39. Ahmed, N., Rafiq, J. I., & Islam, M. R. (2020). Enhanced human activity recognition based on smartphone sensor data using hybrid feature selection model. *Sensors*, 20(1), 317.
40. Martinez, J., Black, M. J., & Romero, J. (2017). On human motion prediction using recurrent neural networks. In *Proceedings of the IEEE Conference on Computer Vision and Pattern Recognition* (pp. 2891–2900). IEEE.
41. Ferrari, A., Micucci, D., Mobilio, M., & Napoletano, P. (2020). On the personalization of classification models for human activity recognition. *IEEE Access*, 8, 32066–32079.
42. Chen, K., Zhang, D., Yao, L., Guo, B., Yu, Z., & Liu, Y. (2021). Deep learning for sensor-based human activity recognition: Overview, challenges, and opportunities. *ACM Computing Surveys (CSUR)*, 54(4), 1–40.
43. Mekruksavanich, S., & Jitpattanakul, A. (2021). Biometric user identification based on human activity recognition using wearable sensors: An experiment using deep learning models. *Electronics*, 10(3), 308.
44. Chen, L., Liu, X., Peng, L., & Wu, M. (2021). Deep learning based multimodal complex human activity recognition using wearable devices. *Applied Intelligence*, 51(6), 4029–4042.
45. Khan, Z. N., & Ahmad, J. (2021). Attention induced multi-head convolutional neural network for human activity recognition. *Applied Soft Computing*, 110, 107671.
46. Khan, I. U., Afzal, S., & Lee, J. W. (2022). Human activity recognition via hybrid deep learning based model. *Sensors*, 22(1), 323.
47. Li, Y., & Wang, L. (2022). Human activity recognition based on residual network and BiLSTM. *Sensors*, 22(2), 635.

48. Fan, Y. C., Tseng, Y. H., & Wen, C. Y. (2022). A novel deep neural network method for HAR-based team training using body-worn inertial sensors. *Sensors*, 22(21), 8507.
49. Kim, E., Helal, S., & Cook, D. (2009). Human activity recognition and pattern discovery. *IEEE Pervasive Computing*, 9(1), 48–53.
50. Ann, O. C., & Theng, L. B. (2014, November). Human activity recognition: A review. In *2014 IEEE International Conference on Control System, Computing and Engineering (ICCSCE 2014)* (pp. 389–393). IEEE.
51. Bahgat, W. M., Balaha, H. M., AbdulAzeem, Y., & Badawy, M. M. (2021). An optimized transfer learning-based approach for automatic diagnosis of COVID-19 from chest x-ray images. *PeerJ Computer Science*, 7, e555.
52. Shelke, M. S., Deshmukh, P. R., & Shandilya, V. K. (2017). A review on imbalanced data handling using undersampling and oversampling technique. *International Journal of Engineering Research*, 3(4), 444–449.
53. Zheng, Z., Cai, Y., & Li, Y. (2015). Oversampling method for imbalanced classification. *Computing and Informatics*, 34(5), 1017–1037.
54. Balaha, H. M., & Hassan, A. E.-S. (2022a). A variate brain tumor segmentation, optimization, and recognition framework. *Artificial Intelligence Review*, 1–54.
55. Balaha, H. M., Ali, H. A., & Badawy, M. (2021). Automatic recognition of handwritten Arabic characters: A comprehensive review. *Neural Computing and Applications*, 33(7), 3011–3034.
56. Amin, A., Anwar, S., Adnan, A., Nawaz, M., Howard, N., Qadir, J., . . . Hussain, A. (2016). Comparing oversampling techniques to handle the class imbalance problem: A customer churn prediction case study. *IEEE Access*, 4, 7940–7957.
57. Mohammed, R., Rawashdeh, J., & Abdullah, M. (2020, April). Machine learning with oversampling and undersampling techniques: Overview study and experimental results. In *2020 11th International Conference on Information and Communication Systems (ICICS)* (pp. 243–248). IEEE.
58. Abdel Razek, A. A. K., Alksas, A., Shehata, M., AbdelKhalek, A., Abdel Baky, K., El-Baz, A., & Helmy, E. (2021). Clinical applications of artificial intelligence and radiomics in neuro-oncology imaging. *Insights into Imaging*, 12(1), 1–17.
59. Elsharkawy, M., Sharafeldeen, A., Taher, F., Shalaby, A., Soliman, A., Mahmoud, A., . . . Razek, A. A. K. A. (2021). Early assessment of lung function in coronavirus patients using invariant markers from chest X-rays images. *Scientific Reports*, 11(1), 1–11.
60. Elgendy, M., Balaha, H. M., Shehata, M., Alksas, A., Ghoneim, M., Sherif, F., . . . Sallah, M. (2022). Role of imaging and AI in the evaluation of COVID-19 infection: A comprehensive survey. *Frontiers in Bioscience (Landmark Edition)*, 27(9), 276.
61. Alksas, A., Shehata, M., Atef, H., Sherif, F., Yaghi, M., Alhalabi, M., . . . El-Baz, A. (2022, August). A comprehensive non-invasive system for early grading of gliomas. In *2022 26th International Conference on Pattern Recognition (ICPR)* (pp. 4371–4377). IEEE.
62. Balaha, H. M., El-Gendy, E. M., & Saafan, M. M. (2022). A complete framework for accurate recognition and prognosis of COVID-19 patients based on deep transfer learning and feature classification approach. *Artificial Intelligence Review*, 1–46.
63. Shalata, A. T., Shehata, M., Van Bogaert, E., Ali, K. M., Alksas, A., Mahmoud, A., . . . El-Baz, A. (2022). Predicting recurrence of non-muscle-invasive bladder cancer: Current techniques and future trends. *Cancers*, 14(20), 5019.
64. Baghdadi, N. A., Alsayed, S. K., Malki, G. A., Balaha, H. M., & Farghaly Abdelaliem, S. M. (2023). An analysis of burnout among female nurse educators in Saudi Arabia using K-means clustering. *European Journal of Investigation in Health, Psychology and Education*, 13(1), 33–53.
65. Sumithra, V., & Surendran, S. (2015). A review of various linear and non linear dimensionality reduction techniques. *International Journal of Computer Science and Information Technologies*, 6(3), 2354–2360.

66. Jia, W., Sun, M., Lian, J., & Hou, S. (2022). Feature dimensionality reduction: A review. *Complex & Intelligent Systems*, 1–31.
67. Khalid, S., Khalil, T., & Nasreen, S. (2014, August). A survey of feature selection and feature extraction techniques in machine learning. In *2014 Science and Information Conference* (pp. 372–378). IEEE.
68. Tharwat, A. (2016). Principal component analysis-a tutorial. *International Journal of Applied Pattern Recognition*, 3(3), 197–240.
69. Fukunaga, K. (2013). *Introduction to Statistical Pattern Recognition*. Elsevier.
70. Tang, B., Shepherd, M., Milios, E., & Heywood, M. I. (2005, April). Comparing and combining dimension reduction techniques for efficient text clustering. *Proceeding of SIAM International Workshop on Feature Selection for Data Mining*, 17–26.
71. Dasgupta, S. (2000). Experiments with random projection. *Uncertainty in Artificial Intelligence: Proceedings of the Sixteenth Conference (UAI-2000)* (pp. 143–151). UAI.
72. Halko, N., Martinsson, P. G., & Tropp, J. A. (2011). Finding structure with randomness: Probabilistic algorithms for constructing approximate matrix decompositions. *SIAM Review*, 53(2), 217–288.
73. Zhang, J., & Liu, B. (2019). A review on the recent developments of sequence-based protein feature extraction methods. *Current Bioinformatics*, 14(3), 190–199.
74. Latif, A., Rasheed, A., Sajid, U., Ahmed, J., Ali, N., Ratyal, N. I., . . . Khalil, T. (2019). Content-based image retrieval and feature extraction: A comprehensive review. *Mathematical Problems in Engineering*. 2019(9658350), 1–21.
75. Balaha, H. M., El-Gendy, E. M., & Saafan, M. M. (2021). CovH2SD: A COVID-19 detection approach based on Harris Hawks optimization and stacked deep learning. *Expert Systems with Applications*, 186, 115805.
76. Tang, J., Alelyani, S., & Liu, H. (2014). Feature selection for classification: A review. *Data Classification: Algorithms and Applications*, 37.
77. Wasserman, L. (2018). Topological data analysis. *Annual Review of Statistics and Its Application*, 5, 501–532.
78. Zomorodian, A. (2012). Topological data analysis. *Advances in Applied and Computational Topology*, 70, 1–39.
79. Bubenik, P. (2015). Statistical topological data analysis using persistence landscapes. *Journal of Machine Learning Research*, 16(1), 77–102.
80. Munch, E. (2017). A user's guide to topological data analysis. *Journal of Learning Analytics*, 4(2), 47–61.
81. Epstein, C., Carlsson, G., & Edelsbrunner, H. (2011). Topological data analysis. *Inverse Problems*, 27(12), 120201.
82. Balaha, H. M., & Hassan, A. E.-S. (2022b). Skin cancer diagnosis based on deep transfer learning and sparrow search algorithm. *Neural Computing and Applications*, 1–39.
83. Alksas, A., Shehata, M., Saleh, G. A., Shaffie, A., Soliman, A., Ghazal, M., . . . El-Baz, A. (2021). A novel computer-aided diagnostic system for accurate detection and grading of liver tumors. *Scientific Reports*, 11(1), 1–18.
84. Ayyad, S. M., Badawy, M. A., Shehata, M., Alksas, A., Mahmoud, A., Abou El-Ghar, M., . . . El-Baz, A. (2022). A new framework for precise identification of prostatic adenocarcinoma. *Sensors*, 22(5), 1848.
85. Sharafeldeen, A., Elsharkawy, M., Khalifa, F., Soliman, A., Ghazal, M., AlHalabi, M., . . . Sandhu, H. S. (2021). Precise higher-order reflectivity and morphology models for early diagnosis of diabetic retinopathy using OCT images. *Scientific Reports*, 11(1), 1–16.
86. Balaha, M. H., El-Ibiary, M. T., El-Dorf, A. A., El-Shewaikh, S. L., & Balaha, H. M. (2022). Construction and writing flaws of the multiple-choice questions in the published test banks of obstetrics and gynecology: Adoption, caution, or mitigation? *Avicenna Journal of Medicine*, 12(3), 138–147.

87. Shehata, M., Alksas, A., Abouelkheir, R. T., Elmahdy, A., Shaffie, A., Soliman, A., . . . El-Baz, A. (2021, April). A new computer-aided diagnostic (CAD) system for precise identification of renal tumors. In *2021 IEEE 18th International Symposium on Biomedical Imaging (ISBI)* (pp. 1378–1381). IEEE.
88. Alksas, A., Shehata, M., Saleh, G. A., Shaffie, A., Soliman, A., Ghazal, M., . . . El-Baz, A. (2021, January). A novel computer-aided diagnostic system for early assessment of hepatocellular carcinoma. In *2020 25th International Conference on Pattern Recognition (ICPR)* (pp. 10375–10382). IEEE.
89. Farahat, I. S., Sharafeldeen, A., Elsharkawy, M., Soliman, A., Mahmoud, A., Ghazal, M., . . . Aladrousy, W. (2022). The role of 3D CT imaging in the accurate diagnosis of lung function in coronavirus patients. *Diagnostics*, 12(3), 696.
90. Shehata, M., Alksas, A., Abouelkheir, R. T., Elmahdy, A., Shaffie, A., Soliman, A., . . . El-Baz, A. (2021). A comprehensive computer-assisted diagnosis system for early assessment of renal cancer tumors. *Sensors*, 21(14), 4928.
91. Alksas, A., Shehata, M., Atef, H., Sherif, F., Alghamdi, N. S., Ghazal, M., . . . El-Baz, A. (2022). A novel system for precise grading of glioma. *Bioengineering*, 9(10), 532.
92. Elsharkawy, M., Sharafeldeen, A., Soliman, A., Khalifa, F., Ghazal, M., El-Daydamony, E., . . . El-Baz, A. (2022b). *Diabetic Retinopathy Diagnostic CAD System Using 3D-Oct Higher Order Spatial Appearance Model*. IEEE.
93. Fahmy, D., Alksas, A., Elnakib, A., Mahmoud, A., Kandil, H., Khalil, A., . . . El-Baz, A. (2022). The role of radiomics and AI technologies in the segmentation, detection, and management of hepatocellular carcinoma. *Cancers*, 14(24), 6123.
94. Elsharkawy, M., Elrazzaz, M., Sharafeldeen, A., Alhalabi, M., Khalifa, F., Soliman, A., . . . El-Daydamony, E. (2022). The role of different retinal imaging modalities in predicting progression of diabetic retinopathy: A survey. *Sensors*, 22(9), 3490.
95. Batouty, N. M., Saleh, G. A., Sharafeldeen, A., Kandil, H., Mahmoud, A., Shalaby, A., . . . El-Baz, A. (2022). State of the art: Lung cancer staging using updated imaging modalities. *Bioengineering*, 9(10), 493.

7 Application of Texture Analysis in Retinal OCT Imaging

Mukhit Kulmaganbetov and James E. Morgan

7.1 INTRODUCTION

Among the ophthalmic imaging devices, the role of optical coherence tomography (OCT) is progressively increasing due to its unique features and many undiscovered potentials. OCT is a cross-sectional non-invasive imaging technique for biomedical and clinical purposes (Schuman et al. 1995; Drexler and Fujimoto 2008). The benefits of using OCT are the fast acquisition time (seconds and femtoseconds), diagnostic precision, ultrahigh axial resolution, and penetrance (Fujimoto et al. 1995; Fercher 1996; Drexler 2010; Tudor et al. 2014; Morgan et al. 2017).

Recent advancements in the hardware and software development of OCT show a promising future for the further influence of the device in the diagnosis and management of ophthalmic conditions (Drexler 2010; Everett et al. 2021). One of the actively investigating fields in this regard is the application of textural feature analysis of the retinal and choroidal OCT scans (Breher et al. 2020; Tazarjani et al. 2021; Kulmaganbetov et al. 2022).

7.2 OPTICAL COHERENCE TOMOGRAPHY AND RETINA

Due to its ability for detailed visualization of the anterior and posterior segments of a human eye, OCT has been extensively used in the assessment of the morphology and physiology of the eye fundus *in vivo* in the last three decades (Huang et al. 1991; Schuman et al. 1995; Tudor et al. 2014). Among the imaging methods in the arsenal of eye specialists, OCT is the state-of-the-art method for high-resolution imaging techniques (Drexler et al. 2001; Baghaie et al. 2015). Although the widespread clinical use of OCT began at the end of the previous century, the history of technical upgrades is dynamic (Drexler et al. 2001; Drexler and Fujimoto 2008).

7.2.1 OCT Principle and Resolution

As seen in Figure 7.1, the working principle of OCT is based on the beam splitter in the Michelson interferometer and acquires the scans by measuring the optical reflections from backscattered light beams (Huang et al. 1991; Drexler et al. 2001). In this regard, the main idea of OCT operation is similar to the performance of the other medical imaging technology—ultrasonography (Culjat et al. 2010; Marchini

DOI: 10.1201/9780367486082-7

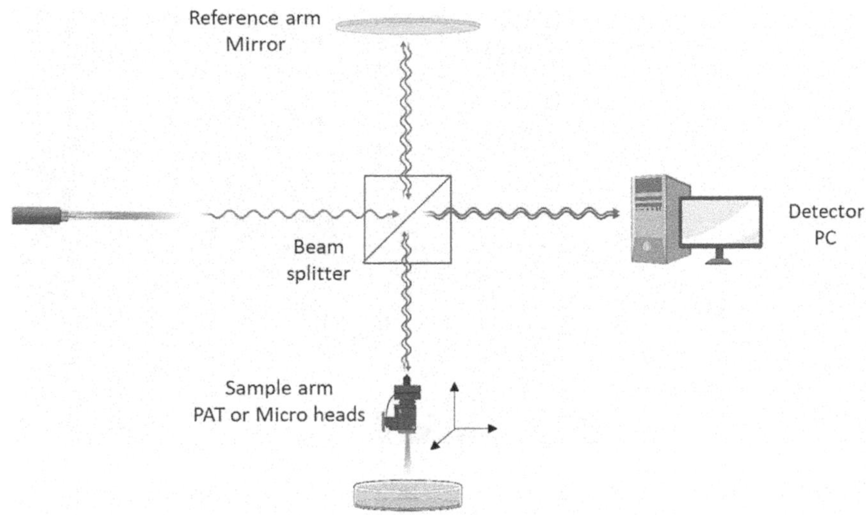

FIGURE 7.1 Principle of optical coherence tomography.

2015). While ultrasound measures sound waves back-reflected from objects of various densities to produce images of the tissues (Chen et al. 2009; Culjat et al. 2010; Gazzard 2015), OCT directly obtains time-encoded signals and determines the quantity of backscattered light (Krauss and Puliafito 1995; Drexler and Fujimoto 2008). This is the working procedure of the first clinical OCTs which analyzed data in the time domain (time domain, TD-OCT) (Pan et al. 1995; Drexler and Fujimoto 2008), which uses a single photodetector and determines the time delay of echo backscattered waves from the sample (Fujimoto et al. 1995).

By contrast, instead of using multiple reference beams from a moving mirror, spectral/Fourier domain OCT (SD-OCT/FD-OCT) can acquire images by multiple detectors (Choma et al. 2003; Drexler and Fujimoto 2008). Therefore, the scanning speed is considerably increased to 27,000 A-scans per second (Gabriele et al. 2011), reaching 100,000 A-scans per second in laboratory-based SD-OCT, which is 200 times faster than TD-OCT (de Amorim Garcia Filho et al. 2013). Furthermore, another advantage of SD-OCT over TD-OCT is sensitivity, which is 20–30 dB higher in the spectral domain device (Choma et al. 2003).

Laser light from a solid-state or superluminescent diode (SLD) source travels to the interferometer and is divided into two high-bandwidth light beams: one directed at the object (sample tissue), with the other going to the reference arm of the OCT (Drexler and Fujimoto 2008). The backscattered light beams from both OCT arms are then combined and registered as an axial A-scan (de Amorim Garcia Filho et al. 2013). Light travelled along with the object is collected as A-scans for further reconstruction to B-scan (Figure 7.2A) (Swanson et al. 1993). Figure 7.2B depicts the fundus photography of the same patient, and the green arrow shows the projection of the B scan along the fovea. OCT produces two- and three-dimensional images with

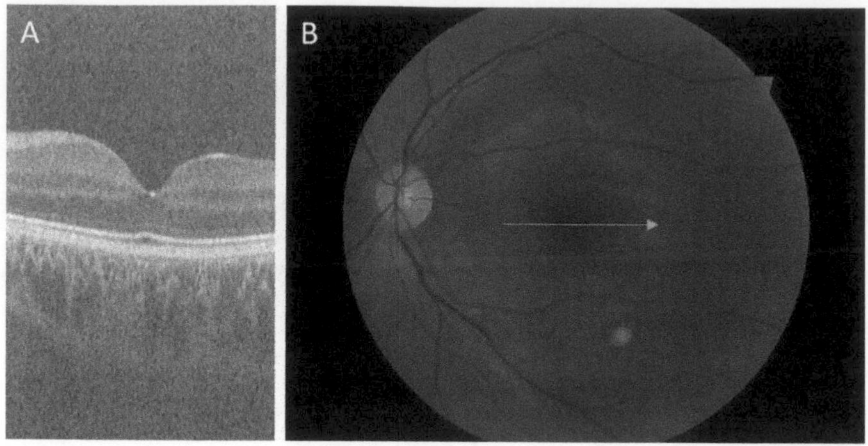

FIGURE 7.2 OCT B-scan through the macula of a healthy retina (a) with the corresponding fundus photo of the same patient's retina (b).

its averaging after a series of scans is accomplished by using the optical sectioning capability of OCT (Huang et al. 1991; Hee et al. 1995).

OCT can provide images with an axial resolution in the range of 1–15 µm (Drexler and Fujimoto 2008), and the coherence length of the light source in OCT determines the spatial resolution of the device (Morgner et al. 2000). Current commercially available OCT devices use a light source centered at a wavelength of approximately 840 nm, and their axial resolution is about 5 µm (de Amorim Garcia Filho et al. 2013). At the same time, laboratory-based OCT with broadband light sources has an axial resolution from 10 µm to 2 µm (Gabriele et al. 2011). Ultrahigh-resolution OCT allows imaging of the biological tissue structure with the axial resolution of 2–3 µm (Drexler et al. 2001). One of the limitations for the lateral resolution of OCT devices (about 20 µm) is the diffraction from the pupil (de Amorim Garcia Filho et al. 2013).

7.2.2 Cell Scattering Contribution

Scatterers of OCT light from the back-reflected retina can be multiple tissue components, including cell organelles, proteins, lipids, and extracellular matrix constituents (Mourant et al. 1998; Wang and Tuchin 2013). Generally, the scattering may depend on the morphological and optical properties of the sample (Ejofodomi 2014; Sun et al. 2020). Light distribution and the spatial variation of the refractive index of the materials influence the scattering and absorption coefficients (Beuthan et al. 1996). While the refractive index of the whole cell equals 1.38, a significant contribution to light scattering belongs to the membranous organelles, the index of refraction of which amounts to 1.48 (Beuthan et al. 1996). For instance, the regions with the mitochondria provide 90% of scattering reflection (Gourley et al. 2005). The membrane of the cell and organelles contains phospholipid polymers, which are the sources of scattering (Beuthan et al. 1996; Mourant et al. 2000; van der Meer et al. 2010).

Besides, according to the Gladstone–Dale equation, the cellular refractive index n is proportional to the local macromolecular mass density (7.1) (Barer and Tkaczyk 1954; Davies et al. 1954; Yi et al. 2016).

$$n = n_o + \rho\alpha, \tag{7.1}$$

Where n_o is the refractive index of water, ρ is the local mass density of macromolecules (g/ml), and α is the refractive index increment (ml/g). For biological materials, this value is approximately 0.17 ml/g (Cherkezyan et al. 2012). Due to the discordant distribution of mass density, the refractive index varies, and it leads to detectable backscattered light due to elastic scattering (Beuthan et al. 1996).

7.2.3 Retinal OCT

The retina comprises a complex multilayered network of nerve cells with a combined thickness of 0.4 mm (Malhotra et al. 2011; Garhart and Lakshminarayanan 2016). Retinal neurons are the sensory part of the visual system that perceives the light and color signals from the external world (Smerdon 2000; Westwood 2009). There are three cell layers in the retina, represented by retinal ganglion cells (RGCs), bipolar cells, and photoreceptors—rods and cones (Sung and Chuang 2010; Tian et al. 2017). The fovea is characterized by a high density of cones and the absence of rods (Schultze 1866). The plexiform layers of the retina consist of axons and dendrites of the corresponding retinal neurons and also amacrine and horizontal cells called interneurons (Boycott et al. 1975; Pascale et al. 2012; Purnyn 2013).

The retinal ganglion cell complex (GCC) extends from the internal limiting membrane (ILM) to the inner nuclear layer (INL) (Kim et al. 2011). The inner plexiform layer (IPL) separates the INL from the retinal ganglion cells layer (GCL) and consists of a tangle of intricately branched and intertwining neuronal processes (Quigley et al. 1989; Tan et al. 2009). RGCs are located in the inner layers of the retina, the thickness of which decreases noticeably toward the periphery. The retinal nerve fiber layer (RNFL) consists of ganglion cell axons that exit the eye to form the optic nerve head (ONH) (Lee et al. 2015).

The sizes of RGC somas vary with retinal eccentricity. Moreover, the most prominent specialization of retinal topography and the majority of ganglion cells are located in the foveal area (Stone and Johnston 1981; Dowling 1987; Wässle and Boycott 1991). To the periphery, on the contrary, the number of RGCs drops sharply. This reflects the fact that in fovea there is an almost one-to-one correspondence between photoreceptors and ganglion cells; on the periphery, there is a strong convergence of photoreceptors to bipolar cells and bipolar cells to one ganglion cell (Perry and Cowey 1985; Curcio et al. 1990). Also, due to the projection of signals to the various layers of the lateral geniculate nucleus (LGN), there are two types of RGCs: parasol cells projects to magnocellular and midget cells—to parvocellular layers of LGN.

Axons of RGCs remain unmyelinated in the retina because of the need to maintain maximum transparency (Yang et al. 2013). In the ONH, they form nerve fascicles that pass through the lamina cribrosa (LC) (Morgan 2004). As they emerge from

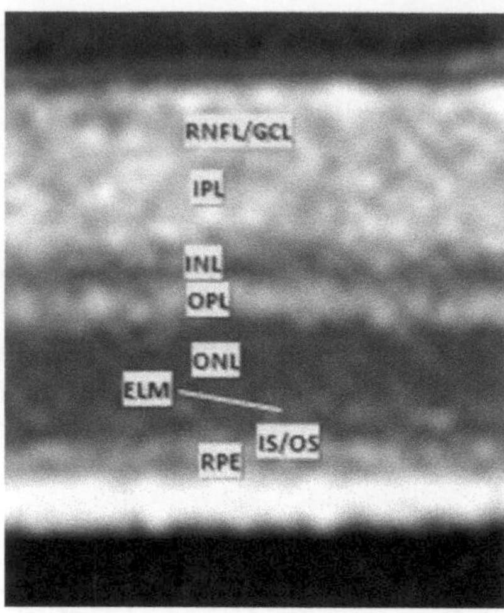

FIGURE 7.3 OCT image of the murine retina. RNFL, retinal nerve fibre layer; GCL, ganglion cell layer; IPL, inner plexiform layer; INL, inner nuclear layer; OPL, outer plexiform layer; ONL, outer nuclear layer; ELM, external limiting membrane; IS/OS, inner and outer segment junction of the photoreceptors; RPE, retinal pigment epithelium.

the LC, these nerve fibers acquire a myelin sheath, which dramatically increases the diameter of the post-laminar optic nerve (Elkington et al. 1990).

LC is a sieve-like structure at the level of the sclera of the eye. In normal eyes, the LC is partly obscured by the axons of the retinal ganglion cells (RGCs), and mostly a small portion of LC is visible using OCT (Morgan-Davies et al. 2004; Chauhan et al. 2013). Due to the pressure differences between the eye and the retrobulbar space, the LC provides structural and functional support to the axons of ganglion cells (Radius 1981; Zeimer and Ogura 1989). During glaucoma, the LC is the main region of injury due to elevated IOP and tissue's physical parameters (Quigley et al. 1989; Quigley 1993; Tian et al. 2017).

Bruch's membrane is the basement of the retinal pigment epithelium (RPE) and connects with the choroid (Benedicto et al. 2017). RPE consists of hexagonal monolayer cells with the apical surface facing the neuronal retinal (Sparrow et al. 2010; Vienola et al. 2020).

The layered structure of the retina is visible in OCT images (Figure 7.3): RNFL and GCL are reflective and are seen as bright colors on a false-color scale, whereas INL and ONL appear hypo-reflective, while IPL and OPL with axons and dendrites of retinal neurons are hyper-reflective (Drexler et al. 2001). The contrast in the reflectivity of biological structures affects the resolution of the retina, and it can be used for the differentiation of the normal and pathologic states of the organ (Toth et al. 1997).

7.2.4 Pathogenesis of Retinal Neurodegeneration

Apoptosis of retinal cells induces the cascade of irreversible functional and morphological alterations in the early etiopathogenesis of neurodegenerative diseases (Adler et al. 1999; Nickells 1999; Remé et al. 2000; Dunaief et al. 2002; Guo et al. 2005; Lo et al. 2011; Almasieh et al. 2012; Ahmad 2017; Bell et al. 2020). The role of membranous organelles, especially mitochondria, is high in cell death, and they generate the optical signal of OCT (Tso et al. 1994; Okamoto and Shaw 2005; Perfettini et al. 2005; Youle and Karbowski 2005; Zhang et al. 2019). Hence, these scatterers may help detect retinal degeneration in pre-clinical stages of retinal neurodegeneration (glaucoma, AD, and AMD).

There are several types of identification of mitochondria-rich regions in the cells. These include a thresholding approach and a texture-based segmentation (Pasternack et al. 2011; Mohammad et al. 2018). Threshold-based technology was used in the study of Pasternack et al., and they concluded that the mitochondria are the likeliest source of the changes in optical scatter in the first three hours of programmed cell death (Pasternack et al. 2010). The reason for these changes might be the structural and functional alterations in mitochondria.

Early diagnosis of RGC apoptosis at the level of subcellular structures may help stop glaucoma and potentially Alzheimer's disease (AD). Research on this subject is many, and many procedures may stop the degeneration of RGC by inhibition (Frank et al. 2001; Shim et al. 2016) or upregulation (Gupta et al. 2014; Binley et al. 2016) of certain genes.

For the purpose of detection of the earliest changes in RGC apoptosis, there were studies of in vivo detection of apoptosis by positron emission tomography (PET) with marker 18F-labelled NST-732 (ApoTrace®) (Aloya et al. 2006), by confocal scanning laser ophthalmoscopy (cSLO) using fluorescent annexin A5 (Ahmad 2017; Yap et al. 2018) and by in vivo confocal neuroimaging (ICON), that allows monitoring the morphology and function of RGC (Prilloff et al. 2010). Also, programmed cell death was diagnosed using the Detection of Apoptotic Retinal Cells (DARC) method, in which fluorescently labelled protein Annexin-A5 binds to the apoptotic membrane phospholipid phosphatidylserine (Cordeiro 2007; Galvao et al. 2013; Yap et al. 2018). However, this technique is unable to detect the number of remaining RGCs. Hence, the proportion of dying cells in healthy compared with glaucomatous retina is unclear. Moreover, DARC requires external ligands and intravenous injections at every patient visit, which may be impractical and unfavorable.

Labelling of externalized phosphatidylserine in apoptosis of photoreceptors was conducted in the study of (Mazzoni et al. 2019). Here, they used Bis(zinc(II)-dipicolylamine (Zn-DPA) with Texas-red (PSVue-550) to label the dying rods and cones. Also, the contribution of RPE apoptosis in the pathogenesis of AMD is significant (Wang et al. 2012), which was characterized by in vivo morphometry and multispectral autofluorescence (Granger et al. 2018).

All the named studies and other approaches to *in vivo* observation of retinal cell apoptosis, however, are dependent on using external ligands, which may be toxic or cause inflammatory reactions. Consequently, there is a great demand for diagnostic methods that can detect and quantify retinal neuron apoptosis on a subcellular level without exogenous ligands to label cells. For visualization in the clinical conditions of the RGC, photoreceptors, and RPE cell degeneration, it is necessary to use ultra-high-resolution OCT imaging. However, OCT is used mainly for the assessment of the morphology of the tissue, its thickness values, and the presence of signs and symptoms of diseases: edema, neovascularization, drusen, tissue detachments, and others (Toth et al. 1997; Unterhuber et al. 2005; Tan et al. 2009; Zweifel et al. 2010; Leuschen et al. 2013).

7.3 OCT TEXTURE

The texture is a measure of regional variations in brightness, intensity, and surface roughness of a small location of an image (Oberholzer et al. 1996; Gossage et al. 2003). The texture of an image demonstrates the spatial variation of pixel intensities (Nailon 2010), and each region in an image may have a constant unique texture. The former is true if a set of local statistics and other structural properties are constant, slightly varying, or approximately periodic (Chi Hau Chen et al. 1993).

Texture analysis (TA) is the characterization of image properties by textural features (Evennett et al. 1993). The aim of this analysis is the recognition of homogeneous regions of an image using textural properties (Davies 2005). This method is frequently used in classification (Bhattacharjee et al. 2011; Anantrasirichai et al. 2013), segmentation (Kajić et al. 2010; González-López et al. 2015), and synthesis (Costantini et al. 2008) of OCT images.

Techniques of pixel variation analysis can be divided into three groups:

- Statistical approaches
- Spectral technologies
- Structural technologies

Statistical technologies for texture analysis are based on histograms of image regions and their moments (Ramola et al. 2020). Examples of these features are coarseness and contrast. The goal of spectral techniques is to detect texture periodicity, orientation, etc. It is based on the power spectrum of a region (Humeau-Heurtier 2019). Pattern primitives and placement rules to describe the texture are used in structural technologies (Gossage et al. 2003).

While in the first-order statistical texture analysis the frequency of a particular gray level at a random image position is measured, co-occurrences between neighboring pixels are calculated in the second-order statistics. Therefore, expanding the number of variables requires higher-order statistical texture analysis (Nailon 2010).

7.3.1 Image Processing

Before the texture analysis, OCT scans undergo a series of procedures, including image registration for the monitoring of the tissue condition over time, noise filtering, or removal. OCT noise management is a debating and simultaneously vastly growing area of research (Baghaie et al. 2015; Qiu et al. 2020; Hu et al. 2021). This is occurring mainly due to the visualization of the microscopic optical tissue parameters, which can be degraded by the instability of fixation of a patient and eye movements. Apart from the laser source features (such as the coherence of the beam, its wavelength, and full width at half-maximum), the OCT system depends on the aperture of the detector, forward, and backward scatters within the coherence length of the beam. Image processing procedures were described in the previous chapters of this book.

7.3.2 Textural Features

7.3.2.1 Gray Level Co-Occurrence Matrix

The gray level co-occurrence matrix (GLCM) is the spatial histogram of the image that shows the relationship between each intensity tone caused by alterations between gray levels i and j. These levels are located at a particular distance d and displaced at a particular angle θ. Using these values, the distributions of image grayscale values probability density functions $s\theta(i, j \mid d, \theta)$ can be quantified (Rogowska et al. 2003; Kulmaganbetov et al. 2020), which can be used for the calculation of the GLCM textural features, including energy or angular second moment, inertia (or contrast), entropy, inverse difference moment, and correlation. The regularity of local grayscale distribution is measured by angular second moment (ASM) by summing the square of each value in the matrices of GLCM (7.2). Inertia (7.3) provides higher weights to $s\theta(i, j \mid d, \theta)$ values that represent regions of high contrast (Gossage et al. 2006). The sum of the multiplication of each $s\theta(i, j \mid d, \theta)$ value by the log of the $s\theta(i, j \mid d, \theta)$ gives us entropy (7.4), where the randomness is evaluated. The inverse difference moment parameter measures the local minimal changes (7.5), and correlation calculates the joint probability of occurrence (7.6). The latter is higher in regions with uniform grayscale values.

$$ASM = \sum_{i=0}^{L-1} \sum_{j=0}^{L-1} [S\theta(i, j|d)]^2 \tag{7.2}$$

$$Inertia = \sum_{i=0}^{L-1} \sum_{j=0}^{L-1} (i-j)^2 \, S\theta(i, j|d) \tag{7.3}$$

$$Entropy = \sum_{i=0}^{L-1} \sum_{j=0}^{L-1} S\theta(i, j|d) \log\left(S\theta(i, j|d)\right) \tag{7.4}$$

$$IDM = \sum_{i=0}^{L-1} \sum_{j=0}^{L-1} \frac{1}{1+(i-j)^2} S\theta(i, j|d) \tag{7.5}$$

$$Correlation = \frac{\sum_{i=0}^{L-1} \sum_{j=0}^{L-1} (i-\mu_x)(j-\mu_y) S\theta(i, j|d)}{\sigma_x \sigma_y} \tag{7.6}$$

Previously, GLCM has been applied for image classification of microscopy (Yogesan et al. 1996), ultrasound (Basset et al. 1993), and OCT images (Gossage et al. 2003) for the classification of normal and pathologic tissues.

7.3.2.2 Gray Level Run-Length Matrix

Gray level run-length matrices (GLRM) compute the number of occurrences of a run with a particular length and gray level in a chosen direction. GLRM properties are calculated by subsampling gray levels in a coarser range of n values and computing the runs of equal levels with lengths from 1 to m along a given direction (Figure 7.4). Then these runs are stored in a run-length $m \times n$ matrix (Podda and Giachetti 2005).

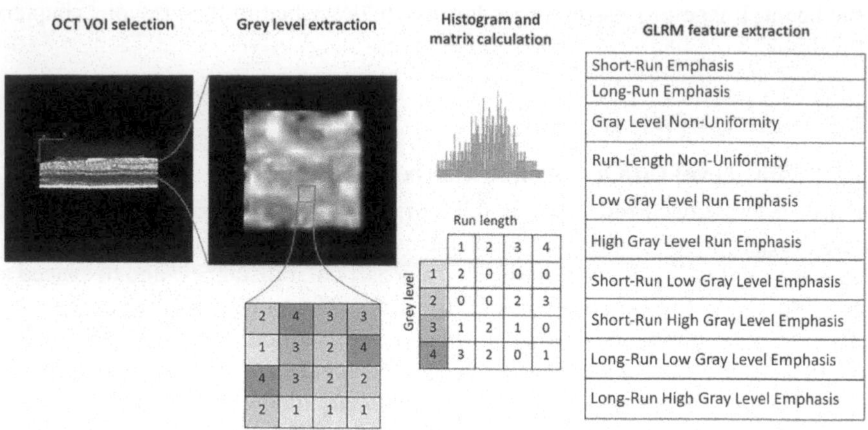

FIGURE 7.4 Demonstration of the GLRM feature extraction. After the VOI selection (column 1), extraction and normalization of voxel gray levels were performed (column 2), as demonstrated by a simplified matrix containing gray levels ranging from 1 to 4. Then, the histogram and GLRM were calculated based on the gray level distribution of voxels (column 3), and GLRM features were derived from the matrix (column 4).

Frequently used GLRM features are short-run emphasis (SRE), long-run emphasis (LRE), run-length non-uniformity (RLNU), gray level non-uniformity (GLNU), and run percentage (RPC) (Galloway 1975; Anantrasirichai et al. 2013). SRE measures the fine texture, while LRE is used for the measurement of coarse texture. Also, GLRM measures the similarity of the gray level values (GNU) and the run lengths (RNU). The number of short runs is measured by RPC. All these five features were extracted from four axes: horizontal, vertical, and two diagonals. Examples of GLRM application for texture analysis can be found in the studies of Podda and Giachetti (2005) for CT scans and Molina et al. (2016) for MR images.

7.3.2.3 Local Binary Pattern

The data of texture orientation and coarseness can be derived from the local binary pattern (LBP) by labelling the pixels of a scan using a threshold of the local pixel neighborhood (Ojala et al. 2002). In a study of Anantrasirichai et al. (2013), LBP properties demonstrated significant benefit for the classification performance with the addition of other textural features.

For the calculation of LBP features, the target window of an image was divided into several sub-regions. The gray value of the neighborhood or reference pixels was then compared to the central or index pixel of a circle. Namely, if the reference pixel is greater or equal to the index pixel, then it is labelled as "1"; if less, then it is labelled as "0." Thus, in LBP, the average sum of image intensity is computed for each local cell (7.7). The results of the local spatial patterns were given in a binary format:

$$\text{LBP}_{PR} = \sum_{P=0}^{P-1} s\left(g_p - g_c\right), \tag{7.7}$$

Where g_c is the gray value of index pixel; g_p, the gray value of reference pixels; and

$$s(x) = 1, \text{ if } x \geq 0$$

$$s(x) = 0, \text{ if } x < 0$$

Sampling points in the cell of this particular size (R = 1 pixel) equal eight (P = 8). This process is followed by the calculation of the local and global (sum of local) histograms (Figure 7.5). There were extracted 59 uniform patterns. If we consider each pattern as a feature, there were selected 59 LBP features.

7.3.2.4 Local Directional Pattern

A non-linear Kirsch compass operator can be used in the detection of textural variation strength in the local directional pattern (LDP), a modified type of LBP (Jabid et al. 2010). Eight sampling points around the index pixel with a certain location (x_c, y_c) correspond to the eight compass directions of the Kirsch kernel.

Given the intensity values of each index (i_c) and reference (i_n) pixels, LDP (7.8) can be calculated for the index pixel:

$$\text{LDP}_k\left(x_c, y_c\right) = \Sigma_{n=0}^{7} \, sign\left(m_n - m_k\right) \times 2^n \tag{7.8}$$

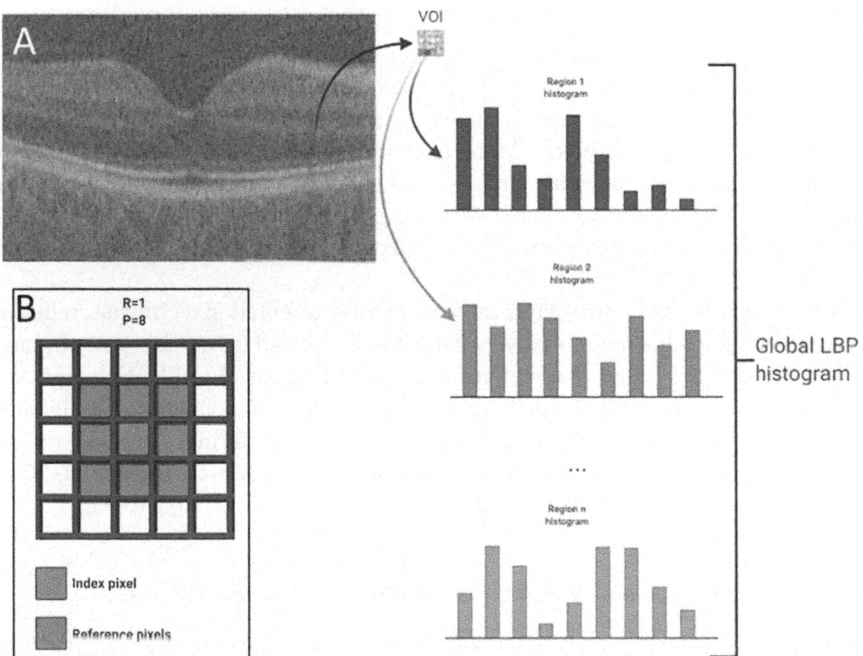

FIGURE 7.5 LBP histogram generation (a) and example of image grayscale matrix with values of R = 1 and P = 8 for a local descriptor (b).

Where m_n are eight corresponding responses of the Kirsch masks and m_k is the kth highest Kirsch activation, $n = 0, 1, \ldots, 7$. If the Kirsch operator of the reference pixel is $> m_k$, then it is assigned 1; otherwise, 0.

7.3.3 Retinal Layers Segmentation Using Texture Analysis

The most common computational method of OCT scan analysis is the segmentation of the retinal layers and local anatomical pathological structures (Rapantzikos et al. 2003; Haeker et al. 2007; Garvin et al. 2009; Baumann et al. 2010; Kajić et al. 2010; Zhang et al. 2012; Lee et al. 2018). Texture analysis–based segmentation is used to establish boundaries between different image regions using the textural features of grayscale and RGB images (Pentland 1984; Mirmehdi et al. 2009).

One of the earliest applications of GLCM textural parameters in the detection of retinal layers was conducted in the study of Baroni et al. (2007) with 79% of segmentation precision for the inner retina; 98.6% of accuracy was achieved in the support vector machine (SVM)–based retinal layer segmentation with the extraction of the GLCM features by Naseri et al. (2010). Similar performance of linear SVM classifier in the early detection of diabetic retinopathy was demonstrated in the recent investigation of Sharafeldeen et al. (2021) when GLCM features with the Markov–Gibbs random field (MGRF) were extracted for the retinal segmentation based on the integration of morphology and reflectivity markers of layers. MGRF was also used in the segmentation of three-dimensional OCT images (Sleman et al. 2021). Earlier in 2009, Mishra et al. (2009) applied a kernel-based algorithm to determine the approximate locations of each retinal layer for further segmentation.

Textural features of the Laplace distribution–based model also can be a useful tool for the extraction of boundary pixels located on the borders between retinal layers for segmentation purposes (Monemian and Rabbani 2020). Kajic et al. (2008) used the unsupervised machine learning tool—Gaussian mixture model and calculation of the Mahanalobis distance in the multidimensional feature space—for the retinal segmentation purposes based on the texture and geometric properties of the OCT images.

The measurement of retinal layer thickness and segmentation techniques relies on the detection of the boundary of layers, which is facilitated by the removal of image noise. The noise has little value with lower-resolution forms of OCT. With high-resolution devices, however, it is likely that an increasing component of this signal is not noise but reflects the subcellular changes in optical scattering. These alterations can be investigated using texture analysis of the OCT images (Gossage et al. 2003; Anantrasirichai et al. 2013; González-López et al. 2015).

7.3.4 Texture Analysis–Based Classification of Retinal OCT Images

The purpose of a classification method in image processing is to categorize different image regions into distinct groups or classes (Dong et al. 2017). The technique provides unique information on the spatial variation of pixels (texture) (Pietikainen 2000) and produces a classification map of the input image. Then each uniform textured region is identified with the texture group or class it belongs to.

Extraction of textural features for the detection and categorization of retinal diseases are the key elements of the majority of recent OCT-related studies (Ghazal et al. 2020; Sandhu et al. 2020; Tazarjani et al. 2021; Schultheiss et al. 2022). Texture-based OCT image classification methods might be further approved for the clinical guidelines as a screening method due to their high accuracy, comparably cheap computational cost, and time efficiency (Eladawi et al. 2018a; Akoushideh et al. 2021; Liu and Aldrich 2022).

7.3.4.1 Glaucoma

Retinal neurodegenerative diseases are the main causative factors of incurable irreversible vision loss. Glaucoma is a group of diseases manifested by the excavation of the optic nerve head and deterioration of the visual field (Quigley and Broman 2006; Quigley 2011; Jonas et al. 2017), and it is often associated with a constant effect of increased intraocular pressure (IOP) (Frankfort et al. 2013). Textural features of the inner plexiform layer (IPL) of the glaucomatous retina have been extracted and used for the SVM-based classification in the study of Anantrasirichai et al. (2013). Retinal nerve fiber layer's (RNFL) texture can also be used for the detection of not only glaucoma but also compressive optic neuropathy, optic neuritis, and non-arteritic anterior ischemic optic neuropathy (Leung et al. 2022).

7.3.4.2 Age-Related Macular Degeneration

Another common type of retinal neurodegenerative condition is age-related macular degeneration (AMD). AMD is a chronic multifactorial progressive disease that affects the macula and leads to loss of central vision (Marshall 1987). The symptoms of AMD are complex and irregular (Bressler et al. 1988). In the early stages of the disease, patients are relatively asymptomatic (Milam et al. 2000). However, as the disease progresses, there may be manifestations such as sudden loss of vision which cannot be corrected with glasses, and the appearance of a gray or dark spot in front of the eye (positive scotoma) (Bressler et al. 1988; Klein 1991). AMD occupies one of the leading positions among the causes of blindness and low vision worldwide (la Cour et al. 2002). OCT is a gold standard method for the detection of AMD, and the application of the machine learning tools may help detect OCT-derived biomarkers of the disease (Elsharkawy et al. 2021).

The source of reflectivity changes in OCT images from eyes with AMD is a topic of active debate. Machine learning has emerged as a powerful method for the identification of subtle optical changes, drusen classification, and the grading of disease severity when applied to OCT images. For instance, multi-scale textural features and shape parameters were extracted from OCT scans to train and segregate different macular conditions: macular hole, macular edema, and AMD (Liu et al. 2011). Apart from the neural retina, the textural signature of the retinal pigment epithelium (RPE) layer in AMD can be used for the early optical detection of dystrophies and intercellular damage of RPE cells (Hanus et al. 2016; Somasundaran et al. 2020). In the study of Haq and Wilk (2018), the most accurate algorithms for the detection of RPE degeneration in neovascular AMD were Naive Bayes and SVM classifiers. Differential diagnosis of AMD from diabetic macular edema (DMO) was achieved by extracting the parameters based on linear configuration pattern (LCP) and using the sequential minimal optimization (SMO) method (Wang et al. 2016).

7.3.4.3 Diabetic Retinopathy

The main ocular complication of diabetes mellitus, one of the four major types of noncommunicable diseases, is diabetic retinopathy (DR). TA-based retinal OCT classification in this field is progressively developing (Sandhu et al. 2018; Shaban et al. 2020; Elsharkawy et al. 2022). Segmentation of retinal layers for the calculation of the retinal layers' thickness values can be performed with the characterization of their tortuosity and reflectivity (ElTanboly et al. 2018).

Another modality of optical coherence tomography (OCT) angiography (OCTA) is extensively used by clinicians for the determination of the disease severity. Linear SVM-based detection of early diabetic vascular alterations was investigated by Eladawi et al. (2018).

The combined application of multiple diagnostic procedures may increase the classification accuracy of TA-based techniques. For instance, the complex use of not only OCT scans but also OCTA images with clinical data (such as the presence of hypertension, hyperlipidemia, level of hemoglobin A1c, etc.) significantly increased the accuracy, sensitivity, and specificity (Sandhu et al. 2020).

7.3.4.4 Neurodegenerative Diseases of the Central Nervous System

The eye as a window to the brain can be used for the purpose of the diagnosis of general neurological diseases. Early *in vivo* detection of this group of the condition has great promise in the future. Texture analysis of the OCT images allowed for categorizing the retinal scans of patients with multiple sclerosis (Nunes et al. 2019; Tazarjani et al. 2021), which causes the inflammation of the central nervous system: brain and spinal cord. Behbehani et al. (2017) segmented the retinal layers of patients with progressive and relapsing remitting types of multiple sclerosis in order to achieve the specific retinal layer predilection and its correlation with disability.

The prevalence of Alzheimer's disease (AD) is steadily increasing with age, and according to "World Alzheimer Report 2015: The Global Impact of Dementia," the incidence of the disease will reach 131.5 million by 2050 (Prince et al. 2015). There is a demand for the broader application of OCT in the diagnosis of early retinal alterations of AD. Ophthalmic manifestations of AD may precede irreversible changes in the brain (Hart et al. 2016; Lim et al. 2016; Czakó et al. 2020), which increases the need for non-invasive, ligand-free detection of the disease. Retinal OCT images of the experimental model of the transgenic AD murine model were acquired in the study of Song et al. (2020) to determine the textural biomarkers of this neurodegenerative disease.

As the role of OCT in the clinical guideline of neurologists is increasing, the machine learning approach to image processing has also begun to integrate. OCT symptoms of central retinal artery occlusions can be pathognomonic symptoms of the acute ischemic event in the brain, whereas the defects in the visual field can indicate the arterial branch occlusion or anterior ischemic optic neuropathy (Schultheiss et al. 2022). Combinative use of OCT and AI-based classification will be increasing in upcoming years as a state-of-the-art, non-invasive, ligand-free *in vivo* technology for early detection of neurodegenerative diseases of the central nervous system.

7.4 LIMITATIONS AND FUTURE DIRECTIONS

OCT is a state-of-the-art technology in clinical medicine, especially in ophthalmology. Nevertheless, in the diagnosis of a human eye, there may be shadows and artifacts, limiting the image quality. Also, the axial resolution of the OCT system might be affected by water absorption in the anterior chamber and the vitreous. Hence, contrast and image penetration depth decrease. In general, the absorption coefficient of water (intraocular fluid and vitreous body) equals 0.04 mm^{-1} for the spectral range of 950–1,130 nm. Therefore, with average axial eye length (24–25 mm), and due to forward–backward light travel, the loss of signal-to-noise ratio amounts to ~9 dB (Hariri et al. 2017). Notwithstanding, this loss in the murine eye is less (~2 dB) and can be ignored as a limiting factor in small animal retinal OCT.

Due to the inhomogeneities in the refractive index of the tissue structures, light reflects back at various angles (Schmitt et al. 1994; Schmitt 1999). There are two groups of debating sides on the nature of speckle: if one cohort thinks this is a source of optical information about the microscopic structures of the tissue, whereas the other group categorizes the speckle as a noise (Schmitt et al. 1999). Hence, in the process of noise filtering, some valuable data can be removed. Consequently, this can lead to misinterpretation of the OCT results. The strategy of efficient noise reduction is one of the goals of the improvement of image processing.

Moreover, averaging the tissue spatial properties of the scans is the signal-carrying speckle together with the signal-degrading one (Schmitt 1999). To avoid this issue, there is a demand for effective speckle reduction techniques. It requires widening the source bandwidth and numerical aperture (NA) of the objective (Schmitt 1999).

The application of machine learning algorithms has been enhanced recently in medicine, especially in the early diagnosis of pathologies. Especially this trend is progressing faster in ophthalmology due to the highly technological basis of the field (Bajwa et al. 2015). Glaucoma, AMD, diabetic retinopathy, and other common socially significant diseases can be detected early with the methods of automatic detection of signs (Abràmoff et al. 2008; Kucur et al. 2018; Zhongyang and Yankui 2019; Bisneto et al. 2020; González-Gonzalo et al. 2020), which also can be served as a screening tool for the tremendous amount of data (Antal and Hajdu 2014; Ting et al. 2017; Grzybowski et al. 2020).

Particularly, supervised machine learning tools demonstrated effectiveness in the performance of binary classification of diseased and healthy participants (Kucur et al. 2018; Bajwa et al. 2019). This type of learning is based on the training of the system from input data (OCT scans) and iteratively improves its operation for the detection of correct output (accurate diagnosis).

After the repetitive training of the program with the increasing amount of input data, the effectiveness of the classifier improves. When new unseen data is imported into the system, the supervised learning tool can identify the correct diagnosis. Hence, in studies with the involvement of machine learning techniques, there should be as much data as possible. Consequently, the validity of the technique is dependent on the size and quality of the training dataset (LeCun et al. 2015).

Although the current available OCT system may not provide cellular and subcellular resolution, analysis of textural parameters has a promising future to measure aspects of cellularity in biological tissues.

ACKNOWLEDGMENTS

The work was supported by the Vice-Chancellor's International Scholarships for Research Excellence (Cardiff University, Cardiff, UK) and the Medical Research Council (G0800547—optophysiological characterization of retinal ganglion cell function by ultrahigh-resolution optical coherence tomography), Tweedie Bequest, University Hospital of Wales.

REFERENCES

Abràmoff, M.D., Niemeijer, M., Suttorp-Schulten, M.S.A., Viergever, M.A., Russell, S.R. and Ginneken, B. 2008. Evaluation of a system for automatic detection of diabetic retinopathy from color fundus photographs in a large population of patients with diabetes. *Diabetes Care* 31(2):193. https://doi.org/10.2337/dc07-1312

Adler, R., Curcio, C., Hicks, D., Price, D. and Wong, F. 1999. Cell death in age-related macular degeneration. *Mol Vis* 3(5):31.

Ahmad, S.S. 2017. An introduction to DARC technology. *Saudi J Ophthalmol* 31(1):38–41. https://doi.org/10.1016/j.sjopt.2016.08.001

Akoushideh, A., Babak Maybodi, M.-N. and Shahbahrami, A. 2021. Features' value range approach to enhance the throughput of texture classification. *IET Image Process* 15:28–46. https://doi.org/10.1049/ipr2.12003

Almasieh, M., Wilson, A.M., Morquette, B., Vargas, J.L.C. and Di Polo, A. 2012. The molecular basis of retinal ganglion cell death in glaucoma. *Prog Retin Eye Res* 31(2):152–181. https://doi.org/10.1016/j.preteyeres.2011.11.002

Aloya, R., Shirvan, A., Grimberg, H., Reshef, A., Levin, G., Kidron, D., Cohen, A. and Ziv, I. 2006. Molecular imaging of cell death in vivo by a novel small molecule probe. *Apoptosis* 11(12):2089. https://doi.org/10.1007/s10495-006-0282-7

Anantrasirichai, N., Achim, A., Morgan, J.E., Erchova, I. and Nicholson, L. 2013. *SVM-based texture classification in optical coherence tomography*. IEEE ISBI, San Francisco, CA, pp. 1332–1335.

Antal, B. and Hajdu, A. 2014. An ensemble-based system for automatic screening of diabetic retinopathy. *Knowl Based Syst* 60:20–27. https://doi.org/10.1016/j.knosys.2013.12.023

Baghaie, A., Yu, Z. and D'Souza, R.M. 2015. State-of-the-art in retinal optical coherence tomography image analysis. *Quant Imaging Med Surg* 5(4):603–617. https://doi.org/10.3978/j.issn.2223-4292.2015.07.02

Bajwa, A., Aman, R. and Reddy, A.K. 2015. A comprehensive review of diagnostic imaging technologies to evaluate the retina and the optic disk. *Int Ophthalmol* 35(5):733–755. https://doi.org/10.1007/s10792-015-0087-1

Bajwa, M.N., Malik, M.I., Siddiqui, S.A., Dengel, A., Shafait, F., Neumeier, W. and Ahmed, S. 2019. Two-stage framework for optic disc localization and glaucoma classification in retinal fundus images using deep learning. *BMC Medical Inform Decis Mak* 19(1):136. https://doi.org/10.1186/s12911-019-0842-8

Barer, R. and Tkaczyk, S. 1954. Refractive index of concentrated protein solutions. *Nature* 173(4409):821–822. https://doi.org/10.1038/173821b0

Baroni, M., Diciotti, S., Evangelisti, A., Fortunato, P. and La Torre, A. 2007. Texture classification of retinal layers in optical coherence tomography. In: Jarm, T., Kramar, P. and Zupanic, A. (eds.), *11th mediterranean conference on medical and biomedical engineering and computing*. IFMBE Proceedings, vol. 16. Springer, Berlin, Heidelberg. https://doi.org/10.1007/978-3-540-73044-6_220

Basset, O., Sun, Z., Mestas, J.L. and Gimenez, G. 1993. Texture analysis of ultrasonic images of the prostate by means of co-occurrence matrices. *Ultrason Imaging* 15(3):218–237. https://doi.org/10.1177/016173469301500303

Baumann, B., Gotzinger, E., Pircher, M., Sattmann, H., Schuutze, C., Schlanitz, F., Ahlers, C., Schmidt-Erfurth, U. and Hitzenberger, C.K. 2010. Segmentation and quantification of retinal lesions in age-related macular degeneration using polarization-sensitive optical coherence tomography. *J Biomed Opt* 15(6):061704. https://doi.org/10.1117/1.3499420

Behbehani, R., Abu Al-Hassan, A., Al-Salahat, A., Sriraman, D., Oakley, J.D. and Alroughani, R. 2017. Optical coherence tomography segmentation analysis in relapsing remitting versus progressive multiple sclerosis. *PLOS One* 12(2):e0172120. https://doi.org/10.1371/journal.pone.0172120

Bell, K., Rosignol, I., Sierra-Filardi, E., Rodriguez-Muela, N., Schmelter, C., Cecconi, F., Grus, F. and Boya, P. 2020. Age related retinal Ganglion cell susceptibility in context of autophagy deficiency. *Cell Death Dis* 6(1):21. https://doi.org/10.1038/s41420-020-0257-4

Benedicto, I., Lehmann, G.L., Ginsberg, M., Nolan, D.J., Bareja, R., Elemento, O., Salfati, Z., Alam, N.M., Prusky, G.T., Llanos, P., Rabbany, S.Y., Maminishkis, A., Miller, S.S., Rafii, S. and Rodriguez-Boulan, E. 2017. Concerted regulation of retinal pigment epithelium basement membrane and barrier function by angiocrine factors. *Nat Commun* 8(1):15374. https://doi.org/10.1038/ncomms15374

Beuthan, J., Minet, O., Helfmann, J., Herrig, M. and Müller, G. 1996. The spatial variation of the refractive index in biological cells. *Phys Med Biol* 41(3):369. https://doi.org/10.1088/0031-9155/41/3/002

Bhattacharjee, M., Ashok, P.C., Rao, K.D., Majumder, S.K., Verma, Y. and Gupta, P.K. 2011. Binary tissue classification studies on resected human breast tissues using optical coherence tomography images. *J Innov Opt Health Sci* 4(1):59–66. https://doi.org/10.1142/s1793545811001083

Binley, K., Ng, W.S., Barde, Y.A., Song, B. and Morgan, J.E. 2016. Brain-derived neurotrophic factor prevents dendritic retraction of adult mouse retinal ganglion cells. *Eur J Neurosci* 44(3):2028–2039. https://doi.org/10.1111/ejn.13295

Bisneto, T.R.V., Carvalho Filho, A.O. and Magalhães, D.M.V. 2020. Generative adversarial network and texture features applied to automatic glaucoma detection. *Appl Soft Comput* 90:106165. https://doi.org/10.1016/j.asoc.2020.106165

Boycott, B.B., Dowling, J.E., Fisher, S.K., Kolb, H. and Laties, A.M. 1975. Interplexiform cells of the mammalian retina and their comparison with catecholamine-containing retinal cells. *Proc R Soc Lond Series B Biolog Sci* 191(1104):353–368. https://doi.org/10.1098/rspb.1975.0133

Breher, K., Terry, L., Bower, T. and Wahl, S. 2020. Choroidal biomarkers: A repeatability and topographical comparison of choroidal thickness and choroidal vascularity index in healthy eyes. *Trans Vis Sci Tech* 9(11):8. https://doi.org/10.1167/tvst.9.11.8

Bressler, N.M., Bressler, S.B. and Fine, S.L. 1988. Age-related macular degeneration. *Surv Ophthalmol* 32(6):375–413. https://doi.org/10.1016/0039-6257(88)90052-5

Chauhan, B.C., O'Leary, N., AlMobarak, F.A., Reis, A.S.C., Yang, H., Sharpe, G.P., Hutchison, D.M., Nicolela, M.T. and Burgoyne, C.F. 2013. Enhanced detection of open-angle glaucoma with an anatomically accurate optical coherence tomography-derived neuroretinal rim parameter. *Ophthalmology* 120(3):535–543. https://doi.org/10.1016/j.ophtha.2012.09.055

Chen, S., Urban, M.W., Pislaru, C., Kinnick, R., Zheng, Y., Yao, A. and Greenleaf, J.F. 2009. Shearwave dispersion ultrasound vibrometry (SDUV) for measuring tissue elasticity and viscosity. *IEEE Trans Ultrason Ferroelectr Freq Control* 56(1):55–62. https://doi.org/10.1109/TUFFC.2009.1005

Cherkezyan, L., Subramanian, H., Stoyneva, V., Rogers, J.D., Yang, S., Damania, D., Taflove, A. and Backman, V. 2012. Targeted alteration of real and imaginary refractive index of biological cells by histological staining. *Opt Letters* 37(10):1601–1603. https://doi.org/10.1364/OL.37.001601

Chen, C.H., Pau, L.F. and Wang, P.S.P. 1993. *Handbook of pattern recognition and computer vision*. World Scientific, Singapore. ISBN 978-981-02-1136-3

Choma, M.A., Sarunic, M.V., Yang, C. and Izatt, J.A. 2003. Sensitivity advantage of swept source and Fourier domain optical coherence tomography. *Opt Express* 11(18):2183–2189. https://doi.org/10.1364/OE.11.002183

Cordeiro, M.F. 2007. DARC: A new method for detecting progressive neuronal death. *Eye* 21(1):S15–S17. https://doi.org/10.1038/sj.eye.6702881

Costantini, R., Sbaiz, L. and Susstrunk, S. 2008. Higher order SVD analysis for dynamic texture synthesis. *IEEE Trans Image Process* 17(1):42–52. https://doi.org/10.1109/TIP.2007.910956

Culjat, M.O., Goldenberg, D., Tewari, P. and Singh, R.S. 2010. A review of tissue substitutes for ultrasound imaging. *Ultrasound Med Biol* 36(6):861–873. https://doi.org/10.1016/j.ultrasmedbio.2010.02.012

Curcio, C.A., Sloan, K.R., Kalina, R.E. and Hendrickson, A.E. 1990. Human photoreceptor topography. *J Comp Neurol* 292(4):497–523. https://doi.org/10.1002/cne.902920402

Czakó, C., Kovács, T., Ungvari, Z., Csiszar, A., Yabluchanskiy, A., Conley, S., Csipo, T., Lipecz, A., Horváth, H., Sándor, G.L., István, L., Logan, T., Nagy, Z.Z. and Kovács, I. 2020. Retinal biomarkers for Alzheimer's disease and vascular cognitive impairment and dementia (VCID): Implication for early diagnosis and prognosis. *GeroScience* 42(6):1499–1525. https://doi.org/10.1007/s11357-020-00252-7

Davies, E.R. 2005. Texture. In: Davies, E.R. (ed.), *Machine vision*, third edition. Morgan Kaufmann, Burlington, pp. 757–779. https://doi.org/10.1016/B978-0-12-206093-9.X5000-X

Davies, H.G., Wilkins, M.H.F., Chayen, J. and La Cour, L.F. 1954. The use of the interference microscope to determine dry mass in living cells and as a quantitative cytochemical method. *Q J Microsc Sci* 95(31):271–304. https://doi.org/10.1242/jcs.s3-95.31.271

de Amorim Garcia Filho, C.A., Yehoshua, Z., Gregori, G., Farah, M.E., Feuer, W. and Rosenfeld, P.J. 2013. Optical coherence tomography A2. In: Wilkinson, C., Hinton, D., Sadda, S., Wiedemann, P. and Ryan, S. (eds.), *Retina*, fifth edition. W.B. Saunders, London, pp. 82–110. https://doi.org/0.1097/IAE.0b013e318285cbd2

Dong, Y., Feng, J., Liang, L., Zheng, L. and Wu, Q. 2017. Multiscale sampling based texture image classification. *IEEE Signal Process Lett* 24(5):614–618. https://doi.org/10.1109/LSP.2017.2670026

Dowling, J.E. 1987. *The retina: An approachable part of the brain.* Belknap Press of Harvard University Press, Cambridge, MA. https://doi.org/10.1002/ajhb.22305

Drexler, W. 2010. Optical coherence tomography. In: Dartt, D.A. (ed.), *Encyclopedia of the eye*. Academic Press, Oxford, pp. 194–204. https://doi.org/10.1136/bjo.58.8.709

Drexler, W. and Fujimoto, J.G. 2008. *Optical coherence tomography: Technology and applications*. Springer-Verlag, Berlin, Heidelberg. https://doi.org/10.1007/978-3-319-06419-2

Drexler, W., Morgner, U., Ghanta, R.K., Kärtner, F.X., Schuman, J.S. and Fujimoto, J.G. 2001. Ultrahigh-resolution ophthalmic optical coherence tomography. *Nat Med* 7:502. https://doi.org/10.1038/86589

Dunaief, J.L., Dentchev, T., Ying, G.S. and Milam, A.H. 2002. The role of apoptosis in age-related macular degeneration. *Arch Ophthalmol* 120(11):1435–1442. https://doi.org/10.1001/archopht.120.11.1435

Ejofodomi, O. 2014. Measurement of optical scattering coefficient of the individual layers of the human urinary bladder using optical coherence tomography. *ISRN Biomed Imaging* 2014:591592. https://doi.org/10.1155/2014/591592

Eladawi, N., Elmogy, M., Fraiwan, L., Pichi, F., Ghazal, M., Aboelfetouh, A., Riad, A., Keynton, R., Schaal, S. and El-Baz, A. 2018. Early diagnosis of diabetic retinopathy in OCTA Images based on local analysis of retinal blood vessels and foveal avascular zone. *24th ICPR* 2018:3886–3891. https://doi.org/10.1109/ICPR.2018.8546250

Eladawi, N., Elmogy, M.M., Ghazal, M., Helmy, O., Aboelfetouh, A., Riad, A., Schaal, S. and El-Baz, A. 2018a. Classification of retinal diseases based on OCT images. *Front Biosci* 23(2):247–264. https://doi.org/10.2741/4589. PMID: 28930545.

Elkington, A.R., Inman, C.B., Steart, P.V. and Weller, R.O. 1990. The structure of the lamina cribrosa of the human eye: An immunocytochemical and electron microscopical study. *Eye* 4:42. https://doi.org/10.1038/eye.1990.5

Elsharkawy, M., Elrazzaz, M., Ghazal, M., Alhalabi, M., Soliman, A., Mahmoud, A., El-Daydamony, E., Atwan, A., Thanos, A., Sandhu, H.S., Giridharan, G. and El-Baz, A. 2021. Role of optical coherence tomography imaging in predicting progression of age-related macular disease: A survey. *Diagnostics* 11(12):2313. https://doi.org/10.3390/diagnostics11122313

Elsharkawy, M., Sharafeldeen, A., Soliman, A., Khalifa, F., Ghazal, M., El-Daydamony, E., Atwan, A., Sandhu, H.S. and El-Baz, A. 2022. A novel computer-aided diagnostic system for early detection of diabetic retinopathy using 3D-OCT higher-order spatial appearance model. *Diagnostics* 12(2):461. https://doi.org/10.3390/diagnostics12020461

ElTanboly, A.H., Palacio, A., Shalaby, A.M., Switala, A.E., Helmy, O., Schaal, S. and El-Baz, A. 2018. An automated approach for early detection of diabetic retinopathy using SD-OCT images. *Front Biosci* 10(2):197–207. https://doi.org/10.2741/e817. PMID: 28930613.

Evennett, P.J., McMahon, J., Mahers, E.G., Joyce, S.C., Reed, M.G., Rood, A.P., Brocklehurst, K.G., Julian, K. and Mills, S.L. 1993. Particle characterisation by microscopy and image analysis. *Analytical Proceedings* 30(5):227–232. https://doi.org/10.1039/AP9933000227

Everett, M., Magazzeni, S., Schmoll, T. and Kempe, M. 2021. Optical coherence tomography: From technology to applications in ophthalmology. *Transl Biophotonics* 3:e202000012. https://doi.org/10.1002/tbio.202000012

Fercher, A. F. 1996. Optical coherence tomography. *J Biomed Opt* 1(2):157–173.

Frank, S., Gaume, B., Bergmann-Leitner, E.S., Leitner, W.W., Robert, E.G., Catez, F., Smith, C.L. and Youle, R.J. 2001. The role of dynamin-related protein 1, a mediator of mitochondrial fission, in apoptosis. *Dev Cell* 1(4):515–525. https://doi.org/10.1016/S1534-5807(01)00055-7

Frankfort, B.J., Khan, A.K., Tse, D.Y., Chung, I., Pang, J.-J., Yang, Z., Gross, R.L. and Wu, S.M. 2013. Elevated intraocular pressure causes inner retinal dysfunction before cell loss in a mouse model of experimental glaucoma. *Invest Ophthalmol Vis Sci* 54(1):762–770. https://doi.org/10.1167/iovs.12-10581

Fujimoto, J.G., Brezinski, M.E., Tearney, G.J., Boppart, S.A., Bouma, B., Hee, M.R., Southern, J.F. and Swanson, E.A. 1995. Optical biopsy and imaging using optical coherence tomography. *Nat Med* 1:970. https://doi.org/10.1038/nm0995-970

Gabriele, M.L., Wollstein, G., Ishikawa, H., Kagemann, L., Xu, J., Folio, L.S. and Schuman, J.S. 2011. Optical coherence tomography: History, current status, and laboratory work. *Invest Ophthalmol Vis Sci* 52(5):2425–2436. https://doi.org/10.1167/iovs.10-6312

Galloway, M.M. 1975. Texture analysis using gray level run lengths. *Comput Graph Image Process* 4(2):172–179. https://doi.org/10.1016/S0146-664X(75)80008-6

Galvao, J., Davis, B.M. and Cordeiro, M.F. 2013. In vivo imaging of retinal ganglion cell apoptosis. *Curr Opin Pharmacol* 13(1):123–127. https://doi.org/10.1016/j.coph.2012.08.007

Garhart, C. and Lakshminarayanan, V. 2016. Anatomy of the eye. In: Chen, J. et al. (eds.), *Handbook of visual display technology*. Springer International Publishing, Cham, pp. 93–104. https://doi.org/10.1007/978-3-319-14346-0_4

Garvin, M.K., Abràmoff, M.D., Wu, X., Russell, S.R., Burns, T.L. and Sonka, M. 2009. Automated 3-D intraretinal layer segmentation of macular spectral-domain optical coherence tomography images. *IEEE Trans Med Imaging* 28(9):1436–1447. https://doi.org/10.1109/TMI.2009.2016958

Gazzard, G. 2015. Angle imaging: Ultrasound biomicroscopy and anterior segment optical coherence tomography. In: Shaarawy, T., Sherwood, M., Hitchings, R. and Crowsto, J. (eds.), *Glaucoma*, second edition. W.B. Saunders, London, pp. 191–200. https://doi.org/10.1016/B978-0-7020-5193-7.00033-9

Ghazal, M., Ali, S.S., Mahmoud, A.H., Shalaby, A.M. and El-Baz, A. 2020. Accurate detection of non-proliferative diabetic retinopathy in optical coherence tomography images using convolutional neural networks. *IEEE Access* 8:34387–34397. https://doi.org/10.1109/ACCESS.2020.2974158

González-Gonzalo, C., Sánchez-Gutiérrez, V., Hernández-Martínez, P., Contreras, I., Lechanteur, Y.T., Domanian, A., van Ginneken, B. and Sánchez, C.I. 2020. Evaluation of a deep learning system for the joint automated detection of diabetic retinopathy and age-related macular degeneration. *Acta Ophthalmologica* 98(4):368–377. https://doi.org/10.1111/aos.14306

González-López, A., Remeseiro, B., Ortega, M., Penedo, M. and Charlón, P. (eds.). 2015. A *texture-based method for choroid segmentation in retinal EDI-OCT images*. Springer International Publishing, Cham. https://doi.org/10.1007/978-3-319-27340-2

Gossage, K.W., Cynthia, M.S., Elizabeth, M.K., Lida, P.H., Alice, L.S., Jeffrey, J.R., Williams, S.K. and Jennifer, K.B. 2006. Texture analysis of speckle in optical coherence tomography images of tissue phantoms. *Phys Med Biol* 51(6):1563–1575. https://doi.org/10.1088/0031-9155/51/6/014

Gossage, K.W., Tkaczyk, T.S., Rodriguez, J.J. and Barton, J.K. 2003. Texture analysis of optical coherence tomography images: Feasibility for tissue classification. *J Biomed Opt* 8(3). https://doi.org/10.1117/1.1577575

Gourley, P.L., Hendricks, J.K., McDonald, A.E., Copeland, R.G., Barrett, K.E., Gourley, C.R., Singh, K.K. and Naviaux, R.K. 2005. Mitochondrial correlation microscopy and nanolaser spectroscopy—new tools for biophotonic detection of cancer in single cells. *Technol Cancer Res Treat* 4(6):585–592. https://doi.org/10.1177/153303460500400602

Granger, C.E., Yang, Q., Song, H., Saito, K., Nozato, K., Latchney, L.R., Leonard, B.T., Chung, M.M., Williams, D.R. and Rossi, E.A. 2018. Human retinal pigment epithelium: In vivo cell morphometry, multispectral autofluorescence, and relationship to cone mosaic. *Invest Ophthalmol Vis Sci* 59(15):5705–5716. https://doi.org/10.1167/iovs.18-24677

Grzybowski, A., Brona, P., Lim, G., Ruamviboonsuk, P., Tan, G.S.W., Abramoff, M. and Ting, D.S.W. 2020. Artificial intelligence for diabetic retinopathy screening: A review. *Eye* 34(3):451–460. https://doi.org/10.1038/s41433-019-0566-0

Guo, L., Moss, S.E., Alexander, R.A., Ali, R.S., Fitzke, F.W. and Cordeiro, M.F. 2005. Retinal ganglion cell apoptosis in glaucoma is related to intraocular pressure and IOP-induced effects on extracellular matrix. *Invest Ophthalmol Vis Sci* 46(1):175–182. https://doi.org/10.1167/iovs.04-0832

Gupta, V., You, Y., Li, J., Gupta, V., Golzan, M., Klistorner, A., van den Buuse, M. and Graham, S. 2014. BDNF impairment is associated with age-related changes in the inner retina and exacerbates experimental glaucoma. *Biochim Biophys Acta Mol Basis Dis* 1842(9):1567–1578. https://doi.org/10.1016/j.bbadis.2014.05.026

Haeker, M., Sonka, M., Kardon, R., Shah, V.A., Wu, X. and Abràmoff, M.D. 2007. Automated segmentation of intraretinal layers from macular optical coherence tomography images. *Proc SPIE 6512, Medical Imag 2007 Image Proc* 651214. https://doi.org/10.1117/12.710231

Hariri, A., Fatima, A., Mohammadian, N., Mahmoodkalayeh, S., Ansari, M.A., Bely, N. and Avanaki, M.R.N. 2017. Development of low-cost photoacoustic imaging systems using very low-energy pulsed laser diodes. *J Biomed Opt* 22(7):075001. https://doi.org/10.1117/1.JBO.22.7.075001

Hanus, J., Anderson, C., Sarraf, D., Ma, J. and Wang, S. 2016. Retinal pigment epithelial cell necroptosis in response to sodium iodate. *Cell Death Disc* 2:16054. https://doi.org/10.1038/cddiscovery.2016.54

Hart, N.J., Koronyo, Y., Black, K.L. and Koronyo-Hamaoui, M. 2016. Ocular indicators of Alzheimer's: Exploring disease in the retina. *Acta Neuropathologica* 132(6):767–787. https://doi.org/10.1007/s00401-016-1613-6

Haq, A. and Wilk, S. 2018. Detection of wet age-related macular degeneration in OCT images: A case study. In: Gzik, M., Tkacz, E., Paszenda, Z. and Piętka, E. (eds.), *Innovations in Biomedical Engineering. IBE 2017. Advances in Intelligent Systems and Computing*, vol. 623. Springer, Cham. https://doi.org/10.1007/978-3-319-70063-2_5

Hee, M.R., Izatt, J.A., Swanson, E.A., Huang, D., Schuman, J.S., Lin, C.P., Puliafito, C.A. and Fujimoto, J.G. 1995. Optical coherence tomography of the human retina. *Arch Ophthalmol* 113(3):325–332. https://doi.org/10.1001/archopht.1995.01100030081025

Hu, Y., Ren, J., Yang, J., Bai, R. and Liu, J. 2021. Noise reduction by adaptive-SIN filtering for retinal OCT images. *Sci Rep* 11:19498. https://doi.org/10.1038/s41598-021-98832-w

Huang, D., Swanson, E.A., Lin, C.P., Schuman, J.S., Stinson, W.G., Chang, W., Hee, M.R., Flotte, T., Gregory, K., Puliafito, C.A. and Fujimoto, J.G. 1991. Optical coherence tomography. *Science* 254(5035):1178–1181. https://doi.org/10.1126/science.1957169

Humeau-Heurtier, A. 2019. Texture feature extraction methods: A survey. *IEEE Access* 7:8975–9000. https://doi.org/10.1109/ACCESS.2018.2890743

Jabid, T., Kabir, M.H. and Chae, O. 2010. Robust facial expression recognition based on local directional pattern. *ETRI J* 32:784–794. https://doi.org/10.4218/etrij.10.1510.0132

Jonas, J.B., Weber, P., Nagaoka, N. and Ohno-Matsui, K. 2017. Glaucoma in high myopia and parapapillary delta zone. *PLoS One* 12(4):e0175120. https://doi.org/10.1371/journal.pone.0175120

Kajić, V., Považay, B., Hermann, B., Hofer, B., Marshall, D., Rosin, P.L. and Drexler, W. 2010. Robust segmentation of intraretinal layers in the normal human fovea using a novel statistical model based on texture and shape analysis. *Opt Exp* 18(14):14730–14744. https://doi.org/10.1364/OE.18.014730

Kajic, V., Powell, G., Povazay, B., Hermann, B., Hofer, B., Garcia-Sanchez, Y., Marshall, D., Rosin, P.L. and Drexler, W. 2008. Texture analysis and geometry based retinal segmentation for three-dimensional OCT. *Invest Ophthalmol Vis Sci* 49(13):1888.

Kim, N.R., Lee, E.S., Seong, G.J., Kang, S.Y., Kim, J.H., Hong, S. and Kim, C.Y. 2011. Comparing the ganglion cell complex and retinal nerve fibre layer measurements by Fourier domain OCT to detect glaucoma in high myopia. *Br J Ophthalmol* 95(8):1115–1121. https://doi.org/10.1136/bjo.2010.182493

Klein, R. 1991. Age-related eye disease, visual impairment, and driving in the elderly. *Human Factors* 33(5):521–525. https://doi.org/10.1177/001872089103300504

Krauss, J.M. and Puliafito, C.A. 1995. Lasers in ophthalmology. *Lasers Surg Med* 17(2):102–159. https://doi.org/10.1002/lsm.1900170203

Kucur, Ş.S., Holló, G. and Sznitman, R. 2018. A deep learning approach to automatic detection of early glaucoma from visual fields. *PLOS One* 13(11):e0206081. https://doi.org/10.1371/journal.pone.0206081

Kulmaganbetov, M., Albon, J., White, N. and Morgan, J.E. 2020. Texture analysis of OCT phantoms. In: *Biophotonics congress: Biomedical optics 2020 (translational, microscopy, OCT, OTS, BRAIN), OSA technical digest*. Optica Publishing Group, Washington, paper JTu3A.23.

Kulmaganbetov, M., Bevan, R.J., Anantrasirichai, N., Achim, A., Erchova, I., White, N., Albon, J. and Morgan, J.E. 2022. Textural feature analysis of optical coherence tomography phantoms. *Electronics* 11(4):669. https://doi.org/10.3390/electronics11040669

la Cour, M., Kiilgaard, J.F. and Nissen, M.H. 2002. Age-related macular degeneration. *Drugs & Aging* 19(2):101–133. https://doi.org/10.2165/00002512-200219020-00003

LeCun, Y., Bengio, Y. and Hinton, G. 2015. Deep learning. *Nature* 521(7553):436–444. https://doi.org/10.1038/nature14539

Lee, H.-J., Kang, K.E., Chung, H. and Kim, H.C. 2018. Automated segmentation of lesions including subretinal hyperreflective material in neovascular age-related macular degeneration. *Am J Ophthalmol* 191:64–75. https://doi.org/10.1016/j.ajo.2018.04.007

Lee, H.-J., Kim, M.-S., Jo, Y.-J. and Kim, J.-Y. 2015. Ganglion cell-inner plexiform layer thickness in retinal diseases: Repeatability study of spectral-domain optical coherence tomography. *Am J Ophthalmol* 160(2):283–289. https://doi.org/10.1016/j.ajo.2015.05.015

Leung, C.K.S., Lam, A.K.N., Weinreb, R.N., Garway-Heath, D.F., Yu, M., Guo, P.W., Chiu, V.S.M., Wan, K.H.N., Wong, M., Wu, K.Z., Cheung, C.Y.L., Lin, C., Chan, C.K.M., Chan, N.C.Y., Kam, K.W. and Lai, G.W.K. 2022. Diagnostic assessment of glaucoma and non-glaucomatous optic neuropathies via optical texture analysis of the retinal nerve fibre layer. *Nat Biomed Eng* 6:593–604. https://doi.org/10.1038/s41551-021-00813-x

Leuschen, J.N., Schuman, S.G., Winter, K.P., McCall, M.N., Wong, W.T., Chew, E.Y., Hwang, T., Srivastava, S., Sarin, N., Clemons, T., Harrington, M. and Toth, C.A. 2013. Spectral-Domain optical coherence tomography characteristics of intermediate age-related macular degeneration. *Ophthalmology* 120(1):140–150. https://doi.org/10.1016/j.ophtha.2012.07.004

Lim, J.K.H., Li, Q.-X., He, Z., Vingrys, A.J., Wong, V.H.Y., Currier, N., Mullen, J., Bui, B.V. and Nguyen, C.T.O. 2016. The eye as a biomarker for Alzheimer's disease. *Front Neurosci* 10:536–536. https://doi.org/10.3389/fnins.2016.00536

Liu, X. and Aldrich, C. 2022. Deep learning approaches to image texture analysis in material processing. *Metals* 12(2):355. https://doi.org/10.3390/met12020355

Liu, Y.Y., Ishikawa, H., Chen, M., Wollstein, G., Duker, J.S., Fujimoto, J.G., Schuman, J.S. and Rehg, J.M. 2011. Computerized macular pathology diagnosis in spectral domain optical coherence tomography scans based on multiscale texture and shape features. *Invest Ophthalmol Vis Sci* 52(11):8316–8322. https://doi.org/10.1167/iovs.10-7012

Lo, A.C.Y., Woo, T.T.Y., Wong, R.L.M. and Wong, D. 2011. Apoptosis and other cell death mechanisms after retinal detachment: Implications for photoreceptor rescue. *Ophthalmologica* 226:10–17. https://doi.org/10.1159/000328206

Malhotra, A., Minja, F.J., Crum, A. and Burrowes, D. 2011. Ocular anatomy and cross-sectional imaging of the eye. *Seminars in Ultrasound, CT and MRI* 32(1):2–13. https://doi.org/10.1053/j.sult.2010.10.009

Marchini, G. 2015. Ultrasound biomicroscopy. In: Shaarawy, T., Sherwood, M., Hitchings, R. and Crowsto, J. (eds.), *Glaucoma*, second edition. W.B. Saunders, London, pp. 179–190. https://doi.org/10.1016/B978-0-7020-5193-7.00033-9

Marshall, J. 1987. The ageing retina: Physiology or pathology. *Eye* 1(2):282–295. https://doi.org/10.1038/eye.1987.47

Mazzoni, F., Müller, C., DeAssis, J., Lew, D., Leevy, W.M. and Finnemann, S.C. 2019. Non-invasive in vivo fluorescence imaging of apoptotic retinal photoreceptors. *Sci Rep* 9(1):1590. https://doi.org/10.1038/s41598-018-38363-z

Milam, A.H., Curcio, C.A., Cideciyan, A.V., Saxena, S., John, S.K., Kruth, H.S., Malek, G., Heckenlively, J.R., Weleber, R.G. and Jacobson, S.G. 2000. Dominant late-onset retinal degeneration with regional variation of sub-retinal pigment epithelium deposits, retinal function, and photoreceptor degeneration. *Ophthalmol* 107(12):2256–2266. https://doi.org/10.1016/S0161-6420(00)00419-X

Mirmehdi, M., Xie, X. and Suri, J. 2009. *Handbook of texture analysis*. Imperial College Press, London. http://www.iapr.org/docs/newsletter-2009-04.pdf

Mishra, A., Wong, A., Bizheva, K. and Clausi, D.A. 2009. Intra-retinal layer segmentation in optical coherence tomography images. *Opt Express* 17:23719–23728.

Mohammad, N., Schloss, R.S., Berjaud, P. and Boustany, N.N. 2018. Label-free dynamic segmentation and morphological analysis of subcellular optical scatterers. *J Biomed Opt* 23(9):1–11. https://doi.org/10.1117/1.JBO.23.9.096004

Molina, D., Pérez-Beteta, J., Luque, B., Arregui, E., Calvo, M., Borrás, J. M., López, C., Martino, J., Velasquez, C., Asenjo, B., Benavides, M., Herruzo, I., Martínez-González, A., Pérez-Romasanta, L., Arana, E. and Pérez-García, V.M. 2016. Tumour heterogeneity in glioblastoma assessed by MRI texture analysis: A potential marker of survival. *Br J Radiol* 89(1064):20160242. https://doi.org/10.1259/bjr.20160242

Monemian, M. and Rabbani, H. 2020. Mathematical analysis of texture indicators for the segmentation of optical coherence tomography images. *Optik* 219:165227. https://doi.org/10.1016/j.ijleo.2020.165227

Morgan, J.E. 2004. Circulation and axonal transport in the optic nerve. *Eye* 18(11):1089–1095. https://doi.org/10.1038/sj.eye.6701574

Morgan, J.E., Tribble, J., Fergusson, J., White, N. and Erchova, I. 2017. The optical detection of retinal ganglion cell damage. *Eye* 31:199. https://doi.org/10.1038/eye.2016.290

Morgan-Davies, J., Taylor, N., Hill, A.R., Aspinall, P., O'Brien, C.J. and Azuara-Blanco, A. 2004. Three dimensional analysis of the lamina cribrosa in glaucoma. *Br J Ophthalmol* 88(10):1299–1304. https://doi.org/10.1136/bjo.2003.036020

Morgner, U., Drexler, W., Kärtner, F.X., Li, X.D., Pitris, C., Ippen, E.P. and Fujimoto, J.G. 2000. Spectroscopic optical coherence tomography. *Opt Letters* 25(2):111–113. https://doi.org/10.1364/OL.25.000111

Mourant, J.R., Canpolat, M., Brocker, C., Esponda-Ramos, O., Johnson, T.M., Matanock, A., Stetter, K. and Freyer, J.P. 2000. Light scattering from cells: The contribution of the nucleus and the effects of proliferative status. *J Biomed Opt* 5(2). https://doi.org/10.1117/1.429979

Mourant, J.R., Freyer, J.P., Hielscher, A.H., Eick, A.A., Shen, D. and Johnson, T.M. 1998. Mechanisms of light scattering from biological cells relevant to noninvasive optical-tissue diagnostics. *Appl Opt* 37(16):3586–3593. https://doi.org/10.1364/AO.37.003586

Nailon, W.H. 2010. Texture analysis methods for medical image characterisation. In: Mao, Y. (ed.), *Biomedical imaging*. IntechOpen, London. https://doi.org/10.5772/8912

Naseri, A., Pouyan, A.A. and Kavian, N. 2010. An image processing approach to automatic detection of retina layers using texture analysis. *2010 17th Iranian Conference of Biomedical Engineering (ICBME)* 1–4. https://doi.org/10.1109/ICBME.2010.5704951

Nickells, R.W. 1999. Apoptosis of retinal ganglion cells in glaucoma: An update of the molecular pathways involved in cell death. *Surv Ophthalmol* 43:S151–S161. https://doi.org/10.1016/S0039-6257(99)00029-6

Nunes, A., Silva, G., Alves, C., Batista, S., Sousa, L., Castelo-Branco, M., Bernardes, R. 2019. Textural information from the retinal nerve fibre layer in multiple sclerosis. *2019 IEEE 6th Portuguese Meeting on Bioengineering (ENBENG)* 1–4. https://doi.org/10.1109/ENBENG.2019.8692454

Oberholzer, R. and Rateitschak, K.H. 1996. Root cleaning or root smoothing: An in vivo study. *J Clin Periodontol* 23:326–330. https://doi.org/10.1111/j.1600-051X.1996.tb00553.x

Ojala, T., Pietikainen, M. and Maenpaa, T. 2002. Multiresolution gray-scale and rotation invariant texture classification with local binary patterns. *IEEE Trans Pattern Anal Mach Intell* 24(7):971–987. https://doi.org/10.1109/TPAMI.2002.1017623

Okamoto, K. and Shaw, J.M. 2005. Mitochondrial morphology and dynamics in yeast and multicellular eukaryotes. *Annu Rev Genet* 39(1):503–536. https://doi.org/10.1146/annurev.genet.38.072902.093019

Pan, Y., Birngruber, R., Rosperich, J. and Engelhardt, R. 1995. Low-coherence optical tomography in turbid tissue: Theoretical analysis. *Appl Opt* 34(28):6564–6574. https://doi.org/10.1364/AO.34.006564

Pascale, A., Dragob, F. and Govonia, S. 2012. Protecting the retinal neurons from glaucoma: Lowering ocular pressure is not enough. *Pharmacol Res* 66(1):19–32. https://doi.org/10.1016/j.phrs.2012.03.002

Pasternack, R.M., Zheng, J.-Y. and Boustany, N.N. 2010. Optical scatter changes at the onset of apoptosis are spatially associated with mitochondria. *J Biomed Opt* 15(4):040504. https://doi.org/10.1117/1.3467501

Pasternack, R.M., Zheng, J.-Y. and Boustany, N.N. 2011. Detection of mitochondrial fission with orientation-dependent optical Fourier filters. *Cytometry* 79A(2):137–148. https://doi.org/10.1002/cyto.a.21011

Pentland, A.P. 1984. Fractal-based description of natural scenes. *IEEE Trans Pattern Anal Mach Intell* 6(6):661–674. https://doi.org/10.1109/TPAMI.1984.4767591

Perfettini, J.-L., Roumier, T. and Kroemer, G. 2005. Mitochondrial fusion and fission in the control of apoptosis. *Trends in Cell Biol* 15(4):179–183. https://doi.org/10.1016/j.tcb.2005.02.005

Perry, V.H. and Cowey, A. 1985. The ganglion cell and cone distributions in the monkey's retina: Implications for central magnification factors. *Vis Res* 25(12):1795–1810. https://doi.org/10.1016/0042-6989(85)90004-5

Pietikainen, M.K. 2000. *Texture analysis in machine vision*. World Scientific Publishing Co., Inc., Singapore. https://doi.org/10.1142/4483

Podda, B. and Giachetti, A. (eds.). 2005. *Texture analysis of CT images for vascular segmentation: A revised run length approach*. Springer, Berlin, Heidelberg. https://doi.org/10.1007/11553595_111

Prilloff, S., Fan, J., Henrich-Noack, P. and Sabel, B.A. 2010. In vivo confocal neuroimaging (ICON): Non-invasive, functional imaging of the mammalian CNS with cellular resolution. *Eur J Neurosci* 31(3):521–528. https://doi.org/10.1111/j.1460-9568.2010.07078.x

Prince, M.J., Wimo, A., Guerchet, M.M., Ali, G.C., Wu, Y.-T. and Prina, M. 2015. *World Alzheimer report 2015—the global impact of dementia: An analysis of prevalence, incidence, cost and trends*. Alzheimer's Disease International, London. http://www.alz.co.uk/research/world-report-2015

Purnyn, H. 2013. The mammalian retina: Structure and blood supply. *Neurophysiology* 45(3):266–276. https://doi.org/10.1007/s11062-013-9365-6

Qiu, B., Huang, Z., Liu, X., Meng, X., You, Y., Liu, G., Yang, K., Maier, A., Ren, Q. and Lu, Y. 2020. Noise reduction in optical coherence tomography images using a deep neural network with perceptually-sensitive loss function. *Biomed Opt Express* 11(2):817–830. https://doi.org/10.1364/BOE.379551

Quigley, H.A. 1993. Open-angle glaucoma. *N Engl J Med* 328(15):1097–1106. https://doi.org/10.1056/nejm199304153281507

Quigley, H.A. 2011. Glaucoma. *Lancet* 377(9774):1367–1377. https://doi.org/10.1016/S0140-6736(10)61423-7

Quigley, H.A. and Broman, A.T. 2006. The number of people with glaucoma worldwide in 2010 and 2020. *Br J Ophthalmol* 90(3):262. https://doi.org/10.1136/bjo.2005.081224

Quigley, H.A., Dunkelberger, G.R. and Green, W.R. 1989. Retinal ganglion cell atrophy correlated with automated perimetry in human eyes with glaucoma. *Am J Ophthalmol* 107(5):453–464. https://doi.org/10.1016/0002-9394(89)90488-1

Radius, R.L. 1981. Regional specificity in anatomy at the lamina cribrosa. *Arch Ophthalmol* 99(3):478–480. https://doi.org/10.1001/archopht.1981.03930010480020

Ramola, A., Shakya, A.K. and Van Pham, D. 2020. Study of statistical methods for texture analysis and their modern evolutions. *Engineering Reports* 2(4):e12149. https://doi.org/10.1002/eng2.12149

Rapantzikos, K., Zervakis, M. and Balas, K. 2003. Detection and segmentation of drusen deposits on human retina: Potential in the diagnosis of age-related macular degeneration. *Med Image Anal* 7(1):95–108. https://doi.org/10.1016/S1361-8415(02)00093-2

Remé, C.E., Grimm, C., Hafezi, F., Wenzel, A. and Williams, T.P. 2000. Apoptosis in the retina: The silent death of vision. *Physiology* 15(3):120–124. https://doi.org/10.1152/physiologyonline.2000.15.3.120

Rogowska, J., Bryant, C.M. and Brezinski, M.E. 2003. Cartilage thickness measurements from optical coherence tomography. *J Opt Soc Am* 20:357–367.

Sandhu, H.S., Elmogy, M., Sharafeldeen, A.T., Elsharkawy, M., El-Adawy, N., Eltanboly, A., Shalaby, A., Keynton, R. and El-Baz, A. 2020. Automated diagnosis of diabetic retinopathy using clinical biomarkers, optical coherence tomography, and optical coherence tomography angiography. *Am J Ophthalmol* 216:201–206.

Sandhu, H.S., Eltanboly, A., Shalaby, A., Keynton, R., Schaal, S. and El-Baz, A. 2018. Automated diagnosis and grading of diabetic retinopathy using optical coherence tomography. *Invest Ophthalmol Vis Sci* 59(7):3155–3160. https://doi.org/10.1167/iovs.17-23677.

Schmitt, J.M. 1999. Optical coherence tomography (OCT): A review. *IEEE J Sel Top Quantum Electron* 5(4):1205–1215. https://doi.org/10.1109/2944.796348

Schmitt, J.M., Knüttel, A., Yadlowsky, M. and Eckhaus, M.A. 1994. Optical-coherence tomography of a dense tissue: Statistics of attenuation and backscattering. *Phys Med Biol* 39(10):1705. https://doi.org/10.1088/0031-9155/39/10/013

Schmitt, J.M., Xiang, S.H. and Yung, K.M. 1999. Speckle in optical coherence tomography. *J Biomed Opt* 4(1):95–105. https://doi.org/10.1117/1.429925

Schultheiss, M., Wenzel, D.A., Spitzer, M.S., Poli, S., Wilhelm, H., Tonagel, F. and Kelbsch, C. 2022. Die optische Kohärenztomographie in der Differenzialdiagnostik wichtiger neuroophthalmologischer Krankheitsbilder [Optical coherence tomography in the differential diagnostics of important neuro-ophthalmological disease patterns]. *Der Nervenarzt* 93:629–642. https://doi.org/10.1007/s00115-022-01302-5

Schultze, M. 1866. Zur Anatomie und Physiologie der Retina. *Archiv für mikroskopische Anatomie* 2(1):175–286. https://doi.org/10.1007/bf02962033

Schuman, J.S., Hee, M.R., Arya, A.V., Pedut-Kloizman, T., Puliafito, C.A., Fujimoto, J.G. and Swanson, E.A. 1995. Optical coherence tomography: A new tool for glaucoma diagnosis. *Curr Opin Ophthalmol* 6(2):89–95. https://doi.org/10.1097/00055735-199504000-00014

Shaban, M., Ogur, Z., Mahmoud, A., Switala, A., Shalaby, A., Abu Khalifeh, H., Ghazal, M., Fraiwan, L., Giridharan, G., Sandhu, H. and El-Baz, A. 2020. A convolutional neural network for the screening and staging of diabetic retinopathy. *PLOS One* 15(6):e0233514. https://doi.org/10.1371/journal.pone.0233514

Sharafeldeen, A., Elsharkawy, M., Khalifa, F., Soliman, A., Ghazal, M., AlHalabi, M., Yaghi, M., Alrahmawy, M., Elmougy, S., Sandhu, H.S. and El-Baz, A. 2021. Precise higher-order reflectivity and morphology models for early diagnosis of diabetic retinopathy using OCT images. *Sci Rep* 11:4730. https://doi.org/10.1038/s41598-021-83735-7

Shim, M.S., Takihara, Y., Kim, K.-Y., Iwata, T., Yue, B.Y.J.T., Inatani, M., Weinreb, R.N., Perkins, G.A. and Ju, W.-K. 2016. Mitochondrial pathogenic mechanism and degradation in optineurin E50K mutation-mediated retinal ganglion cell degeneration. *Sci Rep* 6:33830. https://doi.org/10.1038/srep33830

Sleman, A.A., Soliman, A., Elsharkawy, M., Giridharan, G., Ghazal, M., Sandhu, H., Schaal, S., Keynton, R., Elmaghraby, A. and El-Baz, A. 2021. A novel 3D segmentation approach for extracting retinal layers from optical coherence tomography images. *Med Phys* 48(4):1584–1595. https://doi.org/10.1002/mp.14720. Epub 2021 February 24. PMID: 33450073.

Smerdon, D. 2000. Anatomy of the eye and orbit. *Curr Anaesth Crit Care* 11(6):286–292. https://doi.org/10.1054/cacc.2000.0296

Somasundaran, S., Constable, I.J., Mellough, C.B. and Carvalho, L.S. 2020. Retinal pigment epithelium and age-related macular degeneration: A review of major disease mechanisms. *Clin Exp Ophthalmol* 48(8):1043–1056. https://doi.org/10.1111/ceo.13834

Song, G., Steelman, Z.A., Finkelstein, S., Yang, Z., Martin, L., Chu, K.K., Farsiu, S., Arshavsky, V.Y. and Wax, A. 2020. Multimodal coherent imaging of retinal biomarkers of Alzheimer's disease in a mouse model. *Sci Rep* 10:7912. https://doi.org/10.1038/s41598-020-64827-2

Sparrow, J.R., Hicks, D. and Hamel, C.P. 2010. The retinal pigment epithelium in health and disease. *Curr Mol Med* 10(9):802–823. https://doi.org/10.2174/156652410793937813

Stone, J. and Johnston, E. 1981. The topography of primate retina: A study of the human, bushbaby, and new- and old-world monkeys. *J Comp Neurol* 196(2):205–223. https://doi.org/10.1002/cne.901960204

Sun, M., Zhang, H., Chen, X. and Zhang, Q. 2020. Quantitative analysis of macular retina using light reflection indices derived from SD-OCT for pituitary adenoma. *J Ophthalmol* 2020:8896114. https://doi.org/10.1155/2020/8896114

Sung, C.-H. and Chuang, J.-Z. 2010. The cell biology of vision. *J Cell Biol* 190(6):953–963. https://doi.org/10.1083/jcb.201006020

Swanson, E.A., Izatt, J.A., Hee, M.R., Huang, D., Lin, C.P., Schuman, J.S., Puliafito, C.A. and Fujimoto, J.G. 1993. In vivo retinal imaging by optical coherence tomography. *Opt Lett* 18(21):1864–1866. https://doi.org/10.1364/OL.18.001864

Tan, O., Chopra, V., Lu, A.T.-H., Schuman, J.S., Ishikawa, H., Wollstein, G., Varma, R. and Huang, D. 2009. Detection of macular ganglion cell loss in glaucoma by Fourier-domain optical coherence tomography. *Ophthalmology* 116(12):2305–2314. https://doi.org/10.1016/j.ophtha.2009.05.025

Tazarjani, H.D., Amini, Z., Kafieh, R., Ashtari, F. and Sadeghi, E. 2021. Retinal OCT texture analysis for differentiating healthy controls from multiple sclerosis (MS) with/without optic neuritis. *Biomed Res Int* 2021:5579018. https://doi.org/10.1155/2021/5579018

Tian, H., Li, L. and Song, F. 2017. Study on the deformations of the lamina cribrosa during glaucoma. *Acta Biomater* 55:340–348. https://doi.org/10.1016/j.actbio.2017.03.028

Ting, D.S.W., Cheung, C.Y.-L., Lim, G., Tan, G.S.W., Quang, N.D., Gan, A., Hamzah, H., Garcia-Franco, R., Yeo, I.Y.S., Lee, S.Y., Wong, E.Y.M., Sabanayagam, C., Baskaran, M., Ibrahim, F., Tan, N.C., Finkelstein, E.A., Lamoureux, E.L., Wong, I.Y., Bressler, N.M., Sivaprasad, S., Varma, R., Jonas, J.B., He, M.G., Cheng, C.-Y., Cheung, G.C.M, Aung, T., Hsu, W., Lee, M.L. and Wong, T.Y. 2017. Development and validation of a deep learning system for diabetic retinopathy and related eye diseases using retinal images from multiethnic populations with diabetes. *JAMA* 318(22):2211–2223. https://doi.org/10.1001/jama.2017.18152

Toth, C.A., Narayan, D.G., Boppart, S.A., Hee, M.R., Fujimoto, J.G., Birngruber, R., Cain, C.P., DiCarlo, C.D. and Roach, W.P. 1997. A comparison of retinal morphology viewed by optical coherence tomography and by light microscopy. *Arch Ophthalmol* 115(11):1425–1428. https://doi.org/10.1001/archopht.1997.01100160595012

Tso, M.O., Zhang, C., Abler, A.S., Chang, C.J., Wong, F., Chang, G.Q. and Lam, T.T. 1994. Apoptosis leads to photoreceptor degeneration in inherited retinal dystrophy of RCS rats. *Invest Ophthalmol Vis Sci* 35(6):2693–2699.

Tudor, D., Kajić, V., Rey, S., Erchova, I., Považay, B., Hofer, B., Powell, K.A., Marshall, D., Rosin, P.L., Drexler, W. and Morgan, J.E. 2014. Non-invasive detection of early retinal neuronal degeneration by ultrahigh resolution optical coherence tomography. *PLOS One* 9(4):e93916. https://doi.org/10.1371/journal.pone.0093916

Unterhuber, A., Považay, B., Hermann, B., Sattmann, H., Chavez-Pirson, A. and Drexler, W. 2005. In vivo retinal optical coherence tomography at 1040 nm—enhanced penetration into the choroid. *Opt Exp* 13(9):3252–3258. https://doi.org/10.1364/opex.13.003252

van der Meer, F.J., Faber, D.J., Aalders, M.C.G., Poot, A.A., Vermes, I. and van Leeuwen, T.G. 2010. Apoptosis- and necrosis-induced changes in light attenuation measured by optical coherence tomography. *Lasers Med Sci* 25(2):259–267. https://doi.org/10.1007/s10103-009-0723-y

Vienola, K.V., Zhang, M., Snyder, V.C., Sahel, J.-A., Dansingani, K.K. and Rossi, E.A. 2020. Microstructure of the retinal pigment epithelium near-infrared autofluorescence in healthy young eyes and in patients with AMD. *Sci Rep* 10(1):9561. https://doi.org/10.1038/s41598-020-66581-x

Wang, R.K. and Tuchin, V.V. 2013. Optical coherence tomography: Light scattering and imaging enhancement. In: Tuchin, V.V. (ed.), *Handbook of coherent-domain optical methods: Biomedical diagnostics, environmental monitoring, and materials science*. Springer, New York, NY, pp. 665–742. https://doi.org/10.1007/978-1-4614-5176-1_16

Wang, Y., Shen, D., Wang, V.M., Yu, C.-R., Wang, R.-X., Tuo, J. and Chan, C.-C. 2012. Enhanced apoptosis in retinal pigment epithelium under inflammatory stimuli and oxidative stress. *Apoptosis Int J Prog Cell Death* 17(11):1144–1155. https://doi.org/10.1007/s10495-012-0750-1

Wang, Y., Zhang, Y., Yao, Z., Zhao, R. and Zhou, F. 2016. Machine learning based detection of age-related macular degeneration (AMD) and diabetic macular edema (DME) from optical coherence tomography (OCT) images. *Biomed Opt Express* 7(12):4928–4940. https://doi.org/10.1364/BOE.7.004928

Wässle, H. and Boycott, B.B. 1991. Functional architecture of the mammalian retina. *Phys Rev* 71(2):447–480. https://doi.org/10.1152/physrev.1991.71.2.447

Westwood, D.A. 2009. Visual pathways for perception and action. In: Binder, M.D. et al. (eds.), *Encyclopedia of neuroscience*. Springer, Berlin, Heidelberg, pp. 4324–4327. https://doi.org/10.1007/978-3-540-29678-2_6362

Yang, X., Zou, H., Jung, G., Richard, G., Linke, S.J., Ader, M. and Bartsch, U. 2013. Nonneuronal control of the differential distribution of myelin along retinal ganglion cell axons in the mouse. *Invest Ophthalmol Vis Sci* 54(13):7819–7827. https://doi.org/10.1167/iovs.13-12596

Yap, T.E., Donna, P., Almonte, M.T. and Cordeiro, M.F. 2018. Real-time imaging of retinal ganglion cell apoptosis. *Cells* 7(6):60. https://doi.org/10.3390/cells7060060

Yi, J., Puyang, Z., Feng, L., Duan, L., Liang, P., Backman, V., Liu, X. and Zhang, H.F. 2016. Optical detection of early damage in retinal ganglion cells in a mouse model of partial optic nerve crush injury. *Invest Ophthalmol Vis Sci* 57(13):5665–5671. https://doi.org/10.1167/iovs.16-19955

Yogesan, K., Jørgensen, T., Albregtsen, F., Tveter, K.J. and Danielsen, H.E. 1996. Entropy-based texture analysis of chromatin structure in advanced prostate cancer. *Cytometry* 24(3):268–276. https://doi.org/10.1002/(SICI)1097–0320(19960701)24:3<268::AID-CYTO10>3.0.CO;2-O

Youle, R.J. and Karbowski, M. 2005. Mitochondrial fission in apoptosis. *Nat Rev Mol Cell Biol* 6:657. https://doi.org/10.1038/nrm1697

Zeimer, R.C. and Ogura, Y. 1989. The relation between glaucomatous damage and optic nerve head mechanical compliance. *Arch Ophthalmol* 107(8):1232–1234. https://doi.org/10.1001/archopht.1989.01070020298042

Zhang, L., Lee, K., Niemeijer, M., Mullins, R.F., Sonka, M. and Abràmoff, M.D. 2012. Automated segmentation of the choroid from clinical SD-OCT. *Invest Ophthalmol Vis Sci* 53(12):7510–7519. https://doi.org/10.1167/iovs.12-10311

Zhang, L.-Q., Cui, H., Yu, Y.-B., Shi, H.-Q., Zhou, Y. and Liu, M.-J. 2019. MicroRNA-141–3p inhibits retinal neovascularization and retinal ganglion cell apoptosis in glaucoma mice through the inactivation of Docking protein 5-dependent mitogen-activated protein kinase signaling pathway. *J Cell Phys* 234(6):8873–8887. https://doi.org/10.1002/jcp.27549

Zhongyang, S. and Yankui, S. 2019. Automatic detection of retinal regions using fully convolutional networks for diagnosis of abnormal maculae in optical coherence tomography images. *J Biomed Opt* 24(5):1–9. https://doi.org/10.1117/1.JBO.24.5.056003

Zweifel, S.A., Imamura, Y., Spaide, T.C., Fujiwara, T. and Spaide, R.F. 2010. Prevalence and significance of subretinal drusenoid deposits (reticular pseudodrusen) in age-related macular degeneration. *Ophthalmology* 117(9):1775–1781. https://doi.org/10.1016/j.ophtha.2010.01.027

8 Automation in Pneumonia Detection

Nur Syafiqah Shaharudin, Noraini Hasan, and Nurbaity Sabri

8.1 INTRODUCTION: BACKGROUND STUDY

Pneumonia, a respiratory system disease, is not foreign in Malaysia. Its most recent related disease attack, the coronavirus (COVID-19), scientifically under pneumonia, has taken the lives of so many people around the globe, affecting their respiratory system and causing mortality. Its deadly infections has also taken the lives of people primarily with low immunity the most, such as the elderly and children between 14 years old and earlier. Pneumonia is also recorded at 11.8% as one of the top mortality diseases, after ischemic heart disease, in Malaysia, by the Department of Statistics Malaysia (DOSM), with 4.8% mortality rate among children between 14 years old and infants that are affected (released by the press in 2019) [1]. The infection causes inflammation on one or both lungs, where the alveoli are filled with pus, causing difficulty breathing and other symptoms [2].

The locality of the infections in a restricted area causes difficulty in diagnosing [3] the contamination simply by radiographic images. Other than viral and bacterial causes, pneumonia can also be classified as atypical or fungal. However, the most common attacks are viral and bacterial pneumonia [4]. It happens when the air sacs (alveoli) are inflamed because of the filled-up fluids, which leads to difficulty breathing because of a lack of oxygen and blood transfer to the whole body. Consequently, it leads to shortness of breath, chills, cough, chest pain, and other symptoms [2] or, if unfortunate, may lead to death. On that account, the research was to find an alternative way to classify the identical appearance of X-ray images of viral and bacterial pneumonia. Thus, it benefits doctors and other experts in discovering the infections in a short period and in making a decision immediately to avoid any complications.

Other than that, the goal is to overcome misinterpretation of images and the scarcity of insufficient expert and aid support [5, 6] in managing the inspection of the infection type through radiographic images. In addition, the restricted area limits the vision for image classification because of its identical appearance in an occluded area with bones structure and air space opacifications. Therefore, the system proposed uses an economically affordable [7, 8] shallow learning approach with improved texture features to classify uninfected and pneumonia-infected images, such as viral and bacterial. Thus, the system model aims to alleviate challenges to radiologists, doctors, and other experts in reliable image interpretation and faster diagnosis for the healthcare system.

DOI: 10.1201/9780367486082-8

8.2 DETECTION SYSTEM MATERIALS

To conclude a complete system application, necessary materials are required. Therefore, the data collected was found by a recommended open source, and the tools used for the application production were also able to support scientists, experts, and other researchers for medical application system support.

8.2.1 Dataset Collection

The obtained datasets, X-ray images, were collected from an open-source database site, Kaggle [9]. It is practically an online community space where artificial intelligence (AI) practitioners and experts, as well the beginners, can purposely seek and share their research approach and exchange opinions and findings. The total gathered data, 150 collected radiographic images, was categorized into normal and pneumonia of viral and bacterial origin accordingly. The images gathered, such as in Figure 8.1, are divided into three classes—normal, bacterial pneumonia, and viral pneumonia—and separated for the system training and testing parameter in a ratio of 80:20.

8.2.2 Applied Tools

The system was developed and run using MATLAB R2019a tools. The application also consists of advanced tools, which enable it to utilize imaging and processing techniques, such as in medical, engineering, and experimental processes. It is also well performed in programming, numeric, and media.

8.3 METHODOLOGY

In this new revolutionized era, the modern trend has changed and ushered people and industries into AI. Accordingly, it had also influenced the industries emerging in

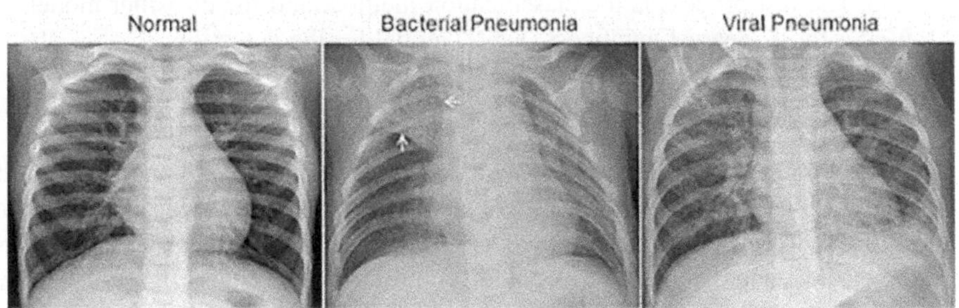

FIGURE 8.1 Radiographic images of normal lungs and lungs with bacterial and viral pneumonia.

Source: Kermany et al. (2018) [10].

technology, overcoming tedious and repetitive tasks in many areas, simultaneously reducing scarcities [5] and complexity. On top of that, many have relied on modernized systems to overcome time constraints and human errors. The related issue, pneumonia, has also been done by other researchers and experts. However, there are still some that have come into inefficient technique. Therefore, the image processing method using the shallow learning (SL) approach was conducted. The research and findings gathered have also accordingly validated the supervised system, using the appropriate techniques, as in Figure 8.2, where it is able to perform the analysis and classification of image datasets, detecting its abnormality from collected normal and pneumonia-infected images accurately.

8.3.1 Technique

The techniques applied for this system approach are conducted successfully. The shallow learning approach can be classified into a few process techniques, which consists of:

- ***Image Acquisition.*** The 150 collected images were obtained in a digitized format (.jpeg) from the open-source Kaggle dataset that comes with the normal and two types of pneumonia, viral and bacterial, images separately.
- ***Preprocessing.*** Each of the uploaded images is resized into 800 × 800 row–column size. It is then denoised and filtered for a clearer and balanced image contrast and intensities, using contrast-limited adaptive histogram equalization (CLAHE) and the average filter. The image segmentation technique used are morphological and thresholding to automatically select the interest region and separate from the other unnecessary objects, such as bones, wires, and others.
- ***Feature Extraction.*** The image values are extracted and normalized using a second-order statistical method feature, the gray level co-occurrence matrix (GLCM) for the image contrast, correlation, energy, and homogeneity.
- ***Classification.*** The extracted features support a better process for classification. The last process is the classification method, and the classifier model used is the K-nearest neighbors (KNN) model for the detection.

8.3.2 Equations

In the development of an application system using the image processing technique, it is essential to use equations. The statistical method structure for image classification is a must for better performance. In this approach, the texture feature GLCM calculated the tabled image values and its contrast combinations value of the image gray level occurrence to bring the informative textures for the process. Thus, it enables the processed images to be classified easily to distinguish the difference of the radiographic images between normal and bacterial and viral pneumonia images. GLCM compares the gray level of the image with the original image pixels, measuring its concentration in a specific area.

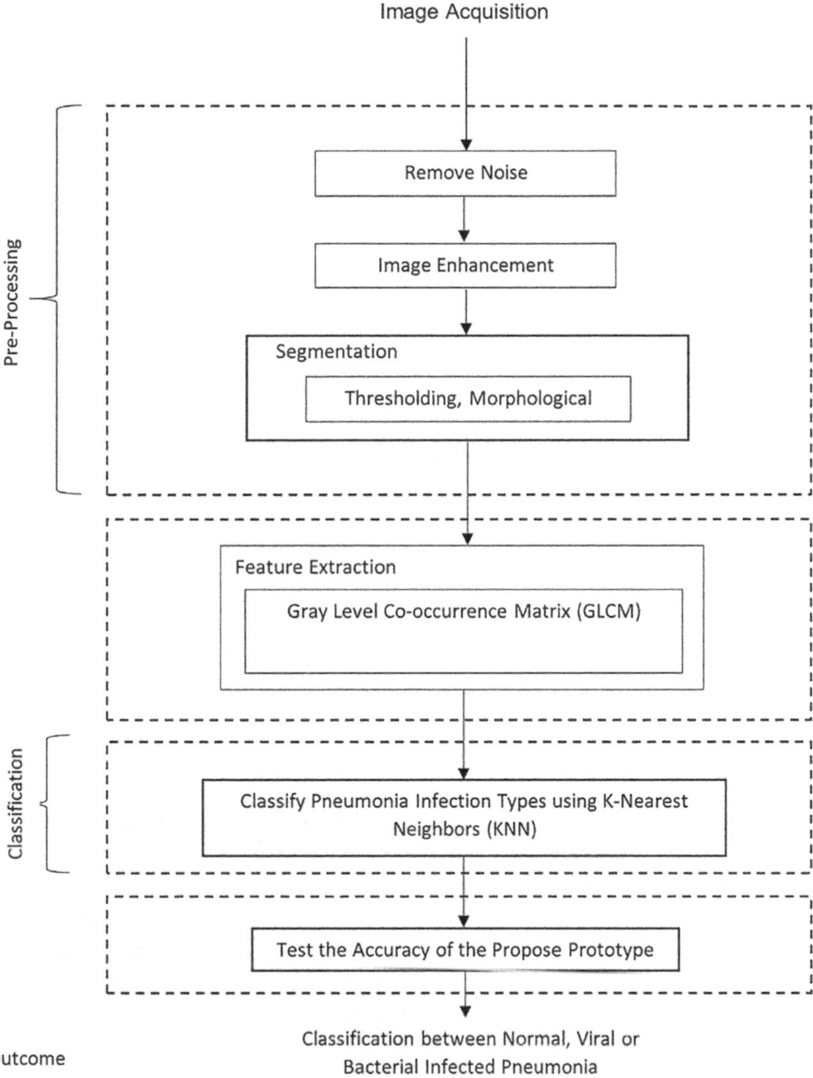

FIGURE 8.2 Shallow learning system framework.

Thus, to find the normalized symmetric GLCM elements, the operation exchanges the gray level values with the images row–column pixel values and combines the matrices of the image co-occurrence with the transposed matrix, referring to Figure 8.3. Further equations (8.1)–(8.4) represent the texture analysis statistics operation, with Pij as the normalized symmetrical of GLCM, and the N representing the number of gray levels of the image.

$$\text{Energy} = \sum\_(\text{i,j}=0)\wedge(\text{N}-1)(\text{P\_ij})\wedge 2 \quad (8.1)$$

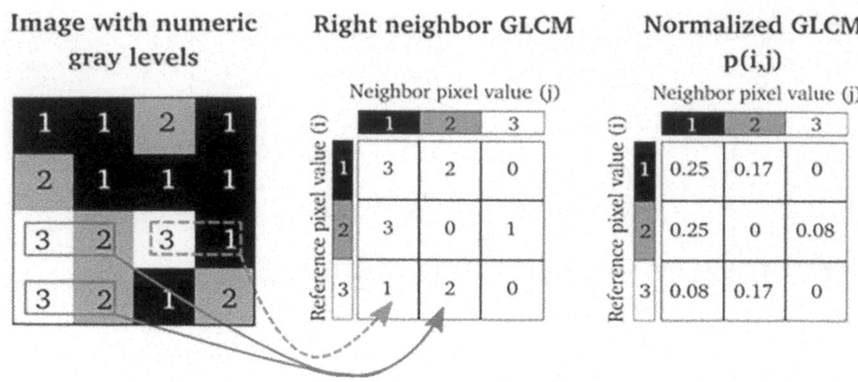

FIGURE 8.3 Illustrated GLCM architecture.

Source: Löfstedt, Brynolfsson, Asklund, Nyholm, and Garpebring (2019) [11].

$$\text{Contrast} = \Sigma\_(i,j=0)^\wedge(N-1)\ P\_ij\ (i-j)^\wedge 2 \tag{8.2}$$

$$\text{Correlation} = \left(\Sigma\left(x\_(i-)x^{-}\right)\left(y\_(i-)y^{-}\right)\right)/\sqrt{}\left(\Sigma(x\_(i-)x^{-})^\wedge 2(y\_(i-)y^{-})^\wedge 2\right) \tag{8.3}$$

$$\text{Homogenetity} = \Sigma\_(i,j=0)^\wedge(N-1)P\_ij/\left(1+(i-j)^\wedge 2\right) \tag{8.4}$$

Apart from that, KNN model is used, using the Euclidean distance method (8.5), for the classification operation process, with its distance value of 4. The algorithm measures the object distance and similarities. It detects the nearest distance of the training images and the test samples. Thus, using cross-validation with sufficient datasets, KNN is also known as one of the most adaptable yet simple models for supervised learning, and it is suitable for image classification. Its performance is almost at its best with the other complex classifiers.

$$\text{distance}(a,b) = \sqrt{}((x\_1-x\_2)^\wedge 2+(y\_1-y\_2)^\wedge 2) \tag{8.5}$$

8.4 RESULTS AND FINDINGS

The classification results of the shallow learning approach efficiently classify the three classes of X-ray images, the two types of pneumonia, bacterial and viral images, also the normal. Accordingly, the validated classifier, KNN, was trained well for all obtained X-ray images, with 120 training images and 30 testing images for the final process. The percentage of classification model accuracy is measured using equation (8.6). Thus, model accuracy was measured using the number of positively (true positive) classified images, TP, divided its value by the total of classified images, then multiplied the results with the percentage (100%) to get its final percentage value.

TABLE 8.1
Image Testing Classification Results

Type of Classes	Total of Tested Images	Classified X-Ray Images	
		True	False
Bacterial	10	9	1
Normal	10	9	1
Viral	10	8	2

Therefore, Table 8.1 indicates the accuracy results for the classification method, with its accuracy of 86.67%.

$$\text{Accuracy} = \text{TP}/(\text{Total Images}) \times 100 \quad (8.6)$$

8.5 CONCLUSIONS

The proposed system, automated pneumonia detection, was developed to overcome misinterpretation and the time constraints in inspecting infections, between viral and bacterial pneumonia, also the normal images. The developed system is supported using a supervised classifier model, KNN with suitable texture analysis, GLCM, and its proper statistical features for better image processes. Therefore, considering the obtained results and system visibility, the developed classification model using a shallow learning approach is applicable to support experts and other researchers in diagnosing medical imaging accurately.

REFERENCES

1. Statistics on Causes of Death, Malaysia, 2020. Department of Statistics Malaysia Official Portal. (2020, November 26). https://www.dosm.gov.my/v1/index.php?r=column%2Fctheme ByCat&cat=401&bul_id=QTU5T0dKQ1g4MHYxd3ZpMzhEMzdRdz09&menu_id=L0pheU43NWJwRWVSZklWdzQ4TlhUUT09
2. Pneumonia. (n.d.). https://www.nhlbi.nih.gov/health-topics/pneumonia
3. Rajpurkar, P. L., Irvin, J. P., Ball, R. N., Zhu, K. W., Yang, B. G., Mehta, H. J., . . . Lungren, M. (2017, December 25). *Deep learning for chest radiograph diagnosis: A retrospective comparison of the CheXNeXt algorithm to practicing radiologists*. https://journals.plos.org/plosmedicine/article/authors?id=10.1371/journal.pmed.1002686
4. Mohd Razali, N. (2017, April 28). *Pneumonia*. Portal MyHEALTH. Retrieved November 6, 2021, from http://www.myhealth.gov.my/en/pneumonia/
5. Rajaraman, S., Candemir, S., Kim, I., Thoma, G., & Antani, S. (2018). Visualization and interpretation of convolutional neural network predictions in detecting pneumonia in pediatric chest radiographs. *Applied Sciences*, 8(10), 1715. https://doi.org/10.3390/app8101715
6. Verma, G., & Prakash, S. (2020). Pneumonia classification using deep learning in healthcare. *International Journal of Innovative Technology and Exploring Engineering Regular Issue*, 9(4), 1715–1723. https://doi.org/10.35940/ijitee.d1599.029420

7. Razzak, M. I., Naz, S., & Zaib, A. (2017). Deep learning for medical image processing: Overview, challenges and the future. *Lecture Notes in Computational Vision and Biomechanics Classification in BioApps*, 323–350. https://doi.org/10.1007/978-3-319-65981-7_12
8. Harris, S. (2018, October). *Predictions for 2019 and beyond*. Purestorage. Retrieved November 6, 2021, from https://s3-eu-west-2.amazonaws.com/signifyresearch/app/uploads/2018/10/16101114/Signify_AI-in-Medical-Imaging-White-Paper.pdf
9. Mooney, P. (2018, March 24). *Chest x-ray images (pneumonia)*. From https://www.kaggle.com/paultimothymooney/chest-xray-pneumonia
10. Kermany, D. S., Goldbaum, M., Cai, W., Valentim, C. C., Liang, H., Baxter, S. L., . . . Zhang, K. (2018). Identifying medical diagnoses and treatable diseases by image-based deep learning. *Cell*, 172(5). doi: 10.1016/j.cell.2018.02.010
11. Löfstedt, T., Brynolfsson, P., Asklund, T., Nyholm, T., & Garpebring, A. (2019). Gray-level invariant Haralick texture features. *PLoS One*, 14(2). doi:10.1371/journal.pone.0212110

9 Texture for Neuroimaging

Ana Nunes, Pedro Serranho, Miguel Castelo-Branco, and Rui Bernardes

9.1 INTRODUCTION

Texture provides essential visual cues about surface properties and is at the core of human vision in the identification of objects and understanding of complex scenes (Julesz 1962) (Bergen and Landy 1991), therefore playing a fundamental role in our daily living. Acknowledging this role, computer vision embraced the concept of visual texture, with research efforts focused on simulating the human perception of texture by computing mathematical representations of image texture that can be efficiently dealt with.

Although fairly intuitive, the concept of image texture is hard to define in an unequivocal fashion. There is no universally agreed-upon definition of this concept, and distinct descriptions can be found across the literature. While some authors describe texture in terms of the qualitative properties that allow discerning visually between different textured regions as "texture as what constitutes a macroscopic region" in Tamura, Mori, and Yamawaki (1978), "the variation of data at scales smaller than the scales of interest" in Petrou and García-Sevilla (2006), or the "properties of an image region which convey tactile information about a corresponding surface" in Matthews (2016), it is pretty common to find definitions of image texture as the (statistical) spatial distribution of the image pixels' gray-level/intensity/brightness values (Haralick, Shanmugam and Dinstein 1973; Tourassi 1999; Castellano, et al. 2004). For the present discussion, we will consider the latter definition—the statistical spatial distribution of pixel values in monochrome images. It then follows that *texture analysis* can be regarded as an umbrella term for different techniques quantifying or characterizing the spatial distribution of the gray levels of an image.

In the upcoming sections of this chapter, we will briefly cover some of the basic concepts and recent developments in texture analysis. We will provide the context for the role and potential of texture-based approaches for applications in the field of neuroimaging.

9.2 HISTORICAL CONTEXT AND CURRENT TRENDS

The earliest studies in texture analysis date back to the 1950s (Kaizer 1956) and 1960s (Julesz 1962), when texture was initially addressed from a psychophysics point of view, with research efforts primarily focused on how the human visual system perceives textures. For approximately two decades, psychophysical analyses of

DOI: 10.1201/9780367486082-9

texture dominated the literature together with the very first practical applications of texture-based approaches in the field of remote sensing (Haralick, Shanmugam and Dinstein 1973; Haralick 1971; Harlow, et al. 1981), namely, in radar (Shanmugan, et al. 1981) and satellite (Darling and Joseph 1968) imagery. In the 1970s, medical imaging applications began to emerge (Ledley 1972), with texture analysis methods being initially discussed and applied in the context of radiographic imaging (Hall, et al. 1971; Chien and Fu 1974; Harlow and Eisenbeis 1973). By the early 2000s, applications of texture analysis in neuroimaging were being regularly reported, notably in neurooncology (Herlidou-Même, et al. 2003; Mahmoud-Ghoneim, et al. 2003), epilepsy (Bernasconi, et al. 2001; Antel, et al. 2003; Bonilha, et al. 2003; Yu, et al. 2001), and schizophrenia (Nedelec, et al. 2004; Saeed and Puri 2002), but also in neurodegenerative disorders like Alzheimer's disease (AD) (Nedelec, et al. 2004; Sayeed, et al. 2001) and multiple sclerosis (MS) (Mathias, Tofts and Losseff 1999).

Strikingly, the initial interest in researching what is perceived as texture has recently been revived under a completely different paradigm: texture-based deep learning. Research efforts are now concentrating on the concept of *learned texture*, that is, the texture elements that are learned by convolutional neural networks (CNNs) during training. CNNs are currently considered the best possible model for human object recognition (Cadieu, et al. 2016; Yamins, et al. 2016; Kubilius, Bracci and Beeck 2016). Different CNNs have been shown to reach remarkably high performances at object detection (Szegedy, et al. 2015), image recognition (He, et al. 2016), and classification (Krivesky, Sutskever and Hinton 2012). The most recent and the widely accepted hypothesis for CNNs competence is grounded on the so-called *shape hypothesis* (Geirhos, et al. 2019). These networks follow a bottom-up approach, aggregating low-level features into increasingly more complex shapes until an object is recognized from a set of pre-defined classes. However, recent empirical findings (Geirhos, et al. 2019; Ballester and Araujo 2016; Brendel and Bethge 2019; Gatys, Ecker and Bethge 2017) have pointed toward the newly proposed *texture hypothesis* (Geirhos, et al. 2019): the idea that CNNs rely more on texture than on the object's shape for object recognition and classification.

Similarly to the *texture renaissance* identified by Kassner and Thornhill more than a decade ago (Kassner and Thornhill 2010), we are perhaps now witnessing yet another inflexion point in the field of texture analysis: the transition from using pre-selected (user-defined) methods to compute texture-based metrics to automatically (deep-)learned texture features. This transition is happening now, as demonstrated by its presence in recent surveys on texture methods (Liu, et al. 2019; Humeau-Heurtier 2019; Ghalati, et al. 2021).

9.3 THE ROLE OF TEXTURE IN NEUROIMAGING

The use of texture analysis as a diagnostic tool is based on the premise that medical images contain a *texture signature* (Tourassi 1999) specific to a given biological process, either normal or pathological. This foundation is established on the grounds of texture importance for human vision and the visual analysis of medical images toward image-based diagnosis. Therefore, texture analysis has a natural application in computer-aided (CAD) diagnosis and radiomics. It can be used as a complementary tool to extract diagnostically valuable quantitative information from medical images—specifically, that the human visual system cannot detect because of the human visual system limitations, for example, spotting subtle differences in intensity, a limitation not suffered by the numerical treatment performed by computers.

The potential of texture-based methods to maximize the information extracted from medical images is broadly recognized. It has been proposed that they should be integrated into the clinical diagnosis decision process after proper validation (Scalco and Rizzo 2017; Varghese, et al. 2019; Eun, et al. 2020) because the quantitative nature of texture-based measures is immune to subjectivity, such as the inter- and intra-observer variability in the visual interpretation of medical images. As such, texture analysis can act as an anchor in image analysis protocols, making them more systematic and robust. This is particularly crucial in longitudinal applications that monitor the normal aging process, neurological disease progression, and response to treatment, to name but a few.

In neuroimaging, texture analysis applications are mostly centred on two medical fields: neurology—the study of neurological and neurodegenerative disorders—and neurooncology. In neurology applications, texture can be an informative instrument used in the quest for disorder-specific biomarkers. In these applications, a set of texture features supports the classification of the analyzed tissue into the healthy/pathological group. However, the suitability of texture-based methods for oncology applications is primarily based on the fact that the tumor structure can be quantified in terms of texture parameters, for example, heterogeneity, which has been associated with tumor aggressiveness (Gay, Baker and Graham 2016; Cyll, et al. 2017), where this heterogeneity is associated with the distinct cell groups affected and the deposition of different aggregates (e.g., amyloid) and elements (e.g., iron).

9.4 TEXTURE METHODS IN NEUROIMAGING

Traditionally, texture analysis methods have been grouped into four classical categories of feature-based methods: statistical, transform-based, model-based, and structural approaches. These categories aggregate methods based on the type of texture information they capture and how it is encoded.

The following lines will cover the most used categories in neuroimaging applications: the statistical, transform-based, and model-based categories. Since they are less frequently applied in the field of neuroimaging, structural methods will be covered in Subsection 9.4.4, along with two other recently proposed categories of feature-based approaches, namely, graph-based and entropy-based methods. In Subsection 9.4.5, learned-based approaches—a new paradigm in texture analysis—will be discussed.

9.4.1 Statistical-Based Methods

Texture features obtained using statistical methods quantify and characterize the local spatial distribution of the gray level values in an image. Statistical texture features are generally considered to be powerful texture descriptors. Globally, this category of methods is the most broadly used across different fields of application, medical imaging (and neuroimaging) included.

A statistical method that stands out for its popularity and range of use is the gray level co-occurrence matrix (GLCM); it tabulates the occurrence of the possible gray level combinations between pixel pairs in a given spatial relation. Despite being one of the earliest developed texture analysis methods (Haralick, Shanmugam and Dinstein 1973), the GLCM is still a benchmark method in comparative studies, namely, in classification-based problems using the aforementioned texture ensemble approaches. Regarding medical imaging applications, it could even be argued that the GLCM method embodies the gold standard to texture analysis.

In neuroimaging, the GLCM has been used in applications following a *texture ensemble* approach, where multiple texture features are computed using different texture analysis methods, in the expectation that a broad range of texture-based metrics will maximize the chances of finding the most descriptive/discriminative ones for the problem at hand. This is a blind search approach, as there is no rationale for using any specific texture analysis method based on the nature of the image's texture under study.

Texture ensemble approaches involving the computation of GLCM metrics have been reported mostly for magnetic resonance imaging (MRI) (Herlidou-Même, et al. 2003; Bonilha, et al. 2003; Yu, et al. 2001; Ortiz-Ramón, et al. 2019; Zhang, et al. 2012; Baskar, Jayanthi and Jayanthi 2019; Rahmim, et al. 2017; Liu, et al. 2020), but also single-photon emission computed tomography (SPECT) (Rahmim, et al. 2017) and optical coherence tomography (OCT) data (Tazarjani, et al. 2021; Nunes, Silva, Duque, et al. 2019; Nunes, et al. 2020). Occasionally, such approaches consist of a simple combination of GLCM-based and histogram features, as in Nedelec et al. (2004), Fujima et al. (2019), Salas-Gonzalez et al. (2009), and Li et al. (2019).

Notably, the GLCM is the texture analysis method most often employed in a standalone fashion, with applications concerning the diagnosis of neurodegenerative disorders such as MS (Mathias, Tofts and Losseff 1999; Nunes, Silva, Alves, et al. 2019), AD (Raut and Dalal 2017; Leandrou, et al. 2020; Lee, Lee and Kim 2020; Lee and Kim 2021; Luk, et al. 2018; Ferreira, et al. 2020), Parkinson's disease (PD) (Sikiö, Holli-Helenius, et al. 2015), and amyotrophic lateral sclerosis (ALS) (Albuquerque, et al. 2016; Maani, et al. 2016; Ishaque, et al. 2018), and the identification of brain tumors (Mahmoud-Ghoneim, et al. 2003; Gazdzinski and Nieman 2014), ischemic stroke (Kassner, et al. 2009; Sikiö, Kölhi, et al. 2015), and epilepsy (Antel, et al. 2003).

As the GLCM considers pairs of image pixels with certain spatial relations, any texture features extracted from it are more discriminative than histogram metrics, which can only provide global texture information on either the whole image or a specific image region. Nonetheless, histogram-based descriptors are also frequently computed in neuroimaging applications, having been applied in the study of epilepsy (Bonilha, et al. 2003; Yu, Mauss, et al. 2001), schizophrenia (Saeed and Puri 2002), PD (Rahmim, et al. 2017) (Betrouni, et al. 2021), AD (Salas-Gonzalez, et al. 2009), and brain tumors (Herlidou-Même, et al. 2003; Fujima, et al. 2019; Vamvakas, et al. 2018; Soltaninejad, et al. 2017). Although first-order statistics are associated with generally poorer performance in image texture discrimination, they are directly computed as part of CAD routines in the clinical setting (Kassner and Thornhill 2010).

The run-length matrix (RLM) (Galloway 1975) is another statistical texture analysis method used in neuroimaging. It allows differentiating between coarse and fine textures based on, respectively, the number of long and short sequences of the same gray level in a particular direction across the image. Despite having been initially proven to be less efficient than other texture analysis methods (Weska, Dyer and Rosenfeld 1976; Conners and Harlow 1980), it is widely applied in neuroimaging texture ensemble implementations (in the study of epilepsy [Bonilha, et al. 2003], ischemic stroke [Ortiz-Ramón, et al. 2019], AD [Baskar, Jayanthi and Jayanthi 2019], PD [Rahmim, et al. 2017; Liu, et al. 2020], MS [Zhang, et al. 2009], and brain tumors [Herlidou-Même, et al. 2003; Vamvakas, et al. 2018]) and much less often individually (Derea, et al. 2019).

Finally, the local binary patterns (LBP) approach (Ojala, Pietikäinen and Harwood 1996), along with its many variations (Hadid, et al. 2015; Brahnam, et al. 2014), analyzes the relationship between a pixel and its neighbors. LBPs are quite popular in some research areas, namely, object and face recognition (Hadid, et al. 2015; Huang, et al. 2011; Sapathy, Jian and Eng 2014). In neuroimaging, LBPs have been applied primarily on texture ensemble approaches (Ortiz-Ramón, et al. 2019; Tazarjani, et al. 2021; Abbasi and Tajeripour 2017), but also in isolation, to detect brain tumors (Kaplan, et al. 2020), PD (Rana, et al. 2017), and early AD (Montagne, et al. 2013), and to distinguish between AD and Lewy body dementia (Oppedal, et al. 2015).

9.4.2 Transform-Based Methods

Transform-based methods analyze images in a different domain: the frequency or scale space, which can be interpreted as texture characteristics. After statistical methods, transform-based approaches are the second most frequently used class of methods in medical imaging applications.

Among transform-based methods, wavelet analysis and similar approaches, for example, the Stockwell transform and Gabor filters, are particularly suited for texture analysis. Wavelet-based methods can provide valuable insight in several image processing (e.g., denoising) and analysis problems, as their representation of the original image contains both spatial and frequency information. As a multiresolution approach that separates high-frequency details from low-frequency, more coarse texture changes, ones, wavelets can be used to complement the local texture information extracted via statistical methods like the application of the GLCM.

Concerning neuroimaging, wavelet features have been computed in combination with the GLCM, RLM, and other statistical methods in a texture ensemble approach to identify ischemic stroke lesions in MRI images (Ortiz-Ramón, et al. 2019). In some approaches aiming to classify brain tumors, wavelet decomposition was applied before the computation of GLCM texture features (Nanthagopal and Rajamony 2013; John 2012; Rezaei and Agahi 2017). Wavelet-based features have also been used in a standalone fashion to identify AD (Feng, Zhang and Chen 2020) and brain tumors (Garg and Singh 2014).

More recently, a more robust version of the complex wavelet transform has been applied in studies of the neuroretina (Nunes, Silva, Duque, et al. 2019; Nunes, et al. 2020) using OCT data, the dual-tree complex wavelet transform, which provides directional selectivity and nearly shift-invariance.

Two other transform-based methods are worth mentioning, both sharing some characteristics with wavelet-based approaches: the Stockwell transform, which is a Gaussian windowed version of the Fourier transform, and Gabor filters, which decompose images using a multichannel filter bank based on the Gabor transform. While the former method has been used to identify oligodendrogliomas (Brown, et al. 2008) and inflammatory lesions in MS (Zhang, et al. 2009), the latter has been used in the identification of AD (Hett, et al. 2018), and the detection, segmentation (Soltaninejad, et al. 2017) (Malviya and Joshi 2014), and classification (Zacharaki, et al. 2009) of brain tumors, and the segmentation of stroke lesions (Subbanna, et al. 2019) from MRI data.

9.4.3 Model-Based Methods

Model-based methods analyze image texture by identifying a mathematical model that reflects the properties of the texture contained in the image, followed by the estimation of model parameters, which are then used as (texture) features. Because estimating these parameters can be computationally complex, model-based methods are not applied as commonly as their statistical and transform-based counterparts.

In neuroimaging applications, the most chosen models to represent texture are either probabilistic (e.g., random field approaches) or geometric (fractal approaches) models. The Markov random field approach, which considers the image's gray levels to be random variables that depend on neighbor pixels, has been applied to MRI brain segmentation (Subbanna, et al. 2019; Tohka, et al. 2010; Held, et al. 1997; Zhang, Brady and Smith 2001; Nie, et al. 2009).

Fractal-based texture analysis is particularly suited for textures with a certain degree of self-similarity, that is, those presenting identical characteristics at different scales. The most popular fractal-based metric is the fractal dimension, which has been associated with texture roughness, as perceived by the human visual system (Pentland 1984). In neuroimaging, fractal features have been extracted from MRI data to aid in the diagnosis of AD (Lahmiri and Boukadoum 2013), brain tumors (Iftekharuddin, Jia and Marsh 2003), and paediatric brain injuries (Dona, et al. 2017). A few texture ensemble implementations have also included fractal analysis in their combo texture feature computation (Baskar, Jayanthi and Jayanthi 2019; Tazarjani, et al. 2021; Soltaninejad, et al. 2017).

9.4.4 Other Feature-Based Methods

While the four classical categories of texture analysis methods (structural, statistical, transform-based, and model-based) are the most consensual and consistent across the literature, at least two other categories of feature-based methods have been proposed: graph-based and entropy-based methods. These two classes of methods were initially presented as separate and dedicated categories of texture analysis in Humeau-Heurtier (2019) and recently revisited in Ghalati et al. (2021). As they both overlap, to some extent, with the four original classes, their recognition and validation as independent categories of texture analysis methods might require additional time and research.

In this subsection, along with the classical category of structural methods, the tentative categories of graph-based and entropy-based methods will be covered. Generally, the methods in these three feature-based categories have been used less frequently in the field of neuroimaging compared to those in the preceding subsection, although occasional applications have been reported.

Structural approaches represent images' texture in terms of *texture primitives* and their placement rules, respectively, the atomic elements that constitute the texture, and the relationships between those elements, which determine how they are arranged in space. These approaches provide highly detailed texture descriptions (Tomita and Tsuji 1990) and are suited for texture synthesis applications, but their applicability in natural image analysis problems—specifically those involving irregular textures—is generally considered to be limited (Castellano, et al. 2004; Humeau-Heurtier 2019; Materka 2004; Chaki and Dey 2020).

In neuroimaging, mathematical morphology, a structural method that uses morphological operators to identify the texture's primitives, has been applied to segment MRI data of the human brain (Senthilkumaran and Kirubakaran 2014; Mendiola-Santibañez, et al. 2014).

In graph-based methods, images are first represented as a graph of nodes from which texture features are extracted, a process which has been framed as an extension/improvement of Markov random field model approaches (Ahmadvand, Yousefi and Shalmani 2017; Gaetano, Scarpa and Sziranyi 2010). In neuroimaging, the region adjacency graph method (Pavlidis 1980) has been applied toward MRI brain segmentation (Ahmadvand, Yousefi and Shalmani 2017).

Entropy-based methods, initially used in time series analysis, were adopted by the image analysis field during the past decade. These methods provide a measure of the image's irregularity by computing entropy metrics directly from the image. In her survey, Humeau-Heurtier (Humeau-Heurtier 2019) argues that this approach is distinct enough from GLCM-based entropy metrics to warrant these methods their own dedicated class instead of falling into the statistical methods category. To the best of the authors' knowledge, neuroimaging applications of entropy-based methods are yet to be reported.

9.4.5 Learned-Based Methods

In opposition to the earlier subsections, which concerned feature-based methods, learned-based methods will be discussed in the next few lines.

Recent texture surveys (Liu, Chen, et al. 2019; Humeau-Heurtier 2019; Ghalati, et al. 2021) have covered yet another emerging category of methods: learned-based approaches. However, more than simply a class of methods, these approaches represent a whole new paradigm in the field of texture analysis: the move toward automatically defining the adequate learned texture features for each context. Rather than selecting the texture-based metrics *a priori*, learned-based approaches infer texture from the datasets under study.

A subset of these methods, vocabulary-based approaches, can be traced back to 2001 (Leung and Malik 2001). In vocabulary-based methods, a dictionary of texture elements is learned from a training set of images by computing and clustering local descriptors. These clusters act as words in a dictionary as they represent the texture patterns present in the image dataset.

In neuroimaging, a vocabulary-based feature computation scheme has been used in conjunction with wavelet-based features to classify medulloblastomas from histopathological images (Galaro, et al. 2011). This type of method has also been applied to MRI data in the analysis of perivascular spaces (González-Castro, et al. 2016) and to discriminate between healthy controls and patients diagnosed with dementia and mild cognitive impairment (MCI) (Bansal, et al. 2020).

More recently, a different avenue of research has been dominating the field: deep-learned texture analysis methods. Deep-learned methods have been integrated into texture analysis research following the remarkable image classification performance results found for deep CNNs like AlexNet (Krivesky, Sutskever and Hinton 2012) and GooLeNet (Szegedy, et al. 2015) networks. Notably, these methods have also been shown to outperform classical texture metrics—specifically, LBP descriptors (Liu, et al. 2016).

Although there might be some practical constraints in applying deep-learned methods to medical imaging problems (deep CNNs are both data-hungry and computationally heavy), several diagnostic applications have recently reported the computation of deep-learned features. In neuroimaging, they have been used to automatically segment brain tumors (Havaei, et al. 2017) and MS lesions (Brosch, et al. 2016). These methods were also used to discriminate among CT brain images of AD patients, patients with brain tumor lesions, and healthy controls (Gao and Hui 2016).

Deep-learned texture analysis methods look particularly promising at present and are unquestionably gaining momentum in current research on texture. A comprehensive review of the timeline and research developments in both vocabulary-based and deep-learned methods can be found in Liu, Chen, et al. (2019).

9.5 TEXTURE APPLICATIONS IN NEUROIMAGING

In the field of neuroimaging, a myriad of texture-based methods has been reported to provide invaluable data in different diagnostic problems, namely, the characterization of brain tumors (Herlidou-Même, et al. 2003; Mahmoud-Ghoneim, et al. 2003; Vamvakas, et al. 2018; Fujima, et al. 2019; Soltaninejad, et al. 2017; Brown, et al. 2008; Zacharaki, et al. 2009), identification of ischemic stroke (Ortiz-Ramón, et al. 2019; Kassner, et al. 2009; Sikiö, Kölhi, et al. 2015), and the detection and monitorization of multiple neurological disorders, including AD (Nedelec, et al. 2004; Zhang, et al. 2012; Baskar, Jayanthi and Jayanthi 2019) (Salas-Gonzalez, et al. 2009; Raut and Dalal 2017; Leandrou, et al. 2020; Lee, Lee and Kim 2020; Lee and Kim 2021; Luk, et al. 2018; Chincarini, et al. 2011), PD (Rahmim, et al. 2017; Liu, et al. 2020; Li, et al. 2019; Sikiö, Holli-Helenius, et al. 2015; Betrouni, et al. 2021), MS (Mathias, Tofts and Losseff 1999; Tazarjani, et al. 2021; Nunes, Silva, Alves, et al. 2019; Zhang, et al. 2009), ALS (Albuquerque, et al. 2016; Maani, et al. 2016; Ishaque, et al. 2018; Elahi, et al. 2020), schizophrenia (Nedelec, et al. 2004; Saeed and Puri 2002; Latha and Kavitha 2019), and epilepsy (Bernasconi, et al. 2001; Antel, et al. 2003; Bonilha, et al. 2003; Yu, et al. 2001). Texture-based approaches have even been employed to help discriminate between some of the disorders previously mentioned

(Nunes, Silva, Duque, et al. 2019) and characterize aging and sex differences in the healthy population (Nunes, et al. 2020). Finally, texture analysis techniques have been applied to images obtained from animal models of disease, namely, to outline damaged tissue in an inflammatory mouse model (Alejo, et al. 2003), characterize the progressive retinal changes caused by AD (Ferreira, et al. 2020), compare brain tumor cellular dynamics (Gazdzinski and Nieman 2014), and predict epilepsy in diseases' rat models (Nedelec, et al. 2004).

The fundamental principle behind all texture-based applications is the *extraction* of texture features (or *texture representation*, as phrased by Liu, Chen, et al. [2019]). Texture features are typically extracted to solve four types of problems: texture classification, texture segmentation, texture synthesis, and shape from texture. To the best of the authors' knowledge, texture synthesis and shape from texture approaches have no applications in neuroimaging so far and will therefore not be addressed in this chapter.

For classification problems, images (or image regions) are classified into one of a set of pre-defined labels based on example cases of those labels. Since texture-based classification is fundamentally a discrimination problem, it has found a natural application in the context of the CAD paradigm. As such, a significant portion of applications of texture analysis in neuroimaging falls into this category. A recent application of texture-based classification in neuroimaging is the work of Lee et al. (Lee, Lee and Kim 2020). Here, the authors computed GLCM features from MRI brain scans to help predict conversion from MCI to AD. They found that texture features predict MCI-to-AD conversion earlier and more accurately than hippocampal volume, traditionally considered a potential biomarker for AD. Other findings across the literature (Bernasconi, et al. 2001; Antel, et al. 2003; Nunes, Silva, Duque, et al. 2019; Leandrou, et al. 2020) have hinted at the potential of texture analysis methods to enhance diagnostic performance, namely, by complementing direct visual inspection and other image analysis techniques, such as volumetry (as earlier).

In texture-based segmentation, images are partitioned into regions with homogeneous texture properties, often to further analyze these regions individually. In neuroimaging, texture-based segmentation has been performed on MRI data since the early 2000s, namely, to segment the cerebellum in healthy controls and schizophrenic patients (Saeed and Puri 2002) and to segment several anatomic structures in an animal model of inflammation and AD patients (Alejo, et al. 2003). In 2001, Kovalev et al. (Kovalev, et al. 2001) proposed 3D co-occurrence descriptors which can be used to segment white matter lesions from MRI data. More recently, Soltinejad et al. (Soltaninejad, et al. 2017) combined texture-based segmentation and classification approaches to delineate brain tumors from fluid-attenuated inversion recovery MRI, while Vamvakas et al. (Vamvakas, et al. 2018) used statistical texture analysis methods to assist in the differentiation of 3D tumor models from diffusion tensor imaging MRI data.

9.5.1 Texture in MRI Applications

MRI is a widely applied imaging technique within the neuroimaging field and the dominating modality when it comes to the use of texture analysis in neuroimaging applications. Texture analysis of MRI neuroimaging data has been extensively applied in the fields of neurooncology and general neurology.

Neurooncology studies using MRI data are among the very first applications of texture analysis in neuroimaging, which included a fractal-based approach for detecting brain tumors (Iftekharuddin, Jia and Marsh 2003), a 3D GLCM application for brain tumor characterization (Mahmoud-Ghoneim, et al. 2003), and a texture ensemble approach for discriminating between different tumor types (Herlidou-Même, et al. 2003). The use of texture analysis to identify, segment, and characterize brain tumors from MRI data remains a frequently used strategy, with an increasing range of available methods over time. Recently applied methods include not only the ever-popular GLCM (Fujima, et al. 2019) but also the LBP (Kaplan, et al. 2020), the RLM (Derea, et al. 2019), and even a deep-learned solution for automatically segmenting brain tumors (Havaei, et al. 2017).

In general neurology, texture analysis applications using MRI data are centred on the diagnosis of ischemic stroke and neurological and neurodegenerative disorders. Recent texture applications in AD, the most studied disorder across MRI-based texture analysis applications, is addressed next.

Raut et al. (Raut and Dalal 2017) and Baskar et al. (Baskar, Jayanthi and Jayanthi 2019) are two recent examples of machine learning texture-based classification problems. In these studies, texture-based features were first computed and then fed to a classifier—an artificial neural network which, after being trained on the computed features, was able to discriminate between healthy controls and different stages of AD progression. While only GLCM-based features were computed in the former study, the latter used a broad set of texture analysis methods in a texture ensemble approach.

Leandrou et al. (Leandrou, et al. 2020) computed several GLCM-based texture features from MRI data of healthy controls, patients diagnosed with MCI and AD, and patients that converted from MCI to AD. The authors focused on the entorhinal cortex, one of the earliest-affected brain locations in AD, and found texture features to be significantly different between groups, surpassing the performance of volumetry-based methods. Furthermore, their results showed that combining texture features with volumetric measurements improved the prediction of conversion from MCI to AD. Similar conclusions regarding the potential role of texture-based metrics as a quantitative MRI-derived biomarker for assessing the progression of AD, specifically when it comes to the prediction of conversion from MCI to AD, were put forward in Lee, Lee, and Kim (2020) and Luk et al. (2018).

In Lee and Kim (2021), four regional GLCM-based features, extracted from MRI data, were found to be correlated with AD pathology in specific brain regions, as measured by tau burden in positron emission tomography (PET). Furthermore, no association could be found between the mean signal intensity of those regions and tau burden, indicating that the GLCM features might be revealing AD-related textural changes in the microstructure of the brain that are not reflected in first-order statistics.

Back in 2004, different texture metrics were computed from MRI data of a cuprizone mouse model to assess the demyelination and remyelination process (Yu, et al. 2004). This study identified texture-based metrics as a potential tool in the longitudinal evaluation of demyelinating disorders, such as MS. Besides this study, to the authors' knowledge, neuroimaging MRI texture analysis on animal models of disease is a rather unexplored avenue of research.

9.5.2 Texture in PET Applications

While texture analysis applications in nuclear medicine neuroimaging do indeed exist (Sayeed, et al. 2001; Rahmim, et al. 2017; Salas-Gonzalez, et al. 2009; Montagne, et al. 2013), they are not nearly as common as in MRI imaging. In this subsection, some of the reported applications of texture-based approaches on PET data will be covered.

One of the earliest applications of texture analysis on PET imaging dates back to 2001 (Sayeed, et al. 2001). In this study, a transform-based method was used to differentiate between AD patients and healthy controls. Specifically, the trace transform—an adaptation of the Radon transform—was used to compute global features that were either sensitive or invariant to rotation, translation, and scaling. The sensitive features were contextualized as anisotropic texture characteristics that may be associated with AD; however, the highest classification accuracy (93%) was achieved when sensitive and invariant features were combined.

More recently, the GLCM method was applied to ^{18}F-florbetapir PET images to identify statistically significant differences between AD, MCI, and healthy control groups (Campbell, Kang and Shokouhi 2017). In this study, the computation of texture features helped streamline the PET preprocessing pipeline, as it allowed the authors to skip the step of region-based intensity normalization—which can be problematic due to variabilities associated with the reference region used. Similarly, in Bouallègue et al. (2019), texture features were used in combination with shape measurements to predict MCI-to-AD conversion, with a performance comparable to that of classical standardized uptake value quantification.

In an animal study, ^{64}Cu-GTSM (thiosemicarbazone) PET imaging was used to assess AD-related changes in copper trafficking and distribution (Torres, et al. 2016). Here, first-order texture features indicated an increased PET signal heterogeneity in a transgenic mouse model of AD, compared to wild-type mice. No reports on neuroimaging PET texture analysis in animal models of disease could be found in the literature.

In neurooncology, one texture analysis application stands out for its recency: a retrospective study that used a texture ensemble approach on ^{11}C-methionine PET data to predict the postoperative prognosis in patients with gliomas (Manabe, et al. 2021). In this study, two features were singled out as particularly significant predictors of patient survival: the GLCM's correlation and the RLM's low gray level run emphasis features.

9.5.3 Texture in OCT Applications

The use of the retina as a window to the brain, which is grounded on the fact that it is an integral and the only visible part of the central nervous system (CNS) (London, Benhar and Schwartz 2013; Svetozarskiy and Kopishinskaya 2015; Czakó, et al. 2020), has been receiving increasing research attention over the past decade.

Unlike the other structures of the CNS, the retina is easily accessible through ophthalmological imaging techniques, such as OCT. Being a widely available, non-invasive imaging technique that allows for fast image acquisition at a low operational cost, OCT constitutes an advantageous alternative to brain imaging modalities like MRI. In the context of neuroimaging, OCT can be used to highlight structural changes in the neuroretina, particularly those brought on by neurodegenerative disorders such as AD (Haan, et al. 2017; Trebbastoni, et al. 2016; Garcia-Martin, et al. 2016), PD (Yu, et al. 2014; Hajee, et al. 2009; Altintaş, et al. 2008), and MS (Tazarjani, et al. 2021; Nunes, Silva, Alves, et al. 2019; Britze and Frederiksen 2018; Varga, et al. 2015; Petzold, et al. 2010).

Traditionally, OCT-based research on neurodegenerative disorders has focused on assessing the potential of retinal layer thickness measurements as biomarkers of neurodegeneration (Haan, et al. 2017; Yu, et al. 2014; Petzold, et al. 2010). More recently, texture-based neurodegeneration studies on OCT data have emerged, with research findings framing image texture as a means to capture complementary information to that provided by thickness measurements (Tazarjani, et al. 2021; Nunes, Silva, Duque, et al. 2019; Varga, et al. 2015). Texture-based approaches have the potential to detect local changes in specific regions of the individual retinal layers, or microstructural alterations spread across the retinal tissue that thickness measurements cannot quantify. In this way, texture analysis can help generate a more comprehensive neurodegeneration retinal map, that is, a disease-specific texture profile (or texture signature) detailing the changes caused by a particular neurodegenerative disorder on the retinal tissue. The establishment of such disease profiles might assist in clarifying the aforementioned conflicting research results and even help discriminate between neurodegenerative disorders, as was explored in Nunes, Silva, Duque, et al. (2019).

The rationale of using the retina as a window to the brain is even further supported by research findings placing visual impairment and changes to the retinal microstructure among the earliest manifestations of AD (Czakó, et al. 2020; Alves, et al. 2019), PD (Turcano, et al. 2019) (Cesareo, et al. 2021), and MS (Borgström, et al. 2020). These research results encourage using the retina to study the onset and progression of neurodegenerative disorders. Furthermore, the use of animal models of disease may help shed light on the changes at the onset of the disease and its progression.

In Ferreira et al. (2020), significant textural differences were found between the multiple retinal layers of wild-type control mice and a transgenic mouse model of AD at the ages of 1 and 2 months old. This study reported a notable increase in the differences between the two mice groups from the first to the second time point, highlighting the potential of texture for tracking the progression of AD during the earliest stages of this neurodegenerative disorder.

9.6 FUTURE PERSPECTIVES FOR TEXTURE IN NEUROIMAGING

The field of texture analysis has grown and evolved extensively over the years. After over six decades since its inception, research efforts remain prolific in this field, with existent texture analysis methods being continuously improved and adapted and new approaches emerging.

The recent shift toward automatic texture feature computation and the increasing research interest in determining how CNNs see images appear to be two of the most promising paths forward for texture analysis research. Deep-learned texture approaches represent an encouraging step toward the automation of medical image interpretation and thus embody a favorable research avenue for neuroimaging.

The growing number of applications of texture-based approaches, over the past two decades, in numerous neurooncology and general neurology research problems and different medical imaging modalities, is a testimony to the potential and suitability of these techniques for neuroimaging.

MRI remains the neuroimaging modality where most texture analysis applications can be found—and will most likely continue to be. While applications centered on AD and brain tumors dominate current research trends, the breadth of neurological disorders where MRI texture analysis studies are now being performed is remarkable, covering pathologies such as, but not restricted to, PD (Betrouni, et al. 2021), ALS (Elahi, et al. 2020), MS (Tazarjani, et al. 2021), and trigeminal neuralgia (Danyluk, et al. 2021).

In the last decade, the use of texture-based approaches in tumor PET image analysis has been receiving some research attention (Buvat, Orlhac and Soussan 2015; Deleu, et al. 2020), given the suitability of some texture metrics (namely, the ones extracted from the GLCM) to quantify tumor heterogeneity. While neurooncology applications have been slower to incorporate texture analysis, it is reasonable to expect an increasing use of these methods in the study of brain tumors in years to come. Amyloid PET imaging remains a less explored path for texture analysis; however, recent research (Campbell, Kang and Shokouhi 2017; Bouallègue, et al. 2019) has framed texture-based metrics as a robust alternative to the traditional standardized uptake value measurements in the diagnosis and prognosis of AD.

In OCT neuroimaging, texture analysis may play an important role in the timely diagnosis of neurodegenerative disorders like AD, which remains both a clinical challenge and a top research priority (Shah, et al. 2016). The role of texture is therefore undisputable at present and seems promising in years to come.

REFERENCES

Abbasi, S., and F. Tajeripour. 2017. "Detection of brain tumor in 3D MRI images using local binary patterns and histogram orientation gradient." *Neurocomputing* 2019: 526–535.

Ahmadvand, A., S. Yousefi, and M. T. M. Shalmani. 2017. "A novel Markov random field model based on region adjacency graph for T1 magnetic resonance imaging brain segmentation." *International Journal of Imaging Systems and Technology* 27 (1): 78–88.

Albuquerque, M. D., L. G. V. Anjos, H. M. T. de Andrade, M. S. de Oliveira, G. Castellano, T. J. R. Rezende, A. Nucci, and M. C. F. Junior. 2016. "MRI texture analysis reveals deep gray nuclei damage in amyotrophic lateral sclerosis." *Journal of Neuroimaging* 26: 201–206.

Alejo, R. P. D., J. Ruiz-Cabello, M. Cortijo, I. Rodriguez, I. Echave, J. Regadera, J. Arrazola, et al. 2003. "Computer-assisted enhanced volumetric segmentation magnetic resonance imaging data using a mixture of artificial neural networks." *Magnetic Resonance Imaging* 21: 901–912.

Altintaş, Ö., P. Işeri, B. Özkan, and Y. Çağlar. 2008. "Correlation between retinal morphological and functional findings and clinical severity in Parkinson's disease." *Documenta Ophthalmologica* 116: 137–146.

Alves, C., L. Jorge, N. Canário, B. Stantiago, I. Santana, J. Castelhano, A. F. Ambrósio, R. Bernardes, and M. Castelo-Branco. 2019. "Interplay between macular retinal changes and white matter integrity in early Alzheimer's disease." *Journal of Alzheimer's Disease* 70: 723–732.

Antel, S. B., D. L. Collins, N. Bernasconi, F. Andermann, R. Shinghal, R. E. Kearny, D. L. Arnold, and A. Bernasconi. 2003. "Automated detection of focal cortical dysplasia lesions using computational models of their MRI characteristics and texture analysis." *NeuroImage* 19: 1748–1759.

Ballester, P., and R. M. Araujo. 2016. "On the performance of GoogLeNet and AlexNet applied to sketches." *30th AAAI Conference on Artificial Intelligence (AAAI-16)*: 1124–1128.

Bansal, D., K. Khanna, R. Chhikara, R. K. Dua, and R. Malhotra. 2020. "Classification of magnetic resonance images using bag of features for detecting dementia." *International Conference on Computational Intelligence and Data Science*: 131–137.

Baskar, D., V. S. Jayanthi, and A. N. Jayanthi. 2019. "An efficient classification approach for detection of Alzheimer's disease from biomedical imaging modalities." *Multimedia Tools and Applications* 78: 12883–12915.

Bergen, J. R., and M. S. Landy. 1991. Computational modeling of visual texture segregation. In *Computational models of visual processing*, edited by M. Landy and J. A. Movshon, Chapter 17, 253–271. MIT Press.

Bernasconi, A., S. B. Antel, D. L. Collins, N. Bernasconi, A. Olivier, F. Dubeau, G. B. Pike, F. Andermann, and D. L. Arnold. 2001. "Texture analysis and morphological processing of magnetic resonance imaging assist detection of focal cortical dysplasia in extra-temporal partial epilepsy." *Annals of Neurology* 49 (6): 770–775.

Betrouni, N., C. Moreau, A.-S. Rolland, N. Carrière, M. Chupin, G. Kuchinski, R. Lopes, R. Viard, L. Defebvre, and D. Devos. 2021. "Texture-based markers from structural imaging correlate with motor handicap in Parkinson's disease." *Scientific Reports* 11: 2724.

Bonilha, L., E. Kobayashi, G. Castellano, G. Coelho, E. Tinois, F. Cendes, and L. M. Li. 2003. "Texture analysis of hippocampal sclerosis." *Epilepsia* 44 (12): 1546–1550.

Borgström, M., A. Tisell, H. Link, E. Wilhem, P. Lundberg, and Y. Huang-Link. 2020. "Retinal thinning and brain atrophy in early MS and CIS." *Acta Neurologica Scandinavica* 142: 418–427.

Bouallègue, F. B., F. Vauchot, D. Mariano-Goulart, and P. Payoux. 2019. "Diagnostic and prognostic value of amyloid PET textural and shape features: Comparison with classical semi-quantitative rating in 760 patients from the ADNI-2 database." *Brain Imaging and Behavior* 13: 111–125.

Brahnam, S., L. C. Jain, L. Nanni, and A. Lumini. 2014. *Local binary patterns: New variants and applications*, vol. 506. Springer.

Brendel, W., and M. Bethge. 2019. "Approximating CNNs with bag-of-local-features models works surprisingly well on ImageNet." arXiv. https://doi.org/10.48550/arxiv.1904.00760

Britze, J., and J. L. Frederiksen. 2018. "Optical coherence tomography in multiple sclerosis." *Eye* 32: 884–888.

Brosch, T., L. Y. W. Tang, Y. Yoo, D. K. B. Li, A. Traboulsee, and R. Tam. 2016. "Deep 3D convolutional encoder networks with shortcuts for multiscale feature integration applied to multiple sclerosis lesion segmentation." *IEEE Transactions on Medical Imaging* 35 (5): 1229–1239.

Brown, R., M. Zlatescu, A. Sijben, G. Roldan, J. Easaw, P. Forsyth, I. Parney, et al. 2008. "The use of magnetic resonance imaging to noninvasively detect genetic signatures in oligodendroglioma." *Clinical Cancer Research* 14 (8): 2357–2362.

Buvat, I., F. Orlhac, and M. Soussan. 2015. "Tumor texture analysis in PET: Where do we stand?" *The Journal of Nuclear Medicine* 56 (11): 1642–1644.

Cadieu, C. F., H. Hong, D. L. K. Yamins, N. Pinto, D. Ardila, E. A. Solomon, N. J. Majaj, and J. J. DiCarlo. 2016. "Deep neural networks rival the representation of primate IT cortex for core visual object recognition." *PLoS Computational Biology* 10 (12): e1003963.

Campbell, D. L., H. Kang, and S. Shokouhi. 2017. "Application of Haralick texture features in brain [18F]-florbetapir positron emission tomography without reference region normalization." *Clinical Interventions in Aging* 12: 2077–2086.

Castellano, G., L. Bonilha, L. M. Li, and F. Cendes. 2004. "Texture analysis of medical images." *Clinical Radiology* 59 (12): 1061–1069.

Cesareo, M., E. Di Marco, C. Giannini, M. Di Marino, D. Aiello, A. Pisani, M. Pierantozzi, N. B. Mercuri, C. Nucci, and R. Mancino. 2021. "The retinal posterior pole in early Parkinson's disease: A fundus perimetry and SD-OCT study." *Clinical Ophthalmology* 15: 4005–4014.

Chaki, J., and N. Dey. 2020. *Texture feature extraction techniques for image recognition.* Springer.

Chien, Y. P., and K.-S. Fu. 1974. "Recognition of x-ray picture patterns." *IEEE Transactions on Systems, Man and Cybernetics* SMC-4 (2): 145–156.

Chincarini, A., P. Bosco, P. Calvini, G. Gemme, M. Esposito, C. Olivieri, L. Rei, et al. 2011. "Local MRI analysis approach in the diagnosis of early and prodromal Alzheimer's disease." *NeuroImage* 58: 469–480.

Conners, R. W., and C. A. Harlow. 1980. "A theoretical comparison of texture algorithms." *IEEE Transactions on Pattern Analysis and Machine Intelligence* PAMI-2 (3): 204–222.

Cyll, K., E. Ersvær, L. Vlatkovic, M. Pradhan, W. Kildal, M. A. Kjær, A. Kleppe, et al. 2017. "Tumour heterogeneity poses a significant challenge to cancer biomarker research." *British Journal of Cancer* 117: 367–375.

Czakó, C., T. Kovács, Z. Ungvari, A. Csiszar, A. Yabluchanskiy, S. Conley, T. Csipo, et al. 2020. "Retinal biomarkers for Alzheimer's disease and vascular cognitive impairment and dementia (VCID): Implication for early diagnosis and prognosis." *GeroScience* 42: 1499–1525.

Danyluk, H., A. Ishaque, D. Ta, Y. H. Yang, B. M. Weatley, S. Kalra, and T. Sankar. 2021. "MRI texture analysis reveals brain abnormalities in medically refractory trigeminal neuralgia." *Frontiers in Neurology* 12 (626504).

Darling, E., and R. D. Joseph. 1968. "Pattern recognition from satellite altitudes." *IEEE Transactions on Systems, Science and Cybernetics* 4 (1): 38–47.

Deleu, A.-L., M. J. Sathekge, A. Maes, B. De Spiegeleer, M. Sathekge, and C. Van de Wiele. 2020. "Characterization of FDG PET images using texture analysis in tumors of the gastro-intestinal tract: A review." *Biomedicines* 8 (304).

Derea, A. S., H. K. Abbas, H. J. Mohamad, and A. A. Al-Zuky. 2019. "Adopting run length features to detect and recognize brain tumor in magnetic resonance images." *2019 First International Conference of Computer and Applied Sciences (CAS)*: 186–192.

Dona, O., M. D. Noseworthy, C. DeMatteo, and J. F. Connolly. 2017. "Fractal analysis of brain blood oxygenation level dependent (BOLD) signals from children with mild traumatic brain injury (mTBI)." *PLOS One* 12 (1): e0169647.

Elahi, G. M. M. E., S. Kalra, L. Zinman, A. Genge, L. Korngut, and Y.-H. Yang. 2020. "Texture classification of MR images of the brain in ALS using M-CoHOG: A multi-center study." *Computerized Medical Imaging and Graphics* 79 (101659).

Eun, N. L., D. Kang, E. J. Son, J. S. Park, J. H. Youk, J.-A. Kim, and H. M. Gweon. 2020. "Texture analysis with 3.0-T MRI for association of response to neoadjuvant chemotherapy in breast cancer." *Radiology* 294: 31–41.

Feng, J., S.-W. Zhang, and L. Chen. 2020. "Identification of Alzheimer's disease based on wavelet transformation energy feature of the structural MRI image and NN classifier." *Artificial Intelligence in Medicine* 108 (101940).

Ferreira, H., J. Martins, A. Nunes, P.I. Moreira, M. Castelo-Branco, A. F. Ambrósio, and R. Bernardes. 2020. "Characterization of the retinal changes of the 3×Tg-AD mouse model of Alzheimer's disease." *Health and Technology* 10: 875–883.

Fujima, N., A. Homma, T. Harada, Y. Shimizu, S. Kano, K. K. Tha, T. Mizumachi, R. Li, K. Kudo, and H. Shirato. 2019. "The utility of MRI histogram and texture analysis for the prediction of histological diagnosis in head and neck malignancies." *Cancer Imaging* 19 (5).

Gaetano, R., G. Scarpa, and T. Sziranyi. 2010. "Graph-based analysis of textured images for hierarchical segmentation." *Proceedings of the British Machine Vision Conference 2010*, 74.1–74.11. https://doi.org/10.5244/c.24.74

Galaro, J., A. R. Judkins, D. Ellison, J. Baccon, and A. Madabhushi. 2011. "An integrated texton and bag of words classifier for identifying anaplastic medulloblastomas." *33rd Annual International Conference of the IEEE EMBS*: 3443–3446.

Galloway, M. M. 1975. "Texture analysis using gray level run lengths." *Computer Graphics and Image Processing* 4 (2): 172–179.

Gao, X. W., and R. Hui. 2016. "A deep learning based approach to classification of CT brain images." *SAI Computing Conference (SAI)*: 28–31.

Garcia-Martin, E., M. P. Bambo, M. L. Marques, M. Satue, S. Otin, J. M. Larrosa, V. Polo, and L. E. Pablo. 2016. "Ganglion cell layer measurements correlate with disease severity in patients with Alzheimer's disease." *Acta Ophthalmologica* 94: e454–e459.

Garg, S., and E. N. Singh. 2014. "MRI brain image quantification using wavelets for tumor detection." *International Journal of Mechanical Engineering and Information Technology* 2 (7): 692–699.

Gatys, L. A., A. S. Ecker, and M. Bethge. 2017. "Texture and art with deep neural networks." *Current Opinion in Neurobiology* 46: 178–186.

Gay, L., A.-M. Baker, and T. A. Graham. 2016. "Tumour cell heterogeneity." *F1000Research* 29 (5): 238.

Gazdzinski, L. M., and B. J. Nieman. 2014. "Cellular imaging and texture analysis distinguish differences in cellular dynamics in mouse brain tumors." *Magnetic Resonance in Medicine* 71: 1531–1541.

Geirhos, R., P. Rubisch, C. Michaelis, M. Bethge, F. A. Wichmann, and W. Brendel. 2019. "ImageNet-trained CNNs are biased towards texture; increasing shape bias improves accuracy and robustness." arXiv. https://doi.org/10.48550/arxiv.1811.12231

Ghalati, M. K., A. Nunes, H. Ferreira, P. Serranho, and R. Bernardes. 2021. "Texture analysis and its applications in biomedical imaging: A survey." *IEEE Reviews in Biomedical Engineering* 15 (in press).

González-Castro, V., M. del C. V. Hernández, P. A. Armitage, and J. M. Wardlaw. 2016. "Automatic rating of perivascular spaces in brain MRI using bag of visual words." Edited by A. Campilho and F. Karray. *International Conference on Image Analysis and Recognition (ICIAR 2016)* 9730: 642–649.

Haan, J. D., F. D. Verbraak, P. J. Visser, and F. H. Bouwman. 2017. "Retinal thickness in Alzheimer's disease: A systematic review and meta-analysis." *Alzheimer's & Dementia: Diagnosis, Assessment & Disease Monitoring* 6: 162–170.

Hadid, A., J. Ylioinas, M. Ghahramani, and A. Taleb-Ahmed. 2015. Gender and texture classification: A comparative analysis using 13 variants of local binary patterns. *Pattern Recognition Letters* 68: 231–238.

Hajee, M. E., W. F. March, D. R. Lazzaro, A. H. Wolintz, E. M. Shrier, S. Glazman, and I. G. Bodis-Wollner. 2009. "Inner retinal layer thinning in Parkinson disease." *Archives of Ophthalmology* 127 (6): 737–741.

Hall, E. L., R. P. Kruger, S. J. Dwyer, D. L. Hall, R. W. McLaren, and G. S. Lodwick. 1971. "A survey of preprocessing and feature extraction techniques for radiographic images." *IEEE Transactions on Computers* C-20 (9): 1032–1044.

Haralick, R. M. 1971. "On a texture-context feature extraction algorithm for remotely sensed imagery." *IEEE Conference on Decision and Control*: 650–657.

Haralick, R. M., K. Shanmugam, and I. H. Dinstein. 1973. "Textural features for image classification." *IEEE Transactions on Systems, Man, and Cybernetics* SMC-3 (6): 610–621.

Harlow, C. A., R. W. Conners, M. M. Trivedi, D. A. DiRosa, and R.E. Vasquez-Espinosa. 1981. "Texture analysis and urban land use classification." *Conference Proceedings Southeastcon* 81: 115–119.

Harlow, C. A., and S. A. Eisenbeis. 1973. "The analysis of radiographic images." *IEEE Transactions on Computers* C-22 (7): 678–689.

Havaei, M., A. Davy, D. Warde-Farley, A. Biard, A. Courville, Y. Bengio, C. Pal, P.-M. Jodoin, and H. Larochelle. 2017. "Brain tumor segmentation with deep neural networks." *Medical Image Analysis* 35: 18–31.

He, K., X. Zhang, S. Ren, and J. Sun. 2016. "Deep residual learning for image recognition." *IEEE Conference on Computer Vision and Pattern Recognition (CVPR)*: 770–778.

Held, K., E. R. Kops, B. J. Krause, W. M. Wells, R. Kikinis, and H.-W. Müller-Gärtner. 1997. "Markov random field segmentation of brain MR images." *IEEE Transactions on Medical Imaging* 16 (6): 878–886.

Herlidou-Même, S., J. M. Constans, B. Carsin, P. A. Eliat, D. Olivie, L. Nadal-Desbarats, C. Gondry, E. Le Rumeur, I. Idy-Peretti, and J. D. de Certaines. 2003. "MRI texture analysis on texture test objects, normal brain and intracranial tumours." *Magnetic Resonance Imaging* 21 (9): 989–993.

Hett, K., V.-T. Ta, J. V. Manjón, and P. Coupé. 2018. "Adaptive fusion of texture-based grading for Alzheimer's disease classification." *Computerized Medical Imaging and Graphics* 70: 8–16.

Huang, D., C. Shan, M. Ardabilian, Y. Wang, and L. Chen. 2011. "Local binary patterns and its application to facial image analysis: A survey." *IEEE Transactions on Systems, Man and Cybernetics—Part C: Applications and Reviews* 41 (6): 765–781.

Humeau-Heurtier, A. 2019. "Texture feature extraction methods: A survey." *IEEE Access* 7: 8975–9000.

Iftekharuddin, K. M., W. Jia, and R. Marsh. 2003. "Fractal analysis of tumor in brain MR images." *Machine Vision and Applications* 13: 352–362.

Ishaque, A., R. Maani, J. Satkunam, P. Seres, D. Mah, A. H. Wilman, S. Naik, Y.-H. Yang, and S. Kalra. 2018. "Texture analysis to detect cerebral degeneration in amyotrophic lateral sclerosis." *Canadian Journal of Neurological Sciences* 45: 533–539.

John, P. 2012. "Brain tumor classification using wavelet and texture based neural network." *International Journal of Scientific & Engineering Research* 3 (10).

Julesz, B. 1962. "Visual pattern discrimination." *IEEE Transactions on Information Theory* 8 (2): 84–92.

Kaizer, H. 1956. "A phsychophysical study of visual texture" (PhD thesis). Boston University.

Kaplan, K., Y. Kaya, M. Kuncan, and H. M. Ertunç. 2020. "Brain tumor classification using modified local binary patterns (LBP) feature extraction methods." *Medical Hypotheses* 139 (109696).

Kassner, A., and R. E. Thornhill. 2010. "Texture analysis: A review of neurologic MR imaging applications." *American Journal of Neuroradiology* 31: 809–816.

Kassner, A., F. Liu, R. E. Thornhill, G. Tomlinson, and D. J. Mikulis. 2009. "Prediction of hemorrhagic transformation in acute ischemic stroke using texture analysis of postcontrast T1-weighted MR images." *Journal of Magnetic Resonance Imaging* 30: 933–941.

Kovalev, V. A., F. Kruggel, H.-J. Gertz, and D. Y. von Cramon. 2001. "Three dimensional texture analysis of MRI brain datasets." *IEEE Transactions on Medical Imaging* 20 (5).

Krivesky, A., I. Sutskever, and G. E. Hinton. 2012. "ImageNet classification with deep convolutional neural networks." *Communications of the ACM* 60 (6): 84–90.

Kubilius, J., S. Bracci, and H. P. Op de Beeck. 2016. "Deep neural networks as a computational model for human shape sensitivity." *PLoS Computational Biology* 12 (4): e1004896.

Lahmiri, S., and M. Boukadoum. 2013. "Alzheimer's disease detection in brain magnetic resonance images using multiscale fractal analysis." *ISRN Radiology* 2013, 627303. https://doi.org/10.5402/2013/627303.

Latha, M., and G. Kavitha. 2019. "Segmentation and texture analysis of structural biomarkers using neighborhood-clustering-based level set in MRI of the schizophrenic brain." *Magnetic Resonance Materials in Physics, Biology and Medicine* 31: 483–499.

Leandrou, S., D. Lamnisos, I. Mamais, P. A. Kyriacou, and C. S. Pattichis. 2020. "Assessment of Alzheimer's disease based on texture analysis of the entorhinal cortex." *Frontiers in Aging Neuroscience* 12 (176).

Ledley, R. S. 1972. "Texture problems in biomedical pattern recognition." *IEEE Conference on Decision and Control and 11th Symposium on Adaptive Processes*: 590–595.

Lee, S., and K. W. Kim. 2021. "Associations between texture of T1-weighted magnetic resonance imaging and radiographic pathologies in Alzheimer's disease." *European Journal of Neuroradiology* 28 (3): 735–744.

Lee, S., H. Lee, and K. W. Kim. 2020. "Magnetic resonance imaging texture predicts progression to dementia due to Alzheimer disease earlier than hippocampal volume." *Journal of Psychiatry and Neuroscience* 45 (1): 7–14.

Leung, T., and J. Malik. 2001. "Representing and recognizing the visual appearance of materials using three-dimensional textons." *International Journal of Computer Vision* 43 (1): 29–44.

Li, G., G. Zhai, X. Zhao, H. An, P. Spincemaille, K. M. Gillen, Y. Xu, Y. Wang, D. Huang, and J. Li. 2019. "3D texture analyses within the substantia nigra of Parkinson's disease patients on quantitative susceptibility maps and R2* maps." *NeuroImage* 188: 465–472.

Liu, L., J. Chen, P. Fieguth, G. Zhao, R. Chellappa, and M. Pietikäinen. 2019. "From BoW to CNN: Two decades of texture representation for texture classification." *International Journal of Computer Vision* 127: 74–109.

Liu, L., P. Fieguth, X. Wang, M. Pietikäinen, and D. Hu. 2016. "Evaluation of LBP and deep texture descriptors with a new robustness benchmark." Edited by B. Leibe, J. Matas, N. Sebe and M. Welling. *European Conference on Computer Vision (ECCV 2016)*: 69–86.

Liu, P., H. Wang, S. Zheng, F. Zhang, and X. Zhang. 2020. "Parkinson's disease diagnosis using neostriatum radiomic features based on T2-weighted magnetic resonance imaging." *Frontiers in Neurology* 11 (248).

London, A., I. Benhar, and M. Schwartz. 2013. "The retina as a window to the brain—from eye research to CNS disorders." *Nature Reviews Neurology* 9: 44–53.

Luk, C. C., A. Ishaque, M. Khan, D. Ta, S. Chenji, Y.-H. Yang, D. Eurich, and S. Kalra. 2018. "Alzheimer's disease: 3-dimensional MRI texture for prediction of conversion from mild cognitive impairment." *Alzheimer's & Dementia: Diagnosis, Assessment & Disease Monitoring* 10: 755–763.

Maani, R., Y.-H. Yang, D. Emery, and S. Kalra. 2016. "Cerebral degeneration in amyotrophic lateral sclerosis revealed by 3-dimensional texture analysis." *Frontiers in Neuroscience* 10 (20).

Mahmoud-Ghoneim, D., G. Toussaint, J. Constans, and J. D. de Certaines. 2003. "Three dimensional texture analysis in MRI: A preliminary evaluation in gliomas." *Magnetic Resonance Imaging* 21: 983–987.

Malviya, M. A. M., and A. S. Joshi. 2014. "Review on automatic brain tumor detection based on Gabor wavelet." *International Journal of Engineering Research & Technology* 3 (1).

Manabe, O., S. Yamaguchi, K. Hirata, K. Kobayashi, H. Kobayashi, S. Terasaka, T. Toyonaga, et al. 2021. "Preoperative texture analysis using 11C-methionine positron emission tomography predicts survival after surgery for glioma." *Diagnostics* 11 (189).

Materka, A. 2004. "Texture analysis methodologies for magnetic resonance imaging." *Dialogues in Clinical Neuroscience* 6 (2): 243–250.

Mathias, J. M., P. S. Tofts, and N. A. Losseff. 1999. "Texture analysis of spinal cord pathology in multiple sclerosis." *Magnetic Resonance in Medicine* 42: 929–935.

Matthews, T. 2016. "Semantics of texture" (PhD thesis). University of Southampton.

Mendiola-Santibañez, J. D., I. M. S. Méndez, C. P. Orta, and I. R. T. Villalobos. 2014. "Algorithm for brain extraction on magnetic resonance images T1 using morphological 3D transformations." *Revista Mexicana de Ingeniería Biomédica* 35 (3): 211–222.

Montagne, C., A. Kodewitz, V. Vigneron, V. Geraud, and S. Lelandais. 2013. "3D local binary pattern for PET image classification by SVM—application to early Alzheimer disease diagnosis." *6th International Conference on Bio-Inspired Systems and Signal Processing (BIOSIGNALS 2013)*: 145–150.

Nanthagopal, A. P., and R. S. Rajamony. 2013. "Classification of benign and malignant brain tumor CT images using wavelet texture parameters and neural network classifier." *Journal of Vizualization* 16: 19–28.

Nedelec, J.-F., O. Yu, J. Chambron, and J.-P. Macher. 2004. "Texture analysis of the brain: From animal models to human applications." *Dialogues in Clinical Neuroscience* 6 (2): 227–233.

Nie, J., Z. Xue, T. Liu, G. Young, K. Setayesh, L. Guo, and S. T. C. Wong. 2009. "Automated brain tumor segmentation using spatial accuracy-weighted hidden Markov random field." *Computerized Medical Imaging Graphics* 33 (6): 431–441.

Nunes, A., G. Silva, C. Alves, S. Batista, L. Sousa, M. Castelo-Branco, and R. Bernardes. 2019. *Textural information from the retinal nerve fibre layer in multiple sclerosis*. Portuguese Meeting on Bioengineering (ENBENG).

Nunes, A., G. Silva, C. Duque, C. Januário, I. Santana, A. F. Ambrósio, M. Castelo-Branco, and R. Bernardes. 2019. "Retinal texture biomarkers may help to discriminate between Alzheimer's, Parkinson's, and healthy controls." *PLOS One* 14 (6): e0218826.

Nunes, A., P. Serranho, H. Quental, A. F. Ambrósio, M. Castelo-Branco, and R. Bernardes. 2020. "Sexual dimorphism of the adult human retina assessed by optical coherence tomography." *Health and Technology* 10: 913–924.

Ojala, T., M. Pietikäinen, and D. Harwood. 1996. "A comparative study of texture measures with classification based on feature distributions." *Pattern Recognition* 29 (1): 51–59.

Oppedal, K., T. Edtestol, K. Engan, M. K. Beyer, and D. Aarsland. 2015. "Classifying dementia using local binary patterns from different regions in magnetic resonance images." *International Journal of Biomedical Imaging* 2015 (572567).

Ortiz-Ramón, R., M. del C. V. Hernández, V. González-Castro, S. Makin, P. A. Armitage, B. S. Aribisala, M. E. Bastin, I. J. Deary, J. M. Wardlaw, and D. Moratal. 2019. "Identification of the presence of ischaemic stroke lesions by means of texture analysis on brain magnetic resonance images." *Computerized Medical Imaging and Graphics* 74: 12–24.

Pavlidis, T. 1980. Structural descriptions and graph grammars. In *Pictorial information systems*, edited by S. K. Chang and K. S. Fu, vol. 80. Springer.

Pentland, A. 1984. "Fractal-based description of natural scenes." *IEEE Transactions on Pattern Analysis and Machine Intelligence* PAMI-6 (6): 661–674.

Petrou, M., and P. García-Sevilla. 2006. *Image processing: Dealing with texture*. John Wiley & Sons.

Petzold, A., J. F. de Boer, S. Schippling, P. Vermersch, R. Kardon, A. Green, P. A. Calabresi, and C. Polman. 2010. "Optical coherence tomography in multiple sclerosis: A systematic review and meta-analysis." *Lancet Neurology* 9: 921–932.

Rahmim, A., P. Huang, N. Shenkov, S. Fotouhi, E. Davoodi-Bojd, L. Lu, Z. Mari, H. Soltanian-Zadeh, and V. Sossi. 2017. "Improved prediction of outcome in Parkinson's disease using radiomics analysis of longitudinal DAT SPECT images." *NeuroImage: Clinical* 16: 539–544.

Rana, B., A. Juneja, M. Saxena, S. Gudwani, S. S. Kumaran, M. Behari, and R. K. Agrawal. 2017. "Relevant 3D local binary pattern based features from fused feature descriptor for differential diagnosis of Parkinson's disease using structural MRI." *Biomedical Signal Processing and Control* 34: 134–143.

Raut, A., and V. Dalal. 2017. "A machine learning based approach for early detection of Alzheimer's disease by extracting texture and shape features of the hippocampus region from MRI scans." *International Journal of Advanced Research in Computer and Communication Engineering*: 236–242.

Rezaei, K., and H. Agahi. 2017. "Malignant and benign brain tumor segmentation and classification using SVM with weighted kernel width." *Signal & Image Processing: An International Journal* 8 (2).

Saeed, N., and B. K. Puri. 2002. "Cerebellum segmentation employing texture properties and knowledge based image processing: Applied to normal adult controls and patients." *Magnetic Resonance Imaging* 20: 425–429.

Salas-Gonzalez, D., J. M. Górriz, J. Ramírez, M. López, I. A. Illan, F. Segovia, C. G. Puntonet, and M. Gómez-Río. 2009. "Analysis of SPECT brain images for the diagnosis of Alzheimer's disease using moments and support vector machines." *Neuroscience Letters* 461: 60–64.

Sapathy, A., X. Jian, and H.-L. Eng. 2014. "LBP-based edge-texture features for object recognition." *IEEE Transactions on Image Processing* 23 (5): 1953–1964.

Sayeed, A., M. Petrou, N. Spyrou, A. Kadyrov, and T. Spinks. 2001. "Diagnostic features of Alzheimer's disease extracted from PET sinograms." *Physics in Medicine & Biology* 47 (1): 137–148.

Scalco, E., and G. Rizzo. 2017. "Texture analysis of medical images for radiotherapy applications." *British Journal of Radiology* 90 (1070).

Senthilkumaran, N., and C. Kirubakaran. 2014. "A case study on mathematical morphology segmentation for MRI brain imaging." *International Journal of Computer Science and Information Technologies* 5 (4): 5336–5340.

Shah, H., E. Albanese, C. Duggan, I. Rudan, K. M. Langa, M. C. Carrillo, K. Y. Chan, et al. 2016. "Research priorities to reduce the global burden of dementia by 2025." *Lancet Neurology* 15: 1285–1294.

Shanmugan, K. S., V. Narayanan, V. S. Frost, J. A. Stiles, and J. C. Holtzman. 1981. "Textural features for radar image analysis." *IEEE Transactions on Geoscience and Remote Sensing* GE-19 (3): 153–156.

Sikiö, M., K. K. Holli-Helenius, L. C. V. Harrison, P. Ryymin, H. Ruottinen, T. Saunamäki, H. J. Eskola, I. Elovaara, and P. Dastidar. 2015. "MR image texture in Parkinson's disease: A longitudinal study." *Acta Radiologica* 56 (1): 97–104.

Sikiö, M., P. Kölhi, P. Ryymin, H. J. Eskola, and P. Dastidar. 2015. "MRI texture analysis and diffusion tensor imaging in chronic right hemisphere ischemic stroke." *Journal of Neuroimaging* 25: 614–619.

Soltaninejad, M., G. Yang, T. Lambrou, N. Allinson, T. L. Jones, T. R. Barrick, F. A. Howe, and X. Ye. 2017. "Automated brain tumour detection and segmentation using superpixel-based extremely randomized trees in FLAIR MRI." *International Journal of Computer Assisted Radiology and Surgery* 12: 183–203.

Subbanna, N. K., D. Rajashekar, B. Cheng, G. Thomalla, J. Fiehler, T. Arbel, and N. D. Forket. 2019. "Stroke lesion segmentation in FLAIR MRI datasets using customized Markov random fields." *Frontiers in Neurology* 10 (541).

Svetozarskiy, S. N., and S. V. Kopishinskaya. 2015. "Retinal optical coherence tomography in neurodegenerative diseases (review)." *Sovremennye Tehnologii v Medicine* 7 (1): 116–123.

Szegedy, C., W. Liu, Y. Jia, P. Sermanet, S. Reed, D. Anguelov, D. Erhan, V. Vanhoucke, and A. Rabinovich. 2015. "Going deeper with convolutions." *IEEE Conference on Computer Vision and Pattern Recognition (CVPR),* 1–9. https://doi.org/10.1109/cvpr.2015.7298594

Tamura, H., S. Mori, and T. Yamawaki. 1978. "Textural features corresponding to visual perception." *IEEE Transactions on Systems, Man and Cybernetics* 8 (6): 460–473.

Tazarjani, H. D., Z. Amini, R. Kafieh, F. Ashtari, and E. Sadeghi. 2021. "Retinal OCT texture analysis for differentiating healthy controls from multiple sclerosis (MS) with/without optic neuritis." *BioMed Research International*: 5579018.

Tohka, J., I. D. Dinov, D. W. Shattuck, and A. W. Toga. 2010. "Brain MRI tissue classification based on local Markov random fields." *Magnetic Resonance Imaging* 28: 557–573.

Tomita, F., and S. Tsuji. 1990. *Computer analysis of visual textures*. Springer Science + Business Media.

Torres, J. B., E. M. Anderozzi, J. T. Dunn, M. Siddique, I. Szanda, D. R. Howlett, K. Sunassee, and P. J. Blowera. 2016. "PET imaging of copper trafficking in a mouse model of Alzheimer's disease." *Journal of Nuclear Medicine* 57 (1): 109–114.

Tourassi, G. 1999. "Journey toward computer-aided diagnosis: Role of image texture analysis." *Radiology* 213: 317–320.

Trebbastoni, A., F. D'Antonio, A. Bruscolini, M. Marcellu, M. Cecere, A. Campanelli, L. Imbriano, C. de Lena, and M. Gharbiya. n.d. "Retinal nerve fibre layer thickness changes in Alzheimer's disease: Results from a 12-month prospective case series." *Neuroscience Letters* 629 (2016): 165–170.

Turcano, P., J. J. Chen, B. L. Bureau, and R. Savica. 2019. "Early ophthalmologic features of Parkinson's disease: A review of preceding clinical and diagnostic markers." *Journal of Neurology* 266: 2103–2111.

Vamvakas, A., I. Tsougos, N. Arikidis, E. Kapsalaki, K. Fountas, I. Fezoulidis, and L. Costaridou. 2018. "Exploiting morphology and texture of 3D tumor models in DTI for differentiating glioblastoma multiforme from solitary metastasis." *Biomedical Signal Processing and Control* 43: 159–173.

Varga, B. E., W. Gao, K. L. Laurik, E. Tátrai, M. Simó, G. M. Simfai, and D. C. DeBuc. 2015. "Investigating tissue optical properties and texture descriptors of the retina in patients with multiple sclerosis." *PLOS One* 10 (11): e0143711.

Varghese, B. A., S. Y. Cen, D. H. Hwang, and V. A. Duddalwar. 2019. "Texture analysis of images: What radiologists need to know." *American Journal of Roentgenology* 212 (3): 520–528.

Weska, J. S., C. R. Dyer, and A. Rosenfeld. 1976. "A comparative study of texture measures for terrain classification." *IEEE Transactions on Systems, Man and Cybernetics* SMC-6 (4): 269–285.

Yamins, D. L. K., Ha Hong, E. A. Solomon, C. F. Cadieu, D. Seibert, and J. J. DiCarlo. 2016. "Performance-optimized hierarchical models predict neural responses in higher visual cortex." *Proceedings of the National Academy of Sciences of the United States of America* 111 (23): 8619–8624.

Yu, J.-G., Y.-F. Feng, Y. Xiang, G. Savini, J.-H. Huang, V. Parisi, W.-J. Yang, and X.-A. Fu. 2014. "Retinal nerve fiber layer thickness changes in Parkinson disease: A meta-analysis." *PLOS One* 9 (1): e85718.

Yu, O., Y. Mauss, I. J. Namer, and J. Chambron. 2001. "Existence of contralateral abnormalities revealed by texture analysis in unilateral intractable hippocampal epilepsy." *Magnetic Resonance Imaging* 19 (10): 1305–1310.

Yu, O., J. Steibel, Y. Mauss, B. Guignard, B. Eclancher, J. Chambron, and D. Grucker. 2004. "Remyelination assessment by MRI texture analysis in a cuprizone mouse model." *Magnetic Resonance Imaging* 22: 1139–1144.

Zacharaki, E. I., S. Wang, S. Chawla, D. S. Yoo, R. Wolf, E. R. Melhem, and C. Davatzikos. 2009. "Classification of brain tumor type and grade using MRI texture and shape in a machine learning scheme." *Magnetic Resonance in Medicine* 62 (6): 1609–1618.

Zhang, J., C. Yu, G. Jiang, W. Liu, and L. Tong. 2012. "3D texture analysis on MRI images of Alzheimer's disease." *Brain Imaging and Behavior* 6: 61–69.

Zhang, Y., M. Brady, and S. Smith. 2001. "Segmentation of brain MR images through a hidden Markov Random field model and the expectation-maximization algorithm." *IEEE Transactions on Medical Imaging* 20 (1): 45–57.

Zhang, Y., H. Zhu, J. R. Mitchell, F. Costello, and L. M. Metz. 2009. "T2 MRI texture analysis is a sensitive measure of tissue injury and recovery resulting from acute inflammatory lesions in multiple sclerosis." *NeuroImage* 47: 107–111.

10 A Multimodal MR-Based CAD System for Precise Assessment of Prostatic Adenocarcinoma

Sarah M. Ayyad, Mohamed Shehata, Ahmed Alksas, Mohamed A. Badawy, Ali H. Mahmoud, Mohamed Abou El-Ghar, Mohammed Ghazal, Moumen El-Melegy, Nahla B. Abdel-Hamid, Labib M. Labib, H. Arafat Ali, and Ayman El-Baz

10.1 INTRODUCTION

Prostatic adenocarcinoma (PCa) is a significant public health problem in the world and is the second diagnosed cancer between men in the United States [1, 2]. In 2021, around 249,000 of new patients with PCa and 34,000 deaths were identified in American men [3]. Hence, PCa represents a serious health problem. Detecting PCa at an early stage often permits for a better treatment and suppress death rate.

There are many screening tests for diagnosing PCa, including prostatic-specific antigen (PSA) blood test, digital rectal examination (DRE), and magnetic resonance imaging (MRI). At present, transrectal ultrasound-guided biopsy (TRUS) is the gold standard diagnostic technique, in which an ultrasound probe is inserted through the rectum and produces sound waves to create an image of the prostate gland. Nevertheless, it is an aching procedure, is costly, and has side effects, such as bleeding and contagion. PSA is a non-invasive way of accessing the level of prostate-specific antigen in the blood [1, 4]. A high PSA value indicates that the patient may have PCa. In general, if the PSA level is larger than 4 ng/ml, the patient may be prone to have PCa [5]. However, PSA has a lower sensitivity, as patients with obesity and inflamed prostate may also have a high PSA level. Another screening test that was first learned by urologists is the DRE test, where the doctor inspects the prostate gland manually to feel any enlargement or hardness [5, 6]. However, this test is uncomfortable and misses the majority of PCa.

Today's MRI has become the most significant and influential in an accurate diagnosis of PCa. Hence, MRI-based computer-aided diagnosis (CAD) systems have gained popularity in the last few decades using different modalities of MRI, such as diffusion-weighted MRI (DW-MRI) [7–12], T2-weighted (T2W- MRI) [13–15],

DOI: 10.1201/9780367486082-10

multiparametric MRI (mpMRI) [16–18], and dynamic contrast-enhanced (DCE-MRI) [19–23]. Generally, DW-MRI and T2W-MRI for diagnosing PCa attain higher accuracy results than other modalities of MRI.

Many MRI-based CAD systems have been designed in the domain of automatic detection of PCa using handcrafted-based machine learning (ML) techniques. For example, authors in [24] proposed a new CAD system using a set of 215 texture features applied on T2-weighted MRI. They extensively compared 11 classifiers to evaluate their CAD. The results showed their CAD was comparable to existing other CADs, as it reached to 90% of area under the curve (AUC) using Bayesian networks. In the same context, authors in [25] introduced a systematic framework to find the best-performing classifier using T2-weighted images applying 110 texture features. They used statistical analysis to determine the best classifier and concluded that support vector machine (SVM) is the best classifier with an accuracy of 92%. Many researchers tried to use different modalities on their CAD to achieve the best results [1, 4, 26, 27]. For example, [26], authors proposed a new CAD system that utilized T2W-MRI, high *b*-value DW-MRI, and apparent diffusion coefficient (ADC) map of DW-MRI. They extracted 124 texture features and then used a three-stage feature selection method to identify clinically significant PCa. Using five-fold cross-validation, they reported 0.92 of AUC. In addition to handcrafted-based techniques, many deep learning (DL) techniques have been designed to detect PCa [28–31].

One of the main DL architectures used extensively in biomedical imaging, convolutional neural network (CNN), has made sound advancement in PCa detection [32–36]. In [26], authors developed a stack of individually trained residual CNN. They also proposed a new technique for converting slice-level classification results to patient level. They tested their technique on 427 patients, and results attained 0.87 of AUC at slice level and 0.84 of AUC at patient level. A new attempt to predict the Gleason score groups and detect prostate cancer lesions was introduced in [27]. They used FocalNet, a variant of CNN, to handle the multi-class classification of PCa and characterize lesion aggressiveness. Using five-fold cross-validation, their model achieved 0.81 of AUC.

The literatures described earlier have the following limitations: (1) none of the existing studies integrates both clinical biomarkers and imaging markers; (2) some of the studies are applicable only to the peripheral zone lesions, ignoring the transitional zone; (3) most of the existing techniques to classify PCa depend on the accuracy of the applied automatic segmentation algorithm, which has the potential to provide inaccurate results; (4) none of the existing handcrafted-features CADs integrated functional, shape, and texture features to enhance the diagnostic accuracy of CAD; and (5) most of the DL-based CADs need a large number of well-curated images for training their models. This work intends to address such limitations to provide a novel PCa-CAD system for precise and early identification of PCa. To the best of our knowledge, we are the first research group to combine clinical biomarker, PSA test, with shape, texture, and functional features extracted from prostate and/or lesion toward an accurate diagnosis of PCa. Details of the proposed PCa-CAD system are illustrated in the next section.

10.2 METHODOLOGY

As shown in Figure 10.1, the proposed PCa-CAD system consists of three main steps: (1) data preparation: prostate and/or lesions are segmented from T2W-MR images and DW-MR images obtained at nine different b-values, which are {b 0, b 100, b 200, b 300, b 400, b 500, b 600, b 700, and b 1400 s/mm^2}; (2) feature extraction: extracting functional, texture, and shape features from the segmented prostate and/or lesion from the two modalities of MR images; and (3) performing feature integration of optimal feature sets with PSA clinical biomarker and obtain the final diagnosis using an SVM classification model. The steps of the proposed CAD system are fully illustrated in the following subsections.

10.2.1 Data Preparation

A total of 66 patients diagnosed with PCa were enrolled in this research after providing informed consent. Ethical board approval was received from the Urology and Nephrology Center, Mansoura University, Egypt (approval number R.21.04.1289). Eligible patients had a history of high PSA or clinical suspicion of PCa and had at least one suspicious lesion appear on MR images. Between 2019 and 2021, 36 benign PCa and 30 malignant PCa patients' MRIs were acquired using three Tesla Ingenia Philips scanners using the acquisition parameters described in Table 10.1. The mean age of the dataset was 66 years (range 48–82 years). For a better diagnostic performance, an experienced radiologist with more than 12 years of hands-on experience manually segmented both prostate and lesion(s) from both DW-MR images and T2W-MR images for the entire dataset.

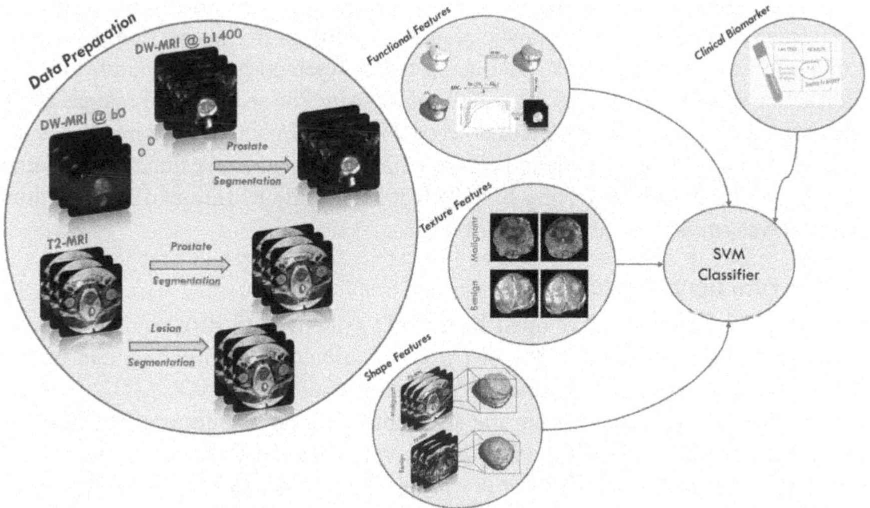

FIGURE 10.1 Pipeline of the developed PCa-CAD for early identification of prostatic adenocarcinoma.

TABLE 10.1
Description of Data Acquisition Protocol

Field of view	200 mm	178 mm	86 mm
Voxel	3 mm	3 mm	3.5 mm
Matrix	68	59	24 slices
Number of signals averaging	4		
Repetition time	4734 ms		
Echo time	88 ms		
Flip angle	90o		
Fat suppression	Spectral attenuated inversion recovery		
Fasting imaging mode	Echo planar imaging		
Folder-over suppression	Oversampling		
Technique	SE		
Water fat shift	Minimum		
Number of b-factors	9		
b-factors	0, 100, 200, 300, 400, 500, 600, 700, 1,400		
Total scan duration	7:58 minute		

10.2.2 Feature Extraction

Feature extraction plays a fundamental role in the machine learning process; specifically speaking, in the medical image analysis domain, it highlights the most discriminating and dominant features from images that can be used afterward as input to the classifier to improve the diagnostic results [37–40]. In this subsection, functional features are extracted from DW-MRI using the segmented prostate gland, followed by extraction of texture features from T2W-MRI using the segmented prostate gland. After that, extraction of shape features from T2W-MRI using the segmented lesion part. Finally, feature integration and classification are conducted. Functional feature extraction was implemented using Matlab R2021b, while texture and shape feature extraction was implemented using Python 3.7.

10.2.2.1 Functional Features

The main purpose of the developed PCa-CAD system is to differentiate between benign and malignant tumors. One of the discriminatory features that can be calculated from DW-MRI is the apparent diffusion coefficient (ADC) [41], which is a quantitative map calculated from the segmented region of interests (ROIs) of the whole prostate gland. DW-MRI depends on evaluating the diffusion of water molecules within tissues. This diffusion is irregular, and the proportion of randomness is positively related to the value of ADC. From an image processing perspective, this randomness can be represented by changes in the gray level values. To produce the ADC map, two DW-MR images are needed. It is estimated at the voxel level as follows:

$$ADC_{(x,y,z)} = \frac{\ln s_0(x,y,z) - \ln s_n(x,y,z)}{b_n - b_0} \quad (10.1)$$

Where s_0 and s_n are the signal intensity obtained at the baseline (b0) and a higher b-value $\in$ {b 100, b 200, b 300, b 400, b 500, b 600, b 700, b 1400}, and n is the higher b-value. The cancerous tumors have higher signal intensity at different b-values, and a smaller ADC value compared with noncancerous tumors [40]. In order to minimize the impact of noise and preserve the continuity of ADC, we applied a continuity correction (smoothing). Furthermore, to account for the variable size of input data and to reduce training time to classify large data volumes, cumulative distribution functions (CDFs) of the ADC values are constructed at the eight different b-values (one CDF for each b-value). After normalization, the ADCs are divided into 100 steps so that all ADC values are kept consistent with a unified size without missing any information, resulting in a CDF of vector size 800 × 1 for each of the prostate and lesion DW-MRI. We used the resulting CDFs after removing the redundant ones to feed to the classifier (122 CDFs for prostate images) to distinguish between benign and malignant images.

10.2.2.2 Texture Features

Texture feature extraction can be defined as capturing the way of distribution of the gray levels over the pixels in the image. These features are captured from T2W-MR images, and specifically from the segmented ROIs of the whole prostate gland. Figure 10.2 reveals an illustrative example to compare benign cases and malignant cases in terms of texture differences. In this chapter, we extracted first-order textural features and second-order textural features as follows:

First Order Texture Features. These texture features explore the frequency distribution in the ROI through a histogram [41–45]. They do not take into account the relationships with near pixel. Our first-order feature set considers 36 features, including mean, median, standard deviation, variance, entropy, kurtosis, skewness, descriptive (mean, variance, Nobs, kurtosis), the number of points in each pin, size of bins, bin width, lower limit, cumulative frequency, CDFs (N = 10); the gray level percentiles were estimated from the 10th to the 100th percentiles with a step of 10%.

Second-Order Texture Features. The features generated from first-order statistics may be prone to noise, so second-order texture features are essential to provide robust diagnostic performance, namely, gray level co-occurrence matrix (GLCM) and gray level run-length matrix (GLRLM) were constructed. GLCM was first described by Shanmugam and Haralick [46] to define the spatial correlation of the grey level values in an image. It is a two-dimensional matrix *G(i, j)*, in which each item *(i, j)* in the matrix

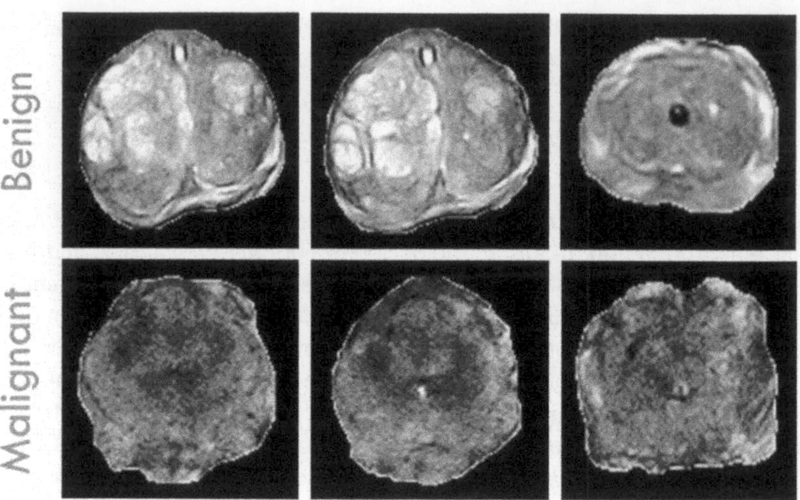

FIGURE 10.2 Some examples of prostatic texture differences between three benign cases (first row) and three malignant cases (second row).

determines the number of times that the various combinations of pixel gray level values happen in an image. This matrix is considered in the zero, $\frac{\pi}{2}$, $\frac{\pi}{4}$, $\frac{3\pi}{4}$ directions, with the change of distance $d \leq \sqrt{2}$. In this chapter, we extracted six features to describe the characteristics and considered all directions, namely, correlation, contrast, angular second moment (ASM), dissimilarity, energy, and homogeneity. On the other hand, GLRLM [42, 47] was used that is based on evaluating runs of gray levels in the image. Here, 16 GLRLM features were used to better collect a variety of features, including gray level non-uniformity normalized, gray level non-uniformity, high gray level run emphasis, gray level variance, long run emphasis, long run high gray level emphasis, long run low gray level emphasis, low gray level run emphasis, run-length non-uniformity, run entropy, run-length non-uniformity normalized, run percentage, run variance, short-run emphasis, short-run low gray level emphasis, and short-run high gray level emphasis.

10.2.2.3 Shape Features

Another set of features, shape features, is also calculated from T2W-MRI and used in the developed PCa-CAD to better describe the shape differences between malignant and benign tumors. Relying on the fact that, usually, malignant tumors have higher growth rate, complex surfaces, and irregular shapes [48–52]. To model the 3D surface of the lesion part, we employed the spectral spherical harmonics (SH) analysis [53], as shown in Figure 10.3. At first, we randomly selected a point to be the point of origin (0,0,0) inside the region of the ROIs. In this coordinate system, the surface of the region can be considered a function of the polar and azimuth angle, f(θ, φ), which can be expressed as a linear set of base functions Yτβ given in the unit sphere. Secondly, the surface of the lesion part is mapped to a unit sphere by the use of our developed attraction–repulsion technique [54–56].

Let α represents the cycle of each of the attraction–repulsion technique, $\mathbf{C}_{\alpha,i}$ signifies the coordinates of the node on the unit sphere, $d_{\alpha,ji} = \mathbf{C}_{\alpha,j} - \mathbf{C}_{\alpha,i}$ signify the vector from node i to node j, and $C_{A,1}$ $C_{A,2}$ are parameters which determine the strength of the attraction–repulsion. After that, the attraction step updates the node's locations according to equation (10.2).

$$\mathbf{C}'_{\alpha+1,i} = \mathbf{C}_{\alpha,l} + C_{A,1} \sum\nolimits_{j \in J_i} \left(\boldsymbol{d}_{\alpha,ji}\, d^2_{\alpha,ji} + C_{A,2} \frac{\boldsymbol{d}_{\alpha,ji}}{d_{\alpha,ji}} \right) \tag{10.2}$$

Where $j = 1, 2, \ldots, J-1, J$, and $i = 1, 2, \ldots, I-1, I$. Assume that C_R is a parameter that manages the shift that occurred in each surface node. Then, the repulsion step expands the spherical mesh to prevent it from deteriorating according to equation (10.3).

$$\mathbf{C}''_{\alpha+1,i} = \mathbf{C}'_{\alpha+1,i} + \frac{C_R}{2I} \sum\nolimits_{j=1; j \neq i}^{I} \frac{\boldsymbol{d}_{\alpha,ji}}{\boldsymbol{d}^2_{\alpha,ji}} \tag{10.3}$$

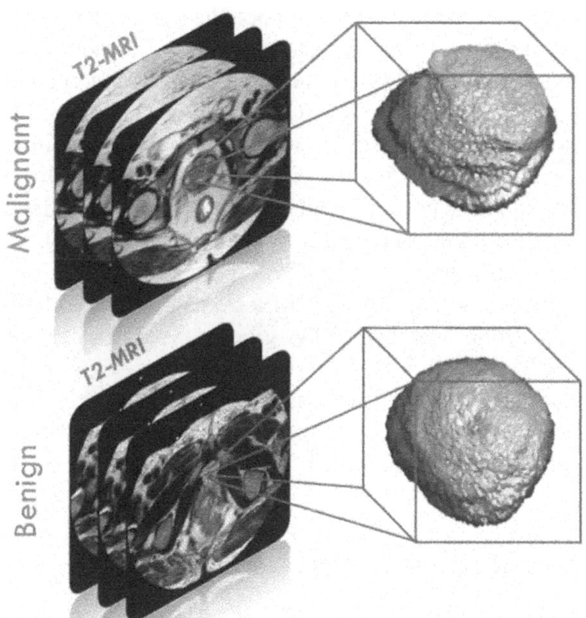

FIGURE 10.3 An illustrative example results from the spectral analysis of spherical harmonics showing the complexity of the surface of a malignant lesion (first row) compared to a benign lesion (second row).

In the last cycle α of the attraction–repulsion technique, each point $\mathbf{C}_i = (x_i, y_i, z_i)$ of the original mesh has been mapped to a corresponding point $\mathbf{C}_{\alpha_f,i} = (\sin\theta_i \cos\varphi_i, \sin\theta_i \sin\varphi_i, \cos\theta_i)$ with polar angle $\theta_i \in [0, \pi]$ and azimuth angle $\varphi_i \in [0, 2\pi]$. Finally, the SH series is generated by solving the isotropic heat equation for the surface, which is a function on the unit sphere, as follows:

$$Y_{\tau\beta} = \begin{cases} c_{\tau\beta}\, G_\tau^{|\beta|} \cos\theta \sin(|\beta|\varphi) & -\tau \le \beta \ge -1 \\ \dfrac{c_{\tau\beta}}{\sqrt{2}} G_\tau^{|\beta|} \cos\theta & \beta = 0 \\ c_{\tau\beta}\, G_\tau^{|\beta|} \cos\theta \cos(|\beta|\varphi) & 1 \le \beta \ge \tau \end{cases} \tag{10.4}$$

Where $c_{\tau\beta}$ is the SH factor and $G_\tau^{|\beta|}$ denotes to the relevant Legendre polynomial of degree τ with order β. In our CAD system, only 85 shape features were automatically extracted for the lesion part to rebuild their shapes. It is noted that benign and healthy lesions are described by a lower-order integration of SH series than cancerous lesions.

10.2.3 Feature Integration and Classification

After estimating the functional discriminating features from ROIs of prostate glands and extracting texture and shape features from ROIs of tumors, we integrated all features with PSA clinical biomarker to enhance the overall diagnostic performance and generalization ability of the constructed model. All these features have unique characteristics. After that, our CAD system continues with a classification stage to obtain the final diagnosis for the input MRI. Toward our goal, we applied four traditional classifiers to get the best possible results, for example, support vector machine (SVM) [57], random forest (RF) [58], decision tree (DT) [59], and linear discriminant analysis (LDA) [60]. SVM is powerful and efficiently performs a linear classification [61–63]. RF is one of the most popular ML classifiers used in CAD systems because of its ability to reduce overfitting without hyperparameter tuning [64–66]. DT conducts the classification model in the form of a tree structure, and it is robust in the presence of noise [67–68]. Moreover, LDA gives a fast classification for the image input. To better explore which integration of feature sets works efficiently, we first evaluate the performances of each feature set individually. Afterward, the feature sets are concatenated to form different feature vectors and utilized the aforementioned classifiers to get the final diagnosis. The description of all feature sets is given in Table 10.2.

10.3 RESULTS AND DISCUSSION

In order to verify the effectiveness of the PCa-CAD system on the aforementioned datasets, evaluation is carried out using four well-known metrics: (1) accuracy, (2) specificity, (3) sensitivity, and (4) AUC. Additional details can be shown in Figure 10.4 and investigated in [69–73]. Different k-fold cross-validation schemas were employed

TABLE 10.2
Details of the Extracted Feature Sets (FS_i)

Feature-Set	Description	N
FS_1 (prostate)	Functional features (8-CDFs)	122
FS_2 (prostate)	Texture features (1st and 2nd order)	58
FS_3 (lesion)	Shape features (SHs)	85
FS_4 (clinical)	PSA	1
FS_5 (combined)	$FS_1 + FS_2 + FS_3 + FS_4$	266

Note: Let *SHs*, *PSA*, and *N* denote spherical harmonics, prostatic-specific antigen, and the number of components for each type of feature set, respectively.

		Actual health condition		
		Malignant	Benign	
Predicted health condition	Malignant	True Positive (TP)	False Positive (FP)	
	Benign	False Negative (FN)	True Negative (TN)	
		Sensitivity = $\frac{TP}{(TP+FN)}$	Specificity = $\frac{TN}{(TN+FP)}$	Accuracy = $\frac{TP+TN}{(TP+FP+FN+TN)}$

FIGURE 10.4 Typical performance metrics for evaluating the diagnostic performance of the designed PCa-CAD.

for better evaluation of the developed PCa-CAD performance, in which *k* was set to 10, 5, and number of patients (leave-one-out cross-validation). All experiments were performed ten times, and the mean and standard deviation for the accuracy, sensitivity, specificity, and AUC were recorded for each feature set. It is worth mentioning that the parameters of the employed classifiers were set by means of a grid search algorithm to explore the optimal set of the four different classifiers. The optimal parameters were found to be as follows: for SVM (kernel function = Gaussian, and kernel scale box = 1), for DT (split criterion = Gini diversity index, and average maximum number of splits = 3), for LDA (discriminant type = diagLinear), and for RF (number of learning cycles = 30, and maximum number of splits = 65). Throughout this study, we have tried to employ functional, texture, and shape feature extraction one time for the whole prostate and another for the ROI of the lesion, only to determine the best segmented image which could be informative to the feature extractor. So it is worth mentioning that the whole prostate gland was more informative in terms of functionality and texture evidenced by a better diagnostic performance in these aspects, while lesion ROIs provided better shape indicators than the whole prostate.

In Tables 10.3 and 10.4, the classification performance was tested by SVM, RF, DT, and LDA classifiers, respectively, using the three validation schemas. The attained results had shown that the overall performance based on feature set FS5 is better than all other feature sets, and this proves the feasibility and the importance of the integration process for the developed PCa-CAD. As shown in Table 10.3, the best results were achieved by the SVM classifier using leave-one-out cross-validation, with 87.88% ± 0.0% of accuracy, 80.00% ± 0.0% of sensitivity, 94.44% ± 0.0% of specificity, and 0.87 ± 0.0 of AUC. This table also depicts that RF classifier attained the second highest performance results using leave-one-out cross-validation with an average of 86.11% ± 0.56% of accuracy, 80.00% ± 2.72% of sensitivity, 91.21% ± 1.91% of specificity, and 0.86 ± 0.01 of AUC. The obtained results demonstrate that the overall performance in the three validation schemas based on SVM classifier is better than all other classifiers, and this indicates that using SVM for classification might improve the diagnostic capabilities. In this regard, SVM was selected in the classification stage for the PCa-CAD system.

Table 10.4 illustrates performance results of all individual feature set that were extracted utilizing an SVM classification model. From this table, we can highlight the merits of using the combined feature set (FS5) over each individual feature set. As shown from that table, the combined feature set (FS5) attained superior values along with the three validation schemas when compared to the individual feature sets. This promotes the integration procedure as a valued addition to the individual feature sets in the developed PCa-CAD. Concerning evaluating the individual features, the best result was attained as expected by the functional features FS1, the second-ranking performance was attained by texture features FS2, while PSA alone achieved the lowest diagnostic performance (please see Table 10.4).

TABLE 10.3

Experimental Results with Regard to Accuracy, Sensitivity, Specificity, and AUC Using Four Optimized Traditional ML Classifiers for the Integrated Feature Set (FS_5)

Classifier	Validation	Accuracy	Sensitivity	Specificity	AUC
SVM	5-fold	85.10 ± 1.04	76.11 ± 2.99	92.59 ± 2.07	0.84 ± 0.01
	10-fold	85.69 ± 1.61	75.93 ± 1.39	93.83 ± 2.54	0.85 ± 0.02
	Leave-one-out	**87.88 ± 0.00**	**80.00 ± 0.00**	94.44 ± 0.00	**0.87 ± 0.00**
RF	5-fold	84.63 ± 1.70	79.52 ± 1.16	88.89 ± 2.97	0.84 ± 0.02
	10-fold	86.15 ± 2.21	78.09 ± 3.92	92.86 ± 2.02	0.85 ± 0.02
	Leave-one-out	86.11 ± 0.56	80.00 ± 2.72	91.21 ± 1.91	0.86 ± 0.01
DT	5-fold	85.61 ± 1.69	**75.7 ± 3.73**	94.44 ± 4.24	0.85 ± 0.02
	10-fold	85.28 ± 1.56	72.38 ± 2.93	96.03 ± 2.02	0.84 ± 0.02
	Leave-one-out	86.36 ± 0.00	73.33 ± 0.00	**97.22 ± 0.00**	0.85 ± 0.00
LDA	5-fold	82.25 ± 1.06	72.85 ± 1.16	90.08 ± 2.02	0.81 ± 0.01
	10-fold	82.57 ± 1.69	73.33 ± 0.00	90.28 ± 3.10	0.82 ± 0.02
	Leave-one-out	81.82 ± 0.00	73.33 ± 0.00	88.89 ± 0.00	0.81 ± 0.00

Note: The SVM classification model outperformed all other classification models.

TABLE 10.4
Experimental Results with Regard to Accuracy, Sensitivity, Specificity, and AUC Using the Optimized SVM Classification Model

Feature Set	Validation	Accuracy	Sensitivity	Specificity	AUC
FS1	5-fold	82.19 ± 1.12	73.33 ± 0	89.58 ± 2.30	0.82 ± 0.01
	10-fold	83.15 ± 1.13	73.7 ± 2.92	89.20 ± 2.05	0.82 ± 0.01
	Leave-one-out	83.33 ± 0.00	76.67 ± 0	88.89 ± 0.00	0.83 ± 0.00
FS2	5-fold	77.88 ± 2.46	57.33 ± 2.91	**95.00 ± 3.89**	0.76 ± 0.02
	10-fold	78.03 ± 2.06	58.00 ± 3.06	**94.72 ± 2.30**	0.76 ± 0.02
	Leave-one-out	78.79 ± 0.00	60.00 ± 0.00	94.44 ± 0.00	0.77 ± 0.00
FS3	5-fold	72.27 ± 2.14	61.66 ± 3.41	81.11 ± 3.24	0.71 ± 0.02
	10-fold	71.97 ± 1.51	61.25 ± 2.86	80.91 ± 1.66	0.71 ± 0.02
	Leave-one-out	74.24 ± 0.00	63.33 ± 0.00	83.33 ± 0.00	0.73 ± 0.00
FS4	5-fold	73.59 ± 1.10	51.9 ± 2.42	91.67 ± 0.00	0.72 ± 0.01
	10-fold	74.24 ± 0.00	53.33 ± 0.00	91.67 ± 0.00	0.73 ± 0.00
	Leave-one-out	74.24 ± 0.00	53.33 ± 0.00	91.67 ± 0.00	0.73 ± 0.00
FS5	5-fold	**85.1 ± 1.04**	**76.11 ± 2.99**	92.59 ± 2.07	**0.84 ± 0.01**
	10-fold	**85.69 ± 1.61**	**75.93 ± 1.39**	93.83 ± 2.54	**0.85 ± 0.02**
	Leave-one-out	**87.88 ± 0.00**	**80.00 ± 0.00**	**94.44 ± 0.00**	**0.87 ± 0.00**

Note: The diagnostic performance of the combined feature set (FS_5) outperformed all other individual features sets (FS_1, FS_2, FS_3, and FS_4) using three different validation schemas.

10.4 CONCLUSION

This chapter introduces a new CAD system for precise and early assessment of prostatic adenocarcinoma. Our dataset consists of 66 biopsy-confirmed patients (malignant [prostatic carcinomas] = 30, and benign [prostatic hyperplasia] = 36) who underwent two types of MRI scans, namely, DW-MRI and T2W-MRI, and provided their consent to participate in this study. The developed PCa-CAD implements the major innovative steps: (1) estimate the best discriminating functional features from DW-MRIs using nine different low (accounts for blood perfusion) and high (accounts for water diffusion) *b*-values, known as voxel-wise ADCs; (2) extract differentiating first and second textural features from T2W-MRI; (3) identify the optimal characterizing shape features based on the spectral analysis of spherical harmonics of T2W-MRI; and (4) concatenate all the aforementioned features with the clinical biomarkers represented by the PSA to obtain the integrated feature set to be used as our final discriminatory features for the optimal diagnosis.

The main advantage of the developed PCa-CAD system is that it integrates imaging features extracted from prostate and/or lesion with clinical biomarkers (PSA), which provides better discriminating feature set that augment the classification performance over any stand-alone feature set. This was evidenced by the highest classification accuracy of 87.88% ± 0.00%, sensitivity of 80% ± 0.00%, specificity of 94.44% ± 0.00%, and AUC of 0.8722 ± 0.00, in differentiating benign from malignant prostate lesions. Experiments on in vivo data demonstrated the robustness of the optimized SVM classification model with the integrated features. Further work will include increasing the dataset, especially the malignant group, in order to investigate the abilities of the PCa-CAD system in specifying the underlying grade of the malignant lesion. Furthermore, a deep learning–based CAD will be designed for the fully automated extraction of discriminatory features.

REFERENCES

1. Lemaitre, G., Martí, R., Rastgoo, M., & Mériaudeau, F. (2017, July). Computer-aided detection for prostate cancer detection based on multi-parametric magnetic resonance imaging. In *2017 39th Annual International Conference of the IEEE Engineering in Medicine and Biology Society (EMBC)* (pp. 3138–3141). IEEE.
2. Pellicer-Valero, O. J., Marenco Jiménez, J. L., Gonzalez-Perez, V., Casanova Ramón-Borja, J. L., Martín García, I., Barrios Benito, M., . . . Martín-Guerrero, J. D. (2022). Deep Learning for fully automatic detection, segmentation, and Gleason Grade estimation of prostate cancer in multiparametric magnetic resonance images. *Scientific Reports*, 12(1), 1–13.
3. Key statistics for prostate cancer. https://www.cancer.org/cancer/prostate-cancer/about/key-statistics.html/, accessed: 2021 December 12.
4. Bleker, J., Kwee, T. C., Dierckx, R. A., de Jong, I. J., Huisman, H., & Yakar, D. (2020). Multiparametric MRI and auto-fixed volume of interest-based radiomics signature for clinically significant peripheral zone prostate cancer. *European Radiology*, 30(3), 1313–1324.
5. Reda, I., Khalil, A., Elmogy, M., Abou El-Fetouh, A., Shalaby, A., Abou El-Ghar, M., . . . El-Baz, A. (2018). Deep learning role in early diagnosis of prostate cancer. *Technology in Cancer Research & Treatment*, 17, 1533034618775530.
6. Wang, S., Burtt, K., Turkbey, B., Choyke, P., & Summers, R. M. (2014). Computer aided-diagnosis of prostate cancer on multiparametric MRI: A technical review of current research. *BioMed Research International*, 2014, Article ID 789561. https://doi.org/10.1155/2014/789561
7. Puech, P., Betrouni, N., Makni, N., Dewalle, A. S., Villers, A., & Lemaitre, L. (2009). Computer-assisted diagnosis of prostate cancer using DCE-MRI data: Design, implementation and preliminary results. *International Journal of Computer Assisted Radiology and Surgery*, 4(1), 1–10.
8. Mazzetti, S., Giannini, V., Russo, F., & Regge, D. (2018). Computer-aided diagnosis of prostate cancer using multi-parametric MRI: Comparison between PUN and Tofts models. *Physics in Medicine & Biology*, 63(9), 095004.
9. Yang, H., Wu, G., Shen, D., & Liao, S. (2021, April). Automatic prostate cancer detection on multi-parametric MRI with hierarchical weakly supervised learning. In *2021 IEEE 18th International Symposium on Biomedical Imaging (ISBI)* (pp. 316–319). IEEE.
10. Shehata, M., Khalifa, F., Soliman, A., Alrefai, R., Abou El-Ghar, M., Dwyer, A. C., . . . El-Baz, A. (2015, April). A novel framework for automatic segmentation of kidney from DW-MRI. In *2015 IEEE 12th International Symposium on Biomedical Imaging (ISBI)* (pp. 951–954). IEEE.
11. Shehata, M., Mahmoud, A., Soliman, A., Khalifa, F., Ghazal, M., Abou El-Ghar, M., . . . El-Baz, A. (2018). 3D kidney segmentation from abdominal diffusion MRI using an appearance-guided deformable boundary. *PLOS One*, 13(7), e0200082.
12. Shehata, M., Khalifa, F., Soliman, A., Takieldeen, A., Abou El-Ghar, M., Shaffie, A., . . . Keynton, R. (2016, April). 3D diffusion MRI-based CAD system for early diagnosis of acute renal rejection. In *2016 IEEE 13th International Symposium on Biomedical Imaging (ISBI)* (pp. 1177–1180). IEEE.

13. Alkadi, R., El-Baz, A., Taher, F., & Werghi, N. (2018). A 2.5 D deep learning-based approach for prostate cancer detection on T2-weighted magnetic resonance imaging. In *Proceedings of the European Conference on Computer Vision (ECCV) Workshops*. ECCV.
14. Alkadi, R., Taher, F., El-Baz, A., & Werghi, N. (2019). A deep learning-based approach for the detection and localization of prostate cancer in T2 magnetic resonance images. *Journal of Digital Imaging*, 32(5), 793–807.
15. Naglah, A., Khalifa, F., Khaled, R., Abdel Razek, A. A. K., Ghazal, M., Giridharan, G., & El-Baz, A. (2021). Novel MRI-based cad system for early detection of thyroid cancer using multi-input CNN. *Sensors*, 21(11), 3878.
16. Alkadi, R., Taher, F., El-Baz, A., & Werghi, N. (2018). Early diagnosis and staging of prostate cancer using magnetic resonance imaging: State of the art and perspectives. *Prostate Cancer Imaging*, 165–188.
17. Reda, I., Shalaby, A., Elmogy, M., Abou Elfotouh, A., Khalifa, F., Abou El-Ghar, M., . . . El-Baz, A. (2017). A comprehensive non-invasive framework for diagnosing prostate cancer. *Computers in Biology and Medicine*, 81, 148–158.
18. Reda, I., Shalaby, A., Abou El-Ghar, M., Khalifa, F., Elmogy, M., Aboulfotouh, A., . . . Keynton, R. (2016, April). A new NMF-autoencoder based CAD system for early diagnosis of prostate cancer. In *2016 IEEE 13th International Symposium on Biomedical Imaging (ISBI)* (pp. 1237–1240). IEEE.
19. Chang, C. Y., Srinivasan, K., Hu, H. Y., Tsai, Y. S., Sharma, V., & Agarwal, P. (2020). SFFS-SVM based prostate carcinoma diagnosis in DCE-MRI via ACM segmentation. *Multidimensional Systems and Signal Processing*, 31(2), 689–710.
20. Khalifa, F., Soliman, A., El-Baz, A., Abou El-Ghar, M., El-Diasty, T., Gimel'farb, G., . . . Dwyer, A. C. (2014). Models and methods for analyzing DCE-MRI: A review. *Medical Physics*, 41(12), 124301.
21. El-Baz, A., Fahmi, R., Yuksel, S., Farag, A. A., Miller, W., El-Ghar, M. A., & Eldiasty, T. (2006, October). A new CAD system for the evaluation of kidney diseases using DCE-MRI. In *International Conference on Medical Image Computing and Computer-Assisted Intervention* (pp. 446–453). Springer.
22. Khalifa, F., Abou El-Ghar, M., Abdollahi, B., Frieboes, H. B., El-Diasty, T., & El-Baz, A. (2013). A comprehensive non-invasive framework for automated evaluation of acute renal transplant rejection using DCE-MRI. *NMR in Biomedicine*, 26(11), 1460–1470.
23. El-Baz, A., Farag, A., Fahmi, R., Yuksela, S., El-Ghar, M. A., & Eldiasty, T. (2006, August). Image analysis of renal DCE MRI for the detection of acute renal rejection. In *18th International Conference on Pattern Recognition (ICPR'06)* (Vol. 3, pp. 822–825). IEEE.
24. Rampun, A., Zheng, L., Malcolm, P., Tiddeman, B., & Zwiggelaar, R. (2016). Computer-aided detection of prostate cancer in T2-weighted MRI within the peripheral zone. *Physics in Medicine & Biology*, 61(13), 4796.
25. Varghese, B., Chen, F., Hwang, D., Palmer, S. L., De Castro Abreu, A. L., Ukimura, O., . . . Pandey, G. (2020, September). Objective risk stratification of prostate cancer using machine learning and radiomics applied to multiparametric magnetic resonance images. In *Proceedings of the 11th ACM International Conference on Bioinformatics, Computational Biology and Health Informatics* (pp. 1–10). ACM.
26. Kwak, J. T., Xu, S., Wood, B. J., Turkbey, B., Choyke, P. L., Pinto, P. A., . . . Summers, R. M. (2015). Automated prostate cancer detection using T2-weighted and high-b-value diffusion-weighted magnetic resonance imaging. *Medical Physics*, 42(5), 2368–2378.
27. Lay, N. S., Tsehay, Y., Greer, M. D., Turkbey, B., Kwak, J. T., Choyke, P. L., . . . Summers, R. M. (2017). Detection of prostate cancer in multiparametric MRI using random forest with instance weighting. *Journal of Medical Imaging*, 4(2), 024506.

28. Yoo, S., Gujrathi, I., Haider, M. A., & Khalvati, F. (2019). Prostate cancer detection using deep convolutional neural networks. *Scientific Reports*, 9(1), 1–10.
29. Cao, R., Bajgiran, A. M., Mirak, S. A., Shakeri, S., Zhong, X., Enzmann, D., . . . Sung, K. (2019). Joint prostate cancer detection and Gleason score prediction in mp-MRI via FocalNet. *IEEE Transactions on Medical Imaging*, 38(11), 2496–2506.
30. Tsehay, Y. K., Lay, N. S., Roth, H. R., Wang, X., Kwak, J. T., Turkbey, B. I., . . . Summers, R. M. (2017, March). Convolutional neural network based deep-learning architecture for prostate cancer detection on multiparametric magnetic resonance images. In *Medical Imaging 2017: Computer-Aided Diagnosis* (Vol. 10134, pp. 20–30). SPIE.
31. Ishioka, J., Matsuoka, Y., Uehara, S., Yasuda, Y., Kijima, T., Yoshida, S., . . . Fujii, Y. (2018). Computer-aided diagnosis of prostate cancer on magnetic resonance imaging using a convolutional neural network algorithm. *BJU International*, 122(3), 411–417.
32. Ayyad, S. M., Shehata, M., Shalaby, A., El-Ghar, A., Ghazal, M., El-Melegy, M., . . . El-Baz, A. (2021). Role of AI and histopathological images in detecting prostate cancer: A survey. *Sensors*, 21(8), 2586.
33. Reda, I., Ayinde, B. O., Elmogy, M., Shalaby, A., El-Melegy, M., Abou El-Ghar, M., . . . El-Baz, A. (2018, April). A new CNN-based system for early diagnosis of prostate cancer. In *2018 IEEE 15th International Symposium on Biomedical Imaging (ISBI 2018)* (pp. 207–210). IEEE.
34. Abdeltawab, H., Shehata, M., Shalaby, A., Khalifa, F., Mahmoud, A., El-Ghar, M. A., . . . El-Baz, A. (2019). A novel CNN-based CAD system for early assessment of transplanted kidney dysfunction. *Scientific Reports*, 9(1), 1–11.
35. Haweel, R., Shalaby, A., Mahmoud, A., Seada, N., Ghoniemy, S., Ghazal, M., . . . El-Baz, A. (2021). A robust DWT–CNN-based CAD system for early diagnosis of autism using task-based fMRI. *Medical Physics*, 48(5), 2315–2326.
36. Hammouda, K., Khalifa, F., Soliman, A., Abdeltawab, H., Ghazal, M., Abou El-Ghar, M., . . . El-Baz, A. (2020, April). A 3D CNN with a learnable adaptive shape prior for accurate segmentation of bladder wall using MR images. In *2020 IEEE 17th International Symposium on Biomedical Imaging (ISBI)* (pp. 935–938). IEEE.
37. Hosseini-Asl, E., Keynton, R., & El-Baz, A. (2016, September). Alzheimer's disease diagnostics by adaptation of 3D convolutional network. In *2016 IEEE International Conference on Image Processing (ICIP)* (pp. 126–130). IEEE.
38. Haweel, R., Shalaby, A., Mahmoud, A., Ghazal, M., Seada, N., Ghoniemy, S., . . . El-Baz, A. (2021, April). A novel DWT-based discriminant features extraction from task-based fMRI: An ASD diagnosis study using CNN. In *2021 IEEE 18th International Symposium on Biomedical Imaging (ISBI)* (pp. 196–199). IEEE.
39. El-Gamal, F. E. Z. A., Elmogy, M. M., Ghazal, M., Atwan, A., Barnes, G. N., Casanova, M. F., . . . El-Baz, A. S. (2017, September). A novel CAD system for local and global early diagnosis of Alzheimer's disease based on PIB-PET scans. In *2017 IEEE International Conference on Image Processing (ICIP)* (pp. 3270–3274). IEEE.
40. Ayyad, S. M., Badawy, M. A., Shehata, M., Alksas, A., Mahmoud, A., Abou El-Ghar, M., . . . El-Baz, A. (2022). A new framework for precise identification of prostatic adenocarcinoma. *Sensors*, 22(5), 1848.
41. Le Bihan, D. (2013). Apparent diffusion coefficient and beyond: What diffusion MR imaging can tell us about tissue structure. *Radiology*, 268(2), 318–322.
42. Alksas, A., Shehata, M., Saleh, G. A., Shaffie, A., Soliman, A., Ghazal, M., . . . El-Baz, A. (2021). A novel computer-aided diagnostic system for accurate detection and grading of liver tumors. *Scientific Reports*, 11(1), 1–18.
43. Shehata, M., Alksas, A., Abouelkheir, R. T., Elmahdy, A., Shaffie, A., Soliman, A., . . . El-Baz, A. (2021, April). A new computer-aided diagnostic (CAD) system for precise identification of renal tumors. In *2021 IEEE 18th International Symposium on Biomedical Imaging (ISBI)* (pp. 1378–1381). IEEE.

44. Alksas, A., Shehata, M., Saleh, G. A., Shaffie, A., Soliman, A., Ghazal, M., . . . El-Baz, A. (2021, January). A novel computer-aided diagnostic system for early assessment of hepatocellular carcinoma. In *2020 25th International Conference on Pattern Recognition (ICPR)* (pp. 10375–10382). IEEE.
45. Shehata, M., Alksas, A., Abouelkheir, R. T., Elmahdy, A., Shaffie, A., Soliman, A., . . . El-Baz, A. (2021). A comprehensive computer-assisted diagnosis system for early assessment of renal cancer tumors. *Sensors*, 21(14), 4928.
46. Haralick, R. M., Shanmugam, K., & Dinstein, I. H. (1973). Textural features for image classification. *IEEE Transactions on Systems, Man, and Cybernetics*, 6, 610–621.
47. Loh, H. H., Leu, J. G., & Luo, R. C. (1988). The analysis of natural textures using run length features. *IEEE Transactions on Industrial Electronics*, 35(2), 323–328.
48. Shaffie, A., Soliman, A., Ghazal, M., Taher, F., Dunlap, N., Wang, B., . . . El-Baz, A. (2017, September). A new framework for incorporating appearance and shape features of lung nodules for precise diagnosis of lung cancer. In *2017 IEEE International Conference on Image Processing (ICIP)* (pp. 1372–1376). IEEE.
49. Khalifa, F., Beache, G. M., Elnakib, A., Sliman, H., Gimel'farb, G., Welch, K. C., & El-Baz, A. (2013, April). A new shape-based framework for the left ventricle wall segmentation from cardiac first-pass perfusion MRI. In *2013 IEEE 10th International Symposium on Biomedical Imaging* (pp. 41–44). IEEE.
50. Hosseini-Asl, E., Gimel'farb, G., & El-Baz, A. (2016). Alzheimer's disease diagnostics by a deeply supervised adaptable 3D convolutional network. *arXiv preprint* arXiv:1607.00556.
51. Ismail, M., Barnes, G., Nitzken, M., Switala, A., Shalaby, A., Hosseini-Asl, E., . . . El-Baz, A. (2017, September). A new deep-learning approach for early detection of shape variations in autism using structural MRI. In *2017 IEEE International Conference on Image Processing (ICIP)* (pp. 1057–1061). IEEE.
52. El-Baz, A., Nitzken, M., Vanbogaert, E., Gimel'farb, G., Falk, R., & El-Ghar, M. A. (2011, March). A novel shape-based diagnostic approach for early diagnosis of lung nodules. In *2011 IEEE International Symposium on Biomedical Imaging: From Nano to Macro* (pp. 137–140). IEEE.
53. El-Baz, A., Nitzken, M., Khalifa, F., Elnakib, A., Gimel'farb, G., Falk, R., & El-Ghar, M. A. (2011, July). 3D shape analysis for early diagnosis of malignant lung nodules. In *Biennial International Conference on Information Processing in Medical Imaging* (pp. 772–783). Springer.
54. Nitzken, M. J., Casanova, M. F., Gimel'farb, G., Inanc, T., Zurada, J. M., & El-Baz, A. (2014). Shape analysis of the human brain: A brief survey. *IEEE Journal of Biomedical and Health Informatics*, 18(4), 1337–1354.
55. Nitzken, M., Casanova, M. F., Gimel'farb, G., Elnakib, A., Khalifa, F., Switala, A., & El-Baz, A. (2011, September). 3D shape analysis of the brain cortex with application to dyslexia. In *2011 18th IEEE International Conference on Image Processing* (pp. 2657–2660). IEEE.
56. Shaffie, A., Soliman, A., Eledkawy, A., van Berkel, V., & El-Baz, A. (2022). Computer-assisted image processing system for early assessment of lung nodule malignancy. *Cancers*, 14(5), 1117.
57. Soentpiet, R. (1999). *Advances in Kernel Methods: Support Vector Learning*. MIT Press.
58. Breiman, L. (2001). Random forests. *Machine Learning*, 45(1), 5–32.
59. Dietterich, T. G., & Kong, E. B. (1995). *Machine Learning Bias, Statistical Bias, and Statistical Variance of Decision Tree Algorithms* (pp. 1–13). Technical Report, Department of Computer Science, Oregon State University.
60. Balakrishnama, S., & Ganapathiraju, A. (1998). Linear discriminant analysis-a brief tutorial. *Institute for Signal and Information Processing*, 18, 1–8.

61. Farag, A. A., Mohamed, R. M., & El-Baz, A. (2005). A unified framework for map estimation in remote sensing image segmentation. *IEEE Transactions on Geoscience and Remote Sensing*, 43(7), 1617–1634.
62. Sandhu, H. S., Eladawi, N., Elmogy, M., Keynton, R., Helmy, O., Schaal, S., & El-Baz, A. (2018). Automated diabetic retinopathy detection using optical coherence tomography angiography: A pilot study. *British Journal of Ophthalmology*, 102(11), 1564–1569.
63. Dekhil, O., Hajjdiab, H., Shalaby, A., Ali, M. T., Ayinde, B., Switala, A., . . . El-Baz, A. (2018). Using resting state functional MRI to build a personalized autism diagnosis system. *PLOS One*, 13(10), e0206351.
64. Haweel, R., Shalaby, A. M., Mahmoud, A. H., Ghazal, M., Seada, N., Ghoniemy, S., . . . El-Baz, A. (2021). A novel grading system for autism severity level using task-based functional MRI: A response to speech study. *IEEE Access*, 9, 100570–100582.
65. Farahat, I. S., Sharafeldeen, A., Elsharkawy, M., Soliman, A., Mahmoud, A., Ghazal, M., . . . El-Baz, A. (2022). The role of 3D CT imaging in the accurate diagnosis of lung function in coronavirus patients. *Diagnostics*, 12(3), 696.
66. Elmogy, M., Khalil, A., Shalaby, A., Mahmoud, A., Ghazal, M., & El-Baz, A. (2019, December). Chronic wound healing assessment system based on color and texture analysis. In *2019 IEEE International Conference on Imaging Systems and Techniques (IST)* (pp. 1–5). IEEE.
67. Ayyad, S. M., Saleh, A. I., & Labib, L. M. (2018). Classification techniques in gene expression microarray data. *International Journal of Computer Science and Mobile Computing*, 7(11), 52–56.
68. Khalifa, F., Soliman, A., Elmaghraby, A., Gimel'farb, G., & El-Baz, A. (2017). 3D kidney segmentation from abdominal images using spatial-appearance models. *Computational and Mathematical Methods in Medicine*, 2017, Article ID 9818506. https://doi.org/10.1155/2017/9818506.
69. El-Baz, A., Beache, G. M., Gimel'farb, G., Suzuki, K., Okada, K., Elnakib, A., . . . Abdollahi, B. (2013). Computer-aided diagnosis systems for lung cancer: Challenges and methodologies. *International Journal of Biomedical Imaging*.
70. El-Baz, A., Elnakib, A., El-Ghar, A., Gimel'farb, G., Falk, R., & Farag, A. (2013). Automatic detection of 2D and 3D lung nodules in chest spiral CT scans. *International Journal of Biomedical Imaging*, 2013, 517632.
71. Elsharkawy, M., Sharafeldeen, A., Soliman, A., Khalifa, F., Ghazal, M., El-Daydamony, E., . . . El-Baz, A. (2022). A novel computer-aided diagnostic system for early detection of diabetic retinopathy using 3D-OCT higher-order spatial appearance model. *Diagnostics*, 12(2), 461.
72. Yasser, I., Khalifa, F., Abdeltawab, H., Ghazal, M., Sandhu, H. S., & El-Baz, A. (2022). Automated diagnosis of optical coherence tomography angiography (OCTA) based on machine learning techniques. *Sensors*, 22(6), 2342.
73. Sharafeldeen, A., Elsharkawy, M., Khaled, R., Shaffie, A., Khalifa, F., Soliman, A., . . . El-Baz, A. (2022). Texture and shape analysis of diffusion-weighted imaging for thyroid nodules classification using machine learning. *Medical Physics*, 49(2), 988–999.

11 Texture Analysis in Cancer Prognosis

Valerio Nardone, Alfonso Reginelli, Roberta Grassi, Giuliana Giacobbe, and Salvatore Cappabianca

11.1 TEXTURE ANALYSIS

Over the past decade, a new technique, called texture analysis (TA), has been developed to help radiologists quantify and identify parameters related to a tumor's heterogeneity that cannot be appreciated by simple visual assessment.[1–8]

TA refers to quantitative measurements of the histogram, distribution, and/or relationship of pixel intensities or gray scales and makes use of several mathematical models that are able to extract these quantitative parameters from regions of interest (ROIs) of selected volumes. These are then statistically correlated to several clinical or biological end points.[9]

There are different methods of TA, including statistical-based, model-based, and transform-based methods. The most utilized form of TA is first-order statistics, which evaluates the distribution (frequency of occurrence, not the spatial relationship) of gray levels in a pixel intensity histogram. Commonly used first-order statistics include mean, SD, threshold, minimum, maximum, skewness, kurtosis, and entropy. Second-order statistics analyze texture in specific direction and length and can be derived from a run-length matrix or a co-occurrence matrix. Higher-order statistics evaluate location and relationships between thre or more pixels and can be derived from neighborhood gray-tone difference matrices.[10]

More recently, a dynamic TA approach has also been developed, which analyzes changes in TA characteristics in subsequent imaging evaluations. This approach is termed delta radiomics.[11,12]

Through this method, it is possible to study TA changes after therapy (usually chemotherapy or radiotherapy) or shortly after the start of therapy, evaluating early the response to treatment. This also allows to avoid aggressive or ineffective treatment strategies.

The goal of radiomics and TA is to construct, through selected features, a standardized prognostic model that can determine clinical outcomes.

There are several applications of radiomics and TA in oncology. These include the ability to discriminate a benign from a malignant lesion and the degree of malignancy, although the greatest impact of TA in oncology is currently represented by the prediction of response to various anticancer strategies and prognosis stratification.

In this chapter, we will discuss the impact of TA in response prediction and prognosis stratification, focusing on breast cancer, lung cancer, liver cancer, gastric

DOI: 10.1201/9780367486082-11

cancer, and rectal cancer, as these diseases have been best described in the literature. We will finally provide a brief summary regarding the texture analysis of other cancers.

11.2 TEXTURE ANALYSIS AND PROGNOSIS: FOCUS ON BREAST CANCER

Breast cancer (BC) represents the most common malignancy in women.[13]

Early detection with mammography screening has been shown to have a large impact on survival. While ultrasonography (US) remains a supportive method to mammography in screening and represents the most accurate method in women of childbearing age with an abundant representation of the glandular component, magnetic resonance imaging (MRI) has assumed in the last decade an increasingly impressive role in high-risk women, in primary staging as well as in the evaluation of therapeutic effect and monitoring of recurrences.

The prognosis of BC is based on the expression of several immunohistochemical biomarkers, including estrogen receptor (ER), progesterone receptor (PR), human epidermal growth factor receptor 2 (HER2), and Ki-67.

Radiomics, integrated with clinical information, has been shown to have a strong clinical impact on BR, especially in the detection of malignant lesions, in the discrimination of tumor grade, as well as in the identification of prognostic factors, such as response to neoadjuvant chemotherapy (NAC) and risk of recurrence.[14–16] In fact, TA has also proved to be able to offer a matrix of potential data to define the biological characteristics of tumors and increasingly for the development of precision medicine.

Although biopsy of the suspected breast lesion is known to be the gold standard for breast cancer characterization to date, it evaluates only the sampled section of a heterogeneous tumor. Radiogenomics, on the other hand, can potentially evaluate the entire tumor burden, with the potential to provide a non-invasive diagnosis.

In addition, the correlation between radiomics and histologic type, as well as between TA and tumor grade, has also been described.

Specifically, two studies found a correlation between entropy calculated on MRI and differentiation between lobular and ductal carcinoma.[17,18] In contrast, other authors have investigated the potential of TA in predicting breast cancer molecular subtypes or tumor grade.[19]

Braman et al.[20] developed an MRI-based model that was able to identify different subtypes of HER2+ breast cancer patients. This model could also predict response to neoadjuvant HER2 targeted therapy.

An important field of investigation also in BR is represented by the evaluation of the response to neoadjuvant chemotherapy, which represents the most used preoperative strategy. Moreover, one should not forget that about half of the patients achieve a pathological complete response (pCR). Several authors have therefore correlated texture features, extracted mainly from MRI, to pCR.[21–23]

Specifically, one study examined a novel combined radiomic approach (intra-tumoral and peri-tumoral) for the prediction of pCR on pretreatment BC dynamic contrast-enhanced magnetic resonance imaging (DCE-MRI). They demonstrated that, especially, TA features extracted from the peri-tumor region can predict the success of pretreatment pCR.[24]

Other researchers have analyzed the risk of positive sentinel or axillary lymph nodes using different approaches. Chai et al.[25] demonstrated that TA features, extracted from 3T DCE-MRI, showed an accuracy of 0.86 and an AUC of 0.91 for the prediction of axillary lymph nodes. Liu et al.[26] also performed a prospective study to predict the same end point; they showed that the combined model (TA and clinical information) had better accuracy (AUC of 0.763). According to these studies, in the near future, TA could support clinical decision-making, avoiding invasive procedures.

Finally, TA was tested in predicting recurrence, predicting disease-free survival (DFS), and the Ki-67 proliferation index.[27–29]

11.3 TEXTURE ANALYSIS AND PROGNOSIS: FOCUS ON LUNG CANCER

Lung cancer (LC) represents one of the most common malignancies and the leading cause of cancer-related death.[30] Surgery or stereotactic radiotherapy (for ineligible patients) is recommended for patients in the early stage of the disease, while a combination of different strategies, such as surgery, radiotherapy, chemotherapy, and immunotherapy, is used in locally advanced or metastatic disease.

Medical imaging is of paramount importance in all phases of the clinical management of patients with LC: from diagnosis to evaluation of therapeutic efficacy and follow-up.

Computed tomography (CT) and positron emission tomography (PET) are usually performed, as these techniques have been widely used in patients with LC, whereas US and MRI are not currently used in the clinical management of such patients.

We will focus on TA end points that include assessment of tumor response and prediction of prognosis.

Assessment of tumor response in LC is usually based on RECIST criteria.[31]

Radiomics is indeed making progress in the field of predicting response to chemotherapy. Two studies in particular have described a potential role for radiomics in predicting response to therapy, based on CT and PET images, respectively.[32,33] Specifically, in the latter study[33], different parameters such as contrast, coarseness, and busyness showed good statistical correlation with RECIST evaluation and outperformed SUV-based parameters after chemoradiotherapy.

In addition, machine learning models have been developed for prognosis prediction and in particular for the analysis of overall survival (OS) and progression-free survival.

Grove et al.[34] found measures of heterogeneity, like spiculation and entropy gradients, to be strong prognostic indicators of OS in patients with early-stage lung cancer. Song et al.[35] describe wavelet features that correlate with OS, while Fried et al.[36] found textural features correlate not just to OS but also to locoregional and distant metastasis control. Different PET-based TA parameters, such as higher contrast, higher tumor heterogeneity, and higher tumor dissimilarity, have also been associated with OS.[37]

D-TA features have also been tested to predict OS with promising results. Paul et al.[38] have shown that some dynamic parameters, such as reduction in mean Hounsfield unit (HU), obtained during radiotherapy correlated with OS.

Because visualization of heterogeneity has been linked to tumor aggressiveness, it represents a negative prognostic factor in lung cancer. Analyzing heterogeneity can be considered difficult for radiologists; the use of TA features allows for a more objective method of quantifying these important prognostic factors.

Other authors, however, have studied the prediction of locoregional or distant recurrence after treatment. Pyka et al.[39] investigated the correlation of TA features in lung cancer patients treated with radiotherapy and found that entropy and correlation outperformed SUV metrics in the prediction of recurrences. Mattonen et al.[40], contrarily, compared the radiologist and the TA features in the correct prediction of recurrence after stereotactic radiotherapy and found that TA features had a lower classification error rate (24% versus 35%).

Another study [41], however, applied CT to the penumbral region, that is, the region extending 10 mm from the tumor surface. The authors found a significant correlation between radiomic features extracted from PET images and prediction of disease recurrence, bringing to light the future role of peri-tumor region analysis as a parameter related to immunotherapy response.

More recently, many researchers have become interested in the potential ability of radiomics to estimate the presence or absence of clinically relevant mutations, a field now referred to as radiogenomics. For example, Aerts et al.[42] applied this concept to LC, showing that data extracted from pretreatment CT scans could non-invasively estimate EGFR mutation status and thus tumor response to gefitinib therapy. They found a mixture of volume, texture, and gradient features could estimate the presence of the mutation. Yipp et al.[43], in a proof-of-concept study on PET images, found that radiomic features could detect EGFR mutation status but instead were unable to find a feature related to KRAS mutations. Finally, others found that tumor location[44] and pleural effusion[45] could be related to ALK gene mutation.

In conclusion, TA in LC has, to date, demonstrated modest prognostic capabilities. Future research should aim at optimizing and standardizing TA features, work on feature selection and model development, in order to improve this approach. Currently, several prospective observational studies are accruing lung cancer patients with different end points, including the development of radiomics models based on CT images to diagnose malignant nodules early.

11.4 TEXTURE ANALYSIS AND PROGNOSIS: FOCUS ON GASTRIC CANCER

Esophageal, esophago-gastric, and gastric cancers (GC) are leading causes of cancer morbidity and death. For patients with potentially resectable disease, multimodality treatment is recommended as it provides the best chance of survival. However, quality of life may be adversely affected by therapy, and in addition, more than 50%

of patients who present have a diagnosis of stage IV disease that precludes curative treatment.[46]

Recent genomic analyses have highlighted genetic heterogeneity as an underlying cause of outcome differences and heterogeneity in response to therapy in GC. In particular, several studies have also suggested that greater tumor heterogeneity is associated with poorer response and outcome.[47, 48]

Therefore, improved patient stratification remains a key challenge for patients with upper gastrointestinal tract cancers.

Indeed, radiomic signatures provide predictive information of the underlying tumor biology and behavior and can be used alone or with other patient-related data (e.g., pathological or genomic data); this allows for improved tumor phenotyping, prediction of treatment response, and prognosis. Radiomic signatures in gastric cancer can be obtained for all cross-sectional imaging modalities, including CT, PET, and MRI.

Among the main applications of CT for GC, it is worth mentioning the evaluation of the staging parameter T, and therefore the assessment of the degree of parietal infiltration, as well as the lymph node involvement, which are important negative prognostic factors with a strong impact on clinical decisions and patient survival. In fact, many studies have focused on the creation and validation of machine learning or deep learning models or algorithms able to predict the state of lymph node disease in the preoperative phase.[49–51] Others, however, have used extrapolated quantitative characteristics to predict survival after radical surgery.[52]

It is also known in the literature that the tumor immune microenvironment plays a critical role in cancer progression, metastasis, and therapeutic response. Current assessment of immune infiltration typically requires tissue samples obtained after surgery. Given the dynamic nature of the immune response, it would be useful to assess the tumor immune microenvironment by a non-invasive means that would allow longitudinal assessment of immune infiltration during the course of treatment. Una serie di studi ha confermato i valori prognostici e potenzialmente predittivi delle cellule immunitarie infiltranti il tumore. In fact, innovative approaches have used radiomics for a non-invasive assessment of the immune microenvironment, correlating the TA features with the Treg cell infiltration or the HER2 expression, la cui espressione correla con una sopravvivenza più scarsa del paziente.[53,54]

In addition to adenocarcinoma, radiomics models have also been created and validated by Zhang L. et al. in order to distinguish between benign and malignant gastrointestinal stromal tumors (GISTs), providing assistance in the preoperative diagnosis of GISTs; moreover, in another study, the same authors developed and validated a radiomic prediction model using CT to preoperatively predict Ki-67 nucleoprotein expression in GISTs. Higher Ki-67 expression correlates with increased tumor size, higher mitotic rate, higher risk of malignancy, and worse prognosis.[55,56]

Zhou et al., instead, investigated PET/CT features that contribute to prognosis prediction in patients with primary gastric diffuse large B-cell lymphoma (PG-DLBCL).[57]

However, these studies, although promising, still need clinical validation.

11.5 TEXTURE ANALYSIS AND PROGNOSIS: FOCUS ON LIVER CANCER

Liver lesions are extremely common in oncology. The liver is a major site of metastasis, and hepatocellular carcinoma (HCC) is the most common primary tumor, representing the second leading cause of death in cancer patients and the first in patients with cirrhosis.[58]

In recent years, many studies have focused on the search for possible applications of radiomics and, in particular, TA in the study of liver injury: from early diagnosis to post-treatment evaluation and prognosis prediction.

In a multi-center study[59], MRTA was able to distinguish HCC from benign liver lesions with a sensitivity of 85% and specificity of 84%, outperforming the subjective qualitative assessments of experienced radiologists.

In a retrospective study[60] including 46 patients with HCC, instead, MRTA was able to differentiate low-grade from high-grade HCC on arterial phase imaging, demonstrating a high potential of radiomics in predicting histologic grade. Another srudy[61], however, evaluated by radiomics the risk of recurrence in patients with early-stage HCC.

CEUS-based deep learning radiomic models were also tested in order to predict early HCC recurrence after ablation and to stratify patients into high- and low-risk subgroups for late recurrence.[62]

Other advanced prediction models have evaluated more complex radiomic features, which, integrated with clinical and laboratory data as well as genomic data, have shown promising results. In particular, radiomics models based on MRI data have been designed in which combining some quantitative data (correlation, inverse difference moment, cluster prominence, uniformity, and GLCM energy) with laboratory data such as AFP, ferritin, and CEA in the preoperative as well as macro-vascular invasion and tumor size has been able to predict the five-year survival of patients with HCC.[63]

Radiomics studies have been performed in order to predict treatment response. Transarterial chemoembolization (TACE), transarterial radioembolization (TARE), and high-frequency ultrasound (HIFU) are minimally invasive methods used to treat a variety of liver malignancies. A retrospective study of 89 subjects with HCC undergoing contrast-enhanced MRI before and one week after TACE/HIFU demonstrated that patients who had a complete response had greater uniformity and energy but lower entropy and skewness than the group with a noncomplete response.[64]

Liver metastases (LM) represent another important chapter in the hepatic field, representing the most frequent lesions (approximately 18–40 times more than primary liver tumors). Therefore, several studies over the years have focused on hepatic secondarisms.[65]

In contrast, a machine learning–based radiomic model attempted to predict the presence of metachronous LM from colorectal cancer (CRC) by studying microvascular changes in healthy liver parenchyma. It was shown that the combined model (AUC 95%) was able to more accurately predict the development of secondarisms in the 24 months after diagnosis, compared with the clinical model (AUC 71%) and the radiomic model (AUC 86%). TA also attempted to predict the presence of synchronous liver metastases in patients with CRC.[66]

Finally, TA has been shown to predict response to Bevacizumab, representing also the best predictor of both overall survival and disease-free survival in patients with LM from unresectable CRC. In particular, it has been observed that the uniformity of lesions, which is strictly dependent on angiogenesis, correlates with a worse response to treatment.[67]

Finally, models were developed in order to predict the response of LMs to first-line oxaliplatin-based chemotherapy in patients with CRC.[68,69]

Although in liver cancer the data that have emerged are still sparse and heterogeneous, both entropy and homogeneity have been shown to be the radiomic features with the strongest clinical impact. Indeed, higher entropy at diagnosis and lower homogeneity of LM were associated with better survival and higher chemotherapy response rates. Such indices could predict good vascularization of LM, while more homogeneous tumors could reflect a tighter cellular structure or necrosis, which might purport reduced therapy effectiveness. Decreased entropy and increased homogeneity after chemotherapy were correlated with radiological tumor response. Entropy and homogeneity were also highly predictive of the degree of tumor regression. Similar considerations are possible for skewness: low baseline values and an increase after chemotherapy were associated with response to chemotherapy.[70]

11.6 TEXTURE ANALYSIS AND PROGNOSIS: FOCUS ON RECTAL CANCER

Rectal cancer (RC) accounts for one-third of all colorectal cancers and is one of the leading causes of cancer death in the Western world in both sexes.[71]

Nowadays, neoadjuvant chemo-radiotherapy (nCRT) followed by total meso-rectal excision (TME) represents the gold standard treatment for patients with locally advanced rectal cancer (LARC). In fact, it is known in the literature that this therapeutic strategy increases progression-free survival (PFS), although its effects on overall survival (OS) are yet to be demonstrated.[72–74]

Neoadjuvant therapy, in particular, has been shown to decrease the rate of local recurrence and, in some cases, has resulted in a pathologically complete response (pCR). This concept has led to the development of emerging, more conservative therapeutic strategies, such as "watch-and-wait" or local excision (LE), in order to avoid invasive surgical treatment burdened by the risk of major complications and a possible worsening of the patient's quality of life. On the other hand, neoadjuvant therapy itself is associated with several side effects and is not always beneficial to patients in terms of tumor burden reduction.[75,76]

Currently, the main clinical challenge in RC is to preoperatively diagnose pCR in patients with LARC after nCRT.

Although magnetic resonance imaging (MRI) represents nowadays the most accurate non-invasive imaging modality for primary staging and re-evaluation after treatment in RC, its clinical utility still appears uncertain.[77]

Imaging parameters such as tumor regression grade (TRG) assessment in MRI, tumor volume, and DWI/ADC functional sequences are dependent on the subjective interpretation of the observer; therefore, to date, more sensitive and reliable MRI markers are needed in order to assess efficacy and predict early response to treatment in clinical practice.[78,79]

In this scenario, radiomics has emerged as a promising tool as it is able to extract, from diagnostic images, a large number of quantitative characteristics; moreover, it has shown a significant predictive power with regard to the assessment of early response to therapy.

Several studies have shown promising results in the prediction and early assessment of response to neoadjuvant therapy using data from both MRI and CT images.

In particular, studies in the literature have shown that several radiomic features, extracted from pre-nCRT and post-nCRT T2-weighted images, are associated with pCR.[80–82]

Other studies, however, have reported the predictive value of pCR based on radiomic features extracted from pre-nCRT DWI MRI.[83,84]

However, many of these studies, even the most recent ones, have focused on the possibility of extracting radiomic features from images belonging to a single sequence in order to discriminate patients with pCR from patients with residual disease.

It must also be considered that RC represents a heterogeneous disease, and tumor sensitivity to treatments might be related to intra-tumor heterogeneity.

TA has been suggested as a promising technique to investigate intra-tumor heterogeneity attributed to various factors, such as hypoxia, necrosis, and angiogenesis, potentially related to tumor aggressiveness and patient prognosis.[85]

During nCRT, intra-tumor heterogeneity is dynamic, so radiomic features extracted from single-sequence images (e.g., T2W sequences or pre-nCRT pre-contrast CT) may overlook tumor change during treatment and have inherent limitations.

Therefore, delta radiomics, which is defined as the change in quantitative features in a series of longitudinal images, can detect information about changes in intra-tumor heterogeneity and adapt therapy, representing a new frontier in radiomics.

Furthermore, delta radiomics has been used to predict the occurrence of distant metastases. The identification of patients with higher risk of developing distant metastases within two years (2yDM) represents another topic of great interest for the clinical community, as it could allow a personalized and accurate management of the patient, thus defining a more rigorous clinical and imaging surveillance or even propose more intensive treatments.[86]

In fact, DMs are the leading cause of treatment failure in LARC. Adjuvant chemotherapy is usually used to control DMs. However, as mentioned earlier, not all patients can benefit from adjuvant chemotherapy, and some show worse outcomes after treatment. Therefore, radiomics models capable of predicting postoperative DM have been validated, stratifying patients who could benefit from adjuvant chemotherapy and patients in whom receiving adjuvant chemotherapy resulted in a worse prognosis than not receiving it.

In addition, a correlation was found between texture analysis performed on pre-nCRT MRI images and early disease progression (ePD)[87]; in fact, it was shown that some patients presenting certain texture parameters showed recurrence and/or distant metastases within three months after radical surgery. These results were in line with further studies in the literature, showing that these parameters correlate with greater tumor aggressiveness and decreased sensitivity to treatment.

Inoltre, la radiomica ha mostrato risultati promettenti nella predizione del parametro T stadiativo[88,89], as well as in the assessment of lymph node invasion[90,91], perineural,[92] and extra-mural vascular invasion (EMVI)[93], important negative prognostic factors.

Ultrasonography (US) has also been used to develop a machine learning–based radiomic model to predict tumor deposits (TD) preoperatively. Studies have confirmed that TD-positive patients have more aggressive tumors, with lower disease-free survival and overall survival. However, TDs are only diagnosed after surgical resection by pathological evaluation. Therefore, detecting TDs preoperatively can greatly improve the treatment strategy and patient prognosis.[94,95]

Finally, advanced prediction models were created in which more complex radiomic features were evaluated with promising results. In one study, Chen et al.[96] demonstrated that MRI-based radiomics represents a sophisticated and non-invasive tool to accurately distinguish recurrent lesions (LR) from lesions (non-recurrent) at the site of anastomosis, and in particular, combining multiple sequences in MRI significantly improved its performance. In this study, the ROC curve of model combination indicated an AUC of 0.864 (validation set), with sensitivity and specificity of 81–82% and 75–86%, respectively, suggesting that model combination provides better discrimination performance than individual models ($p > 0.05$). Interestingly, extracting and combining quantitative features from multiple MRI sequences significantly improve model performance, making it more effective.

At a time in history when we are witnessing the development of targeted therapies, knowing the genetic profile of tumors is particularly crucial for personalized treatment. Determination of KRAS mutational status, which is typical of the histological specimen, has been highly recommended as it is considered a rather specific negative biomarker for antibody-based therapies against epidermal growth factor receptor (EGFR). Generally, metastatic CRs with KRAS mutations tend to be resistant to anti-EGFR-targeted therapy. Predicting, therefore, the mutational status preoperatively represents a point of fundamental importance, although at present few studies have focused in this direction.[97]

11.7 OTHER CANCERS

TA has also been applied in many other tumor diseases, such as brain cancer (both primary and metastatic), head and neck cancer, kidney cancer, bladder cancer, prostate cancer, as well as gynecological cancer.

These heterogeneous diseases are usually managed with a combination of different anticancer strategies, such as surgery, radiotherapy, chemotherapy, and immunotherapy. In this context, TA has been applied for different purposes, such as prognosis prediction, correlation with grading or other histological features, differential diagnosis, and response to therapy.

In brain cancer, several radiomics studies have been published in the last decade, mainly in the field of glioma and meningioma grading prediction, prognosis, as well as differential diagnosis between disease progression and radionecrosis in patients treated with radiotherapy.[98–100]

TA approaches have also been used in the field of head and neck cancers, mainly with the aim of predicting prognosis. Studies have focused on the possibility of differentiating between benign and malignant masses by CT; others on the assessment of response to therapy and locoregional control. TA has also been used to predict cervical lymph node metastasis and Hif1-alpha expression in squamous cell carcinoma (SCC) whose positivity is an expression of hypoxic areas.[101–104]

TA approaches have also been tested for kidney cancer. Renal cell carcinomas (RCC) are classified into subtypes, most commonly, clear cell RCC (ccRCC), papillary RCC (pRCC), and chromophobe RCC (cRCC), with different imaging characteristics and risk profiles. In a retrospective study of 61 patients with ccRCC undergoing preoperative magnetic resonance imaging at 1.5T, MRTA on ADC maps correlated with higher cellularity, tumor necrosis, and hemorrhage, and thus higher stage.[9]

In contrast, Kozikowski et al.[105] investigated the role of TA in predicting muscle invasiveness of bladder cancer by analyzing eight studies with a total of 860 patients included. TA showed good diagnostic accuracy.

Multiparametric MRI (mpMRI) imaging has revolutionized the detection, staging, and management of early prostate carcinoma, and MRTA has the potential to standardize image quantification. In a retrospective study, TA features extracted from T2-weighted images and ADC maps showed potential to differentiate noncancerous from cancerous prostate tissue and tumor energy and entropy on ADC maps in the peripheral zone correlated with Gleason score. It is known in the literature that the transition zone is more difficult to identify than the peripheral zone on MRI. MRTA has shown a high accuracy in identifying prostate cancer located in the transition zone compared to conventional radiological evaluation. In addition, MRI has also been shown to be good at assessing disease recurrence after radiotherapy, which is considered the conventional treatment of prostate cancer.[106–108]

Finally, TA approaches have been tested in gynecological cancers with different aims, such as the prediction of prognosis of ovarian cancer patients, and in particular in the evaluation of grading, histological subtype, and differential diagnosis.[109, 110] TA has also been used in preoperative risk stratification in patients with endometrial carcinoma.[111,112] There is currently insufficient evidence on the benefit of TA approaches in this context, despite this field being promising for future clinical practice.

All the aforementioned studies showed interesting results regarding the application of texture in oncology. However, there are still several pitfalls to be solved before TA can be successfully applied in the clinical management of cancer patients, such as standardization of image acquisition and feature extraction, lack of prospective studies. By setting an ambitious goal to address these challenges, radiomics can become a clinical reality in the near future.

BIBLIOGRAFIA

1. Lambin P, Rios-Velazquez E, Leijenaar R, et al. Radiomics: extracting more information from medical images using advanced feature analysis. *Eur J Cancer*. 2012;48(4):441–446. doi:10.1016/j.ejca.2011.11.036
2. Gillies RJ, Kinahan PE, Hricak H. Radiomics: images are more than pictures, they are data. *Radiology*. 2016;278(2):563–577. doi:10.1148/radiol.2015151169
3. Avanzo M, Wei L, Stancanello J, et al. Machine and deep learning methods for radiomics. *Med Phys*. 2020;47(5):e185–e202. doi:10.1002/mp.13678
4. Horvat N, Bates DDB, Petkovska I. Novel imaging techniques of rectal cancer: what do radiomics and radiogenomics have to offer? A literature review. *Abdom Radiol (New York)*. 2019;44(11):3764–3774. doi:10.1007/s00261-019-02042-y
5. Lo Gullo R, Daimiel I, Morris EA, Pinker K. Combining molecular and imaging metrics in cancer: radiogenomics. *Insights Imaging*. 2020;11(1):1. doi:10.1186/s13244-019-0795-6
6. Granata V, Caruso D, Grassi R, et al. Structured reporting of rectal cancer staging and restaging: a consensus proposal. *Cancers (Basel)*. 2021;13(9). doi:10.3390/cancers13092135
7. Granata V, Grassi R, Miele V, et al. Structured reporting of lung cancer staging: a consensus proposal. *Diagnostics (Basel, Switzerland)*. 2021;11(9). doi:10.3390/diagnostics11091569
8. Mazzei MA, Bagnacci G, Gentili F, et al. Structured and shared CT radiological report of gastric cancer: a consensus proposal by the Italian research group for gastric cancer (GIRCG) and the Italian society of medical and interventional radiology (SIRM). *Eur Radiol*. Published online August 2021:1–12. doi:10.1007/s00330-021-08205-0
9. Thomas JV, Abou Elkassem AM, Ganeshan B, Smith AD. MR imaging texture analysis in the abdomen and pelvis. *Magn Reson Imaging Clin N Am*. 2020;28(3):447–456. doi:10.1016/j.mric.2020.03.009
10. Lubner MG, Smith AD, Sandrasegaran K, Sahani DV, Pickhardt PJ. CT Texture analysis: definitions, applications, biologic correlates, and challenges. *Radiogr a Rev Publ Radiol Soc North Am Inc*. 2017;37(5):1483–1503. doi:10.1148/rg.2017170056
11. Nardone V, Reginelli A, Guida C, et al. Delta-radiomics increases multicentre reproducibility: a phantom study. *Med Oncol*. 2020;37(5):38. doi:10.1007/s12032-020-01359-9
12. Boldrini L, Cusumano D, Chiloiro G, et al. Delta radiomics for rectal cancer response prediction with hybrid 0.35 T magnetic resonance-guided radiotherapy (MRgRT): a hypothesis-generating study for an innovative personalized medicine approach. *Radiol Med*. 2019;124(2):145–153. doi:10.1007/s11547-018-0951-y
13. Libson S, Lippman M. A review of clinical aspects of breast cancer. *Int Rev Psychiatry*. 2014;26(1):4–15. doi:10.3109/09540261.2013.852971
14. Tagliafico AS, Piana M, Schenone D, Lai R, Massone AM, Houssami N. Overview of radiomics in breast cancer diagnosis and prognostication. *Breast*. 2020;49:74–80. doi:10.1016/j.breast.2019.10.018
15. Sutton EJ, Onishi N, Fehr DA, et al. A machine learning model that classifies breast cancer pathologic complete response on MRI post-neoadjuvant chemotherapy. *Breast Cancer Res*. 2020;22(1):57. doi:10.1186/s13058-020-01291-w
16. Conti A, Duggento A, Indovina I, Guerrisi M, Toschi N. Radiomics in breast cancer classification and prediction. *Semin Cancer Biol*. 2021;72:238–250. doi:10.1016/j.semcancer.2020.04.002
17. Waugh SA, Purdie CA, Jordan LB, et al. Magnetic resonance imaging texture analysis classification of primary breast cancer. *Eur Radiol*. 2016;26(2):322–330. doi:10.1007/s00330-015-3845-6

18. Holli-Helenius K, Salminen A, Rinta-Kiikka I, et al. MRI texture analysis in differentiating luminal A and luminal B breast cancer molecular subtypes—a feasibility study. *BMC Med Imaging*. 2017;17(1):69. doi:10.1186/s12880-017-0239-z
19. Li Q, Dormer J, Daryani P, Chen D, Zhang Z, Fei B. Radiomics analysis of MRI for predicting molecular subtypes of breast cancer in young women. *Proc SPIE—the Int Soc Opt Eng*. 2019;10950. doi:10.1117/12.2512056
20. Braman N, Prasanna P, Whitney J, et al. Association of peritumoral radiomics with tumor biology and pathologic response to preoperative targeted therapy for HER2 (ERBB2)-positive breast cancer. *JAMA Netw Open*. 2019;2(4):e192561. doi:10.1001/jamanetworkopen.2019.2561
21. Bian T, Wu Z, Lin Q, et al. Radiomic signatures derived from multiparametric MRI for the pretreatment prediction of response to neoadjuvant chemotherapy in breast cancer. *Br J Radiol*. 2020;93(1115):20200287. doi:10.1259/bjr.20200287
22. Bitencourt AGV, Gibbs P, Rossi Saccarelli C, et al. MRI-based machine learning radiomics can predict HER2 expression level and pathologic response after neoadjuvant therapy in HER2 overexpressing breast cancer. *EBioMedicine*. 2020;61:103042. doi:10.1016/j.ebiom.2020.103042
23. Park H, Lim Y, Ko ES, et al. Radiomics signature on magnetic resonance imaging: association with disease-free survival in patients with invasive breast cancer. *Clin Cancer Res an Off J Am Assoc Cancer Res*. 2018;24(19):4705–4714. doi:10.1158/1078-0432.CCR-17-3783
24. Braman NM, Etesami M, Prasanna P, et al. Intratumoral and peritumoral radiomics for the pretreatment prediction of pathological complete response to neoadjuvant chemotherapy based on breast DCE-MRI. *Breast Cancer Res*. 2017;19(1):57. doi:10.1186/s13058-017-0846-1
25. Chai R, Ma H, Xu M, et al. Differentiating axillary lymph node metastasis in invasive breast cancer patients: a comparison of radiomic signatures from multiparametric breast MR sequences. *J Magn Reson Imaging*. 2019;50(4):1125–1132. doi:10.1002/jmri.26701
26. Liu Z, Feng B, Li C, et al. Preoperative prediction of lymphovascular invasion in invasive breast cancer with dynamic contrast-enhanced-MRI-based radiomics. *J Magn Reson Imaging*. 2019;50(3):847–857. doi:10.1002/jmri.26688
27. Chitalia RD, Rowland J, McDonald ES, et al. Imaging phenotypes of breast cancer heterogeneity in preoperative breast dynamic contrast enhanced magnetic resonance imaging (DCE-MRI) scans predict 10-year recurrence. *Clin Cancer Res an Off J Am Assoc Cancer Res*. 2020;26(4):862–869. doi:10.1158/1078-0432.CCR-18-4067
28. Kim S, Kim MJ, Kim E-K, Yoon JH, Park VY. MRI radiomic features: association with disease-free survival in patients with triple-negative breast cancer. *Sci Rep*. 2020;10(1):3750. doi:10.1038/s41598-020-60822-9
29. Zhang Y, Zhu Y, Zhang K, et al. Invasive ductal breast cancer: preoperative predict Ki-67 index based on radiomics of ADC maps. *Radiol Med*. 2020;125(2):109–116. doi:10.1007/s11547-019-01100-1
30. Schabath MB, Cote ML. Cancer progress and priorities: lung cancer. *Cancer Epidemiol Biomarkers Prev a Publ Am Assoc Cancer Res cosponsored by Am Soc Prev Oncol*. 2019;28(10):1563–1579. doi:10.1158/1055-9965.EPI-19-0221
31. Schwartz LH, Litière S, de Vries E, et al. RECIST 1.1-Update and clarification: from the RECIST committee. *Eur J Cancer*. 2016;62:132–137. doi:10.1016/j.ejca.2016.03.081
32. Coroller TP, Agrawal V, Narayan V, et al. Radiomic phenotype features predict pathological response in non-small cell lung cancer. *Radiother Oncol J Eur Soc Ther Radiol Oncol*. 2016;119(3):480–486. doi:10.1016/j.radonc.2016.04.004
33. Cook GJR, Yip C, Siddique M, et al. Are pretreatment 18F-FDG PET tumor textural features in non-small cell lung cancer associated with response and survival after chemoradiotherapy? *J Nucl Med*. 2013;54(1):19–26. doi:10.2967/jnumed.112.107375

34. Grove O, Berglund AE, Schabath MB, et al. Quantitative computed tomographic descriptors associate tumor shape complexity and intratumor heterogeneity with prognosis in lung adenocarcinoma. *PLOS One*. 2015;10(3):e0118261. doi:10.1371/journal.pone.0118261
35. Song J, Liu Z, Zhong W, et al. Non-small cell lung cancer: quantitative phenotypic analysis of CT images as a potential marker of prognosis. *Sci Rep*. 2016;6:38282. doi:10.1038/srep38282
36. Fried DV, Tucker SL, Zhou S, et al. Prognostic value and reproducibility of pretreatment CT texture features in stage III non-small cell lung cancer. *Int J Radiat Oncol Biol Phys*. 2014;90(4):834–842. doi:10.1016/j.ijrobp.2014.07.020
37. Reginelli A, Nardone V, Giacobbe G, et al. Radiomics as a new frontier of imaging for cancer prognosis: a narrative review. *Diagnostics (Basel, Switzerland)*. 2021;11(10). doi:10.3390/diagnostics11101796
38. Paul J, Yang C, Wu H, et al. Early assessment of treatment responses during radiation therapy for lung cancer using quantitative analysis of daily computed tomography. *Int J Radiat Oncol Biol Phys*. 2017;98(2):463–472. doi:10.1016/j.ijrobp.2017.02.032
39. Pyka T, Bundschuh RA, Andratschke N, et al. Textural features in pre-treatment [F18]-FDG-PET/CT are correlated with risk of local recurrence and disease-specific survival in early stage NSCLC patients receiving primary stereotactic radiation therapy. *Radiat Oncol*. 2015;10:100. doi:10.1186/s13014-015-0407-7
40. Mattonen SA, Palma DA, Johnson C, et al. Detection of local cancer recurrence after stereotactic ablative radiation therapy for lung cancer: physician performance versus radiomic assessment. *Int J Radiat Oncol Biol Phys*. 2016;94(5):1121–1128. doi:10.1016/j.ijrobp.2015.12.369
41. Mattonen SA, Davidzon GA, Bakr S, et al. [18F] FDG positron emission tomography (PET) tumor and penumbra imaging features predict recurrence in non-small cell lung cancer. *Tomogr (Ann Arbor, Mich)*. 2019;5(1):145–153. doi:10.18383/j.tom.2018.00026
42. Aerts HJWL, Velazquez ER, Leijenaar RTH, et al. Decoding tumour phenotype by noninvasive imaging using a quantitative radiomics approach. *Nat Commun*. 2014;5:4006. doi:10.1038/ncomms5006
43. Yip SSF, Kim J, Coroller TP, et al. Associations between somatic mutations and metabolic imaging phenotypes in non-small cell lung cancer. *J Nucl Med*. 2017;58(4):569–576. doi:10.2967/jnumed.116.181826
44. Yamamoto S, Korn RL, Oklu R, et al. ALK molecular phenotype in non-small cell lung cancer: CT radiogenomic characterization. *Radiology*. 2014;272(2):568–576. doi:10.1148/radiol.14140789
45. Rizzo S, Petrella F, Buscarino V, et al. CT radiogenomic characterization of EGFR, K-RAS, and ALK mutations in non-small cell lung cancer. *Eur Radiol*. 2016;26(1):32–42. doi:10.1007/s00330-015-3814-0
46. Sah B-R, Owczarczyk K, Siddique M, Cook GJR, Goh V. Radiomics in esophageal and gastric cancer. *Abdom Radiol (New York)*. 2019;44(6):2048–2058. doi:10.1007/s00261-018-1724-8
47. Ichikawa H, Nagahashi M, Shimada Y, et al. Actionable gene-based classification toward precision medicine in gastric cancer. *Genome Med*. 2017;9(1):93. doi:10.1186/s13073-017-0484-3
48. Katona BW, Rustgi AK. Gastric cancer genomics: advances and future directions. *Cell Mol Gastroenterol Hepatol*. 2017;3(2):211–217. doi:10.1016/j.jcmgh.2017.01.003
49. Li J, Dong D, Fang M, et al. Dual-energy CT-based deep learning radiomics can improve lymph node metastasis risk prediction for gastric cancer. *Eur Radiol*. 2020;30(4):2324–2333. doi:10.1007/s00330-019-06621-x
50. Feng Q-X, Liu C, Qi L, et al. An intelligent clinical decision support system for preoperative prediction of lymph node metastasis in gastric cancer. *J Am Coll Radiol*. 2019;16(7):952–960. doi:10.1016/j.jacr.2018.12.017

51. Sun R-J, Fang M-J, Tang L, et al. CT-based deep learning radiomics analysis for evaluation of serosa invasion in advanced gastric cancer. *Eur J Radiol*. 2020;132:109277. doi:10.1016/j.ejrad.2020.109277
52. Li W, Zhang L, Tian C, et al. Prognostic value of computed tomography radiomics features in patients with gastric cancer following curative resection. *Eur Radiol*. 2019;29(6):3079–3089. doi:10.1007/s00330-018-5861-9
53. Wang N, Wang X, Li W, et al. Contrast-enhanced CT parameters of gastric adenocarcinoma: can radiomic features be surrogate biomarkers for HER2 over-expression status? *Cancer Manag Res*. 2020;12:1211–1219. doi:10.2147/CMAR.S230138
54. Wang Y, Yu Y, Han W, et al. CT radiomics for distinction of human epidermal growth factor receptor 2 negative gastric cancer. *Acad Radiol*. 2021;28(3):e86–e92. doi:10.1016/j.acra.2020.02.018
55. Zhang L, Kang L, Li G, et al. Computed tomography-based radiomics model for discriminating the risk stratification of gastrointestinal stromal tumors. *Radiol Med*. 2020;125(5):465–473. doi:10.1007/s11547-020-01138-6
56. Zhang Q-W, Gao Y-J, Zhang R-Y, et al. Personalized CT-based radiomics nomogram preoperative predicting Ki-67 expression in gastrointestinal stromal tumors: a multicenter development and validation cohort. *Clin Transl Med*. 2020;9(1):12. doi:10.1186/s40169-020-0263-4
57. Zhou Y, Ma X-L, Pu L-T, Zhou R-F, Ou X-J, Tian R. Prediction of overall survival and progression-free survival by the (18)F-FDG PET/CT radiomic features in patients with primary gastric diffuse large B-cell lymphoma. *Contrast Media Mol Imaging*. 2019;2019:5963607. doi:10.1155/2019/5963607
58. Siegel RL, Miller KD, Jemal A. Cancer statistics, 2017. *CA Cancer J Clin*. 2017;67(1):7–30. doi:10.3322/caac.21387
59. Stocker D, Marquez HP, Wagner MW, et al. MRI texture analysis for differentiation of malignant and benign hepatocellular tumors in the non-cirrhotic liver. *Heliyon*. 2018;4(11):e00987. doi:10.1016/j.heliyon.2018.e00987
60. Zhou W, Zhang L, Wang K, et al. Malignancy characterization of hepatocellular carcinomas based on texture analysis of contrast-enhanced MR images. *J Magn Reson Imaging*. 2017;45(5):1476–1484. doi:10.1002/jmri.25454
61. Ji G-W, Zhu F-P, Xu Q, et al. Radiomic features at contrast-enhanced CT predict recurrence in early stage hepatocellular carcinoma: a multi-institutional study. *Radiology*. 2020;294(3):568–579. doi:10.1148/radiol.2020191470
62. Liu F, Liu D, Wang K, et al. Deep learning radiomics based on contrast-enhanced ultrasound might optimize curative treatments for very-early or early-stage hepatocellular carcinoma patients. *Liver Cancer*. 2020;9(4):397–413. doi:10.1159/000505694
63. Zhang J, Wang X, Zhang L, et al. Radiomics predict postoperative survival of patients with primary liver cancer with different pathological types. *Ann Transl Med*. 2020;8(13):820. doi:10.21037/atm-19-4668
64. Yu JY, Zhang HP, Tang ZY, et al. Value of texture analysis based on enhanced MRI for predicting an early therapeutic response to transcatheter arterial chemoembolisation combined with high-intensity focused ultrasound treatment in hepatocellular carcinoma. *Clin Radiol*. 2018;73(8):758.e9–758.e18. doi:10.1016/j.crad.2018.04.013
65. Lincke T, Zech CJ. Liver metastases: detection and staging. *Eur J Radiol*. 2017;97:76–82. doi:10.1016/j.ejrad.2017.10.016
66. Taghavi M, Trebeschi S, Simões R, et al. Machine learning-based analysis of CT radiomics model for prediction of colorectal metachronous liver metastases. *Abdom Radiol (New York)*. 2021;46(1):249–256. doi:10.1007/s00261-020-02624-1
67. Ravanelli M, Agazzi GM, Tononcelli E, et al. Texture features of colorectal liver metastases on pretreatment contrast-enhanced CT may predict response and prognosis in patients treated with bevacizumab-containing chemotherapy: a pilot study including comparison with standard chemotherapy. *Radiol Med*. 2019;124(9):877–886. doi:10.1007/s11547-019-01046-4

68. Nakanishi R, Oki E, Hasuda H, et al. Radiomics texture analysis for the identification of colorectal liver metastases sensitive to first-line oxaliplatin-based chemotherapy. *Ann Surg Oncol*. 2021;28(6):2975–2985. doi:10.1245/s10434-020-09581-5
69. Hewitt DB, Pawlik TM, Cloyd JM. Who will benefit? Using radiomics to predict response to oxaliplatin-based chemotherapy in patients with colorectal liver metastases. *Ann Surg Oncol*. 2021;28(6):2931–2933. doi:10.1245/s10434-020-09586-0
70. Fiz F, Viganò L, Gennaro N, et al. Radiomics of liver metastases: a systematic review. *Cancers (Basel)*. 2020;12(10). doi:10.3390/cancers12102881
71. Berardi R, Maccaroni E, Onofri A, et al. Locally advanced rectal cancer: the importance of a multidisciplinary approach. *World J Gastroenterol*. 2014;20(46):17279–17287. doi:10.3748/wjg.v20.i46.17279
72. Zhu Y, Zhou Y, Zhang W, et al. Value of quantitative dynamic contrast-enhanced and diffusion-weighted magnetic resonance imaging in predicting extramural venous invasion in locally advanced gastric cancer and prognostic significance. *Quant Imaging Med Surg*. 2021;11(1):328–340. doi:10.21037/qims-20-246
73. van de Velde CJH, Boelens PG, Borras JM, et al. EURECCA colorectal: multidisciplinary management: European consensus conference colon & rectum. *Eur J Cancer*. 2014;50(1):1.e1–1.e34. doi:10.1016/j.ejca.2013.06.048
74. Bertocchi E, Barugola G, Nicosia L, et al. A comparative analysis between radiation dose intensification and conventional fractionation in neoadjuvant locally advanced rectal cancer: a monocentric prospective observational study. *Radiol Med*. 2020;125(10):990–998. doi:10.1007/s11547-020-01189-9
75. Cianci R, Cristel G, Agostini A, et al. MRI for rectal cancer primary staging and restaging after neoadjuvant chemoradiation therapy: how to do it during daily clinical practice. *Eur J Radiol*. 2020;131:109238. doi:10.1016/j.ejrad.2020.109238
76. Crimì F, Capelli G, Spolverato G, et al. MRI T2-weighted sequences-based texture analysis (TA) as a predictor of response to neoadjuvant chemo-radiotherapy (nCRT) in patients with locally advanced rectal cancer (LARC). *Radiol Med*. 2020;125(12):1216–1224. doi:10.1007/s11547-020-01215-w
77. Fornell-Perez R, Vivas-Escalona V, Aranda-Sanchez J, et al. Primary and post-chemoradiotherapy MRI detection of extramural venous invasion in rectal cancer: the role of diffusion-weighted imaging. *Radiol Med*. 2020;125(6):522–530. doi:10.1007/s11547-020-01137-7
78. Ciolina M, Caruso D, De Santis D, et al. Dynamic contrast-enhanced magnetic resonance imaging in locally advanced rectal cancer: role of perfusion parameters in the assessment of response to treatment. *Radiol Med*. 2019;124(5):331–338. doi:10.1007/s11547-018-0978-0
79. Wan L, Peng W, Zou S, et al. MRI-based delta-radiomics are predictive of pathological complete response after neoadjuvant chemoradiotherapy in locally advanced rectal cancer. *Acad Radiol*. Published online November 2020; 28. doi:10.1016/j.acra.2020.10.026
80. Horvat N, Veeraraghavan H, Khan M, et al. MR imaging of rectal cancer: radiomics analysis to assess treatment response after neoadjuvant therapy. *Radiology*. 2018;287(3):833–843. doi:10.1148/radiol.2018172300
81. Aker M, Ganeshan B, Afaq A, Wan S, Groves AM, Arulampalam T. Magnetic resonance texture analysis in identifying complete pathological response to neoadjuvant treatment in locally advanced rectal cancer. *Dis Colon Rectum*. 2019;62(2):163–170. doi:10.1097/DCR.0000000000001224
82. De Cecco CN, Ganeshan B, Ciolina M, et al. Texture analysis as imaging biomarker of tumoral response to neoadjuvant chemoradiotherapy in rectal cancer patients studied with 3-T magnetic resonance. *Invest Radiol*. 2015;50(4):239–245. doi:10.1097/RLI.0000000000000116

83. Shi L, Zhang Y, Nie K, et al. Machine learning for prediction of chemoradiation therapy response in rectal cancer using pre-treatment and mid-radiation multi-parametric MRI. *Magn Reson Imaging*. 2019;61:33–40. doi:10.1016/j.mri.2019.05.003
84. Nie K, Shi L, Chen Q, et al. Rectal cancer: assessment of neoadjuvant chemoradiation outcome based on radiomics of multiparametric MRI. *Clin Cancer Res an Off J Am Assoc Cancer Res*. 2016;22(21):5256–5264. doi:10.1158/1078-0432.CCR-15-2997
85. O'Connor JPB, Rose CJ, Waterton JC, Carano RAD, Parker GJM, Jackson A. Imaging intratumor heterogeneity: role in therapy response, resistance, and clinical outcome. *Clin Cancer Res an Off J Am Assoc Cancer Res*. 2015;21(2):249–257. doi:10.1158/1078-0432.CCR-14-0990
86. Chiloiro G, Rodriguez-Carnero P, Lenkowicz J, et al. Delta radiomics can predict distant metastasis in locally advanced rectal cancer: the challenge to personalize the cure. *Front Oncol*. 2020;10:595012. doi:10.3389/fonc.2020.595012
87. Nardone V, Reginelli A, Scala F, et al. Magnetic-resonance-imaging texture analysis predicts early progression in rectal cancer patients undergoing neoadjuvant chemoradiation. *Gastroenterol Res Pract*. 2019;2019:8505798. doi:10.1155/2019/8505798
88. Sun Y, Hu P, Wang J, et al. Radiomic features of pretreatment MRI could identify T stage in patients with rectal cancer: preliminary findings. *J Magn Reson Imaging*. Published online February 2018;48. doi:10.1002/jmri.25969
89. Wen D-G, Hu S-X, Li Z-L, et al. Application of automated machine learning based on radiomics features of T2WI and RS-EPI DWI to predict preoperative T staging of rectal cancer. *Sichuan da xue xue bao Yi xue ban = J Sichuan Univ Med Sci Ed*. 2021;52(4):698–705. doi:10.12182/20210460201
90. Yang L, Dong D, Fang M, et al. Can CT-based radiomics signature predict KRAS/NRAS/BRAF mutations in colorectal cancer? *Eur Radiol*. 2018;28(5):2058–2067. doi:10.1007/s00330-017-5146-8
91. Li H-H, Zhu H, Yue L, et al. Feasibility of free-breathing dynamic contrast-enhanced MRI of gastric cancer using a golden-angle radial stack-of-stars VIBE sequence: comparison with the conventional contrast-enhanced breath-hold 3D VIBE sequence. *Eur Radiol*. 2018;28(5):1891–1899. doi:10.1007/s00330-017-5193-1
92. Huang Y, He L, Dong D, et al. Individualized prediction of perineural invasion in colorectal cancer: development and validation of a radiomics prediction model. *Chin J Cancer Res*. 2018;30(1):40–50. doi:10.21147/j.issn.1000-9604.2018.01.05
93. Yu X, Song W, Guo D, et al. Preoperative prediction of extramural venous invasion in rectal cancer: comparison of the diagnostic efficacy of radiomics models and quantitative dynamic contrast-enhanced magnetic resonance imaging. *Front Oncol*. 2020;10:459. doi:10.3389/fonc.2020.00459
94. Chen L-D, Li W, Xian M-F, et al. Preoperative prediction of tumour deposits in rectal cancer by an artificial neural network-based US radiomics model. *Eur Radiol*. 2020;30(4):1969–1979. doi:10.1007/s00330-019-06558-1
95. Gopal P, Lu P, Ayers GD, Herline AJ, Washington MK. Tumor deposits in rectal adenocarcinoma after neoadjuvant chemoradiation are associated with poor prognosis. *Mod Pathol an Off J United States Can Acad Pathol Inc*. 2014;27(9):1281–1287. doi:10.1038/modpathol.2013.239
96. Chen F, Ma X, Li S, et al. MRI-based radiomics of rectal cancer: assessment of the local recurrence at the site of anastomosis. *Acad Radiol*. Published online November 2020;28. doi:10.1016/j.acra.2020.09.024
97. Allegra CJ, Rumble RB, Hamilton SR, et al. Extended RAS gene mutation testing in metastatic colorectal carcinoma to predict response to anti-epidermal growth factor receptor monoclonal antibody therapy: american society of clinical oncology provisional clinical opinion update 2015. *J Clin Oncol Off J Am Soc Clin Oncol*. 2016;34(2):179–185. doi:10.1200/JCO.2015.63.9674

98. Ortiz-Ramón R, Ruiz-España S, Mollá-Olmos E, Moratal D. Glioblastomas and brain metastases differentiation following an MRI texture analysis-based radiomics approach. *Phys Medica PM an Int J Devoted to Appl Phys to Med Biol Off J Ital Assoc Biomed Phys*. 2020;76:44–54. doi:10.1016/j.ejmp.2020.06.016
99. Larroza A, Moratal D, Paredes-Sánchez A, et al. Comment on 'computer-extracted texture features to distinguish cerebral radionecrosis from recurrent brain tumors on multiparametric MRI: a feasibility study'. *AJNR Am J Neuroradiol*. 2017;38(3):E21. doi:10.3174/ajnr.A5071
100. Soni N, Priya S, Bathla G. Texture analysis in cerebral gliomas: a review of the literature. *AJNR Am J Neuroradiol*. 2019;40(6):928–934. doi:10.3174/ajnr.A6075
101. Forghani R, Chatterjee A, Reinhold C, et al. Head and neck squamous cell carcinoma: prediction of cervical lymph node metastasis by dual-energy CT texture analysis with machine learning. *Eur Radiol*. 2019;29(11):6172–6181. doi:10.1007/s00330-019-06159-y
102. Fruehwald-Pallamar J, Hesselink JR, Mafee MF, Holzer-Fruehwald L, Czerny C, Mayerhoefer ME. Texture-based analysis of 100 MR examinations of head and neck tumors—is it possible to discriminate between benign and malignant masses in a multicenter trial? *Rofo*. 2016;188(2):195–202. doi:10.1055/s-0041-106066
103. Meyer H-J, Hamerla G, Höhn AK, Surov A. CT texture analysis-correlations with histopathology parameters in head and neck squamous cell carcinomas. *Front Oncol*. 2019;9:444. doi:10.3389/fonc.2019.00444
104. Kuno H, Qureshi MM, Chapman MN, et al. CT texture analysis potentially predicts local failure in head and neck squamous cell carcinoma treated with chemoradiotherapy. *AJNR Am J Neuroradiol*. 2017;38(12):2334–2340. doi:10.3174/ajnr.A5407
105. Kozikowski M, Suarez-Ibarrola R, Osiecki R, et al. Role of radiomics in the prediction of muscle-invasive bladder cancer: a systematic review and meta-analysis. *Eur Urol Focus*. Published online June 2021;8. doi:10.1016/j.euf.2021.05.005
106. Pan R, Yang X, Shu Z, et al. Application of texture analysis based on T2-weighted magnetic resonance images in discriminating Gleason scores of prostate cancer. *J Xray Sci Technol*. 2020;28(6):1207–1218. doi:10.3233/XST-200695
107. Sidhu HS, Benigno S, Ganeshan B, et al. Textural analysis of multiparametric MRI detects transition zone prostate cancer. *Eur Radiol*. 2017;27(6):2348–2358. doi:10.1007/s00330-016-4579-9
108. Gnep K, Fargeas A, Gutiérrez-Carvajal RE, et al. Haralick textural features on T(2)-weighted MRI are associated with biochemical recurrence following radiotherapy for peripheral zone prostate cancer. *J Magn Reson Imaging*. 2017;45(1):103–117. doi:10.1002/jmri.25335
109. Wei C, Chen Y-L, Li X-X, et al. Diagnostic performance of MR imaging-based features and texture analysis in the differential diagnosis of ovarian thecomas/fibrothecomas and uterine fibroids in the adnexal area. *Acad Radiol*. 2020;27(10):1406–1415. doi:10.1016/j.acra.2019.12.025
110. An H, Wang Y, Wong EMF, et al. CT texture analysis in histological classification of epithelial ovarian carcinoma. *Eur Radiol*. 2021;31(7):5050–5058. doi:10.1007/s00330-020-07565-3
111. Ytre-Hauge S, Salvesen ØO, Krakstad C, Trovik J, Haldorsen IS. Tumour texture features from preoperative CT predict high-risk disease in endometrial cancer. *Clin Radiol*. 2021;76(1):79.e13–79.e20. doi:10.1016/j.crad.2020.07.037
112. Ueno Y, Forghani B, Forghani R, et al. Endometrial carcinoma: MR imaging-based texture model for preoperative risk stratification-a preliminary analysis. *Radiology*. 2017;284(3):748–757. doi:10.1148/radiol.2017161950

Index

Note: Page numbers in *italic* indicate a figure and page numbers in **bold** indicate a table on the corresponding page.

R

S

T

V

W

Milton Keynes UK
Ingram Content Group UK Ltd.
UKHW031532071024
449327UK00005B/107